Sustainable Heritage Management - Design: Notions, Methods and Techniques

Mar Loren-Méndez · Marta García-Casasola ·
Vladan Djokić · Ana Nikezić · Emanuela Sorbo ·
Maria Philokyprou · Konstantinos Sakantamis
Editors

Sustainable Heritage Management - Design: Notions, Methods and Techniques

Springer

Editors
Mar Loren-Méndez
UNESCO Chair on Modern Urban heritage
CREhAR (Creative Research and Education on heritage Assessment and Regeneration)
Department of Architectural History
Theory and Composition
School of Architecture
Universidad de Sevilla
Sevilla, Spain

Marta García-Casasola
UNESCO Chair on Modern Urban heritage
CREhAR (Creative Research and Education on heritage Assessment and Regeneration)
Department of Architectural History
Theory and Composition
School of Architecture
Universidad de Sevilla
Sevilla, Spain

Vladan Djokić
Department of Urbanism
Faculty of Architecture (UB-FA)
University of Belgrade
Belgrade, Serbia

Ana Nikezić
Department of Architecture
Faculty of Architecture (UB-FA)
University of Belgrade
Belgrade, Serbia

Emanuela Sorbo
Department of Architecture and Arts
Università Iuav di Venezia
Venice, Italy

Maria Philokyprou
Department of Architecture
School of Engineering
University of Cyprus
Nicosia, Cyprus

Konstantinos Sakantamis
Faculty of Engineering
School of Architecture
Aristotle University of Thessaloniki
Thessaloniki, Greece

ISBN 978-3-031-81969-8 ISBN 978-3-031-81970-4 (eBook)
https://doi.org/10.1007/978-3-031-81970-4

This book is developed in an European Network of five Universities, which started in the context of their innovative research work developed in an Erasmus+ Project within the EU programme for education, training, youth and sport, entitled: HERSUS Enhancing of Heritage Awareness and Sustainability of Built Environment in Architectural and Urban Design Higher Education. This publication is coordinated by the UNESCO Chair on Modern Urban heritage CREhAR (Creative Research and Education on heritage Assessment and Regeneration).

This Springer imprint is published by the registered company Springer Nature Switzerland AG
The registered company address is: Gewerbestrasse 11, 6330 Cham, Switzerland

Preface

This book is developed as a complementary result of the research work carried out by an European network of five universities, which started in the context of their innovative Erasmus+ project within the EU program for education, training, youth and sport, entitled: *HERSUS Enhancing of Heritage Awareness and Sustainability of Built Environment in Architectural and Urban Design Higher Education.*

After having worked for more than four years sharing knowledge and experience, at the end of the project, the consolidated research partnership was able to go further, to continue researching together, and develop a new result, on which to build the methodology to be applied in sustainable heritage management and design. The specific background of each member of the partnership allows to firstly transfer the basic conceptualization of the field, framing the NOTIONS related to heritage-sustainability binomial; secondly to clarify the general METHODOLOGY, and highlight the specificity of each case; to focus on the identification and monitorization of VALUES for decision making; to intervene through project planning, designing STRATEGIES AND ACTIONS; and finally to implement new TOOLS to improve knowledge and facilitate sustainable heritage management. This book therefore contributes to opening the way to managing heritage from a sustainable perspective, characterizing it from a holistic approach that includes environmental conservation and bioclimatic design. The five parts offer three transversal perspectives: education, practice and research, trying to give an integral and complete answer to students, practitioners and researchers (Fig. 1).

This publication is coordinated by the UNESCO Chair on Modern Urban heritage CREhAR (Creative Research and Education on heritage Assessment and Regeneration), at the School of Architecture, University of Seville (USE), institutions and their research groups that represent a benchmark in research on cultural heritage and sustainability: their projects specifically address innovation in heritage methodologies for the identification, characterization, enhancement and intervention of heritage.

The European universities of the consortium have been collaborating closely, both as authors and editors: University of Seville, Spain; University of Belgrade,

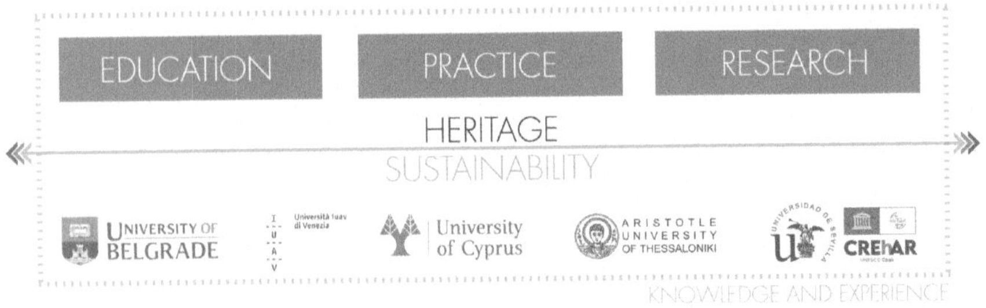

Fig. 1 Five universities working together to share knowledge and experience related to heritage-sustainability binomial through the perspective from education, practice and research. University of Seville (USE) UNESCO Chair CREhAR; University of Belgrade (UB FA) IUAV University of Venice (IUAV), University of Cyprus (UCY), Aristotle University of Thessaloniki (AUTh)

Serbia; IUAV University of Venice, Italy; University of Cyprus, Cyprus; and Aristotle University of Thessaloniki, Greece. Their diverse contributions are therefore developed from a shared place where the universities have built together a process to manage heritage and sustainability, whose results are based on consensus and generate shared knowledge which reinforce the proposal.

The aim of this book is to show the main methodological process to develop good heritage intervention practices from a sustainable perspective. Specific objectives are:

- To understand the specificity within the conceptualization and the identification of heritage types, where the constant change of heritage boundaries unveils emergent types, and redefine preexistent ones.
- To know and manage the heritage methodology to guarantee the results of the intervention.
- To recognize the identification of cultural values as key for project decision making.
- To acknowledge the intervention design strategies and the project decisions, as part and fully integrated within the heritage project management.
- To learn how to design, use and constantly update new tools to improve heritage knowledge.

The five parts, which organize the table of contents of the book, intend to emphasize the key aspects to consider when you are going to intervene in heritage from a sustainable perspective:

PART I. The conceptualization and characterization of heritage: **notions and heritage types**.
PART II. The relevance of the specific **methodologies** related to heritage-sustainability binomial.
PART III. The knowledge and decisions making based on cultural **values**.
PART IV. The development of a project: **design strategies and actions**.
PART V. The **tools**: technologies as sustainable tools for advancing knowledge and heritage management.

SECTIONS

	NOTION AND HERITAGE TYPES	METHODOLOGIES	VALUES	DESIGN STRATEGIES AND ACTIONS	TOOLS
EDUCATION PRACTICE RESEARCH	From the object to the **landscape** **Emerging** heritages: anthropological, modern, industrial... Fomenting heritage perspective from **sustainability**	Holistic and **interdisciplinary** approach Sustainable heritage **management**	Identification and **valuation** Developing critical studies Values and **society**: working with citizens Documenting	**Coevolution** as the key on heritage intervention: working with **temporality** New proceeding to **manage** values and **monitoring** them through to the intervention project	**Creativity**: transforming heritage following a model of work based on knowledge and working with **communities** Energy efficient buildings Mapping **cartographies**

Fig. 2 Diagram to show a summary of the contributions organized in five parts aligned with the main aspects of sustainable heritage management and crossing from the three different perspectives: education, practice and research

These five process-based parts are put in dialog with the three transversal perspectives which run through the university mission, specifically in the field of architecture: EDUCATION, PRACTICE AND RESEARCH. The proposal is organized in response to what has been called **applied research** and **service learning**, where the educational task is transcended in order to contribute hand in hand with professionals to solving the needs of citizens, always working from the basement of the science, implementing the process from the research field. To this end, the contributions address different study cases that allow to verify the correct application of the methodology (Fig. 2).

Regarding its contribution to the literature in the field, this publication offers an international book in the field of heritage and sustainability, providing both conceptual and specific didactic contents and materials, for what is needed to deal with the complex process involved in heritage management. It looks for quality proceedings to project integrated actions on heritage assets, from a transdisciplinary perspective. This book promotes decision making on expert knowledge and cultural values; the proper use of methods, strategies and tools, and reaches agreements through participation of the different agents involved. The central innovation of this contribution rests on combining the processes and methods of the heritage studies with those of the project within the field of Architecture and Urban Planning.

Sevilla, Spain

Mar Loren-Méndez
Marta García-Casasola

Acknowledgements

This book would not have been possible without the work and expertise of the group of researchers involved in the innovative European project *HERSUS Enhancing of Heritage Awareness and Sustainability of the Built Environment in Architectural and Urban Design Higher Education*. An international network of five European universities on architecture, heritage and sustainability. Coordinated by the UNESCO Chair on Modern Urban heritage CREhAR (Creative Research and Education on heritage Assessment and Regeneration), at the School of Architecture, Universidad de Sevilla (USE), the network has continued its research together to make this publication possible, and has expanded to include leading authors in the field of sustainable heritage management and design.

The editors would like to acknowledge the work done by Domingo Galán-Caro as editing coordinator. He contributed to the organization of the documentary process and the communication management. He also participated in the formal review of the different contributions. He is part of the UNESCO Chair on Modern Urban heritage CREhAR (Creative Research and Education on heritage Assessment and Regeneration), at the Universidad de Sevilla, Department of Architectural History, Theory and Composition, School of Architecture, Seville, Spain, dgalan@us.es.

We would like to thank you very much to all the authors.

Contents

Part V Tools

Part I
Notions and Heritage Types

The conceptualization in the field of heritage has been carried out along the twentieth century and the beginning of the twenty-first. This context of constant change unveils emergent types, and redefines pre-existent ones; its classification also allows to identify the specificity of each heritage and therefore points out the shifting perspective from the pristine monument to cultural diversity in the centre of heritage conservation and sustainability.

A genealogical reading through the last century offers a diagnosis to understand heritage from the object to the landscape, from the monument to immaterial significance. It is a nomadic conceptualization, as F. Choay suggests, which must be constantly updated. It is a process of retraced steps to fine-tune different issues.

The permanent revision of the conceptualization forces to review the consolidated types of heritage from different perspectives.

In this case, the approach to historical settlements from sustainability means to consider them as systems, in which it is also relevant the anthropological point of view *apart from being carriers of historical, archaeological, artistic, scientific, social or technical values, are mainly carriers of collective memory* (M. Dousi). It is important to recognize the environment as the stage where life is carried out and where it is possible to look from a heritage perspective, not only to identify relevant monuments but also to find vernacular architecture, anonymous architecture, which should have been valued. This consideration is the key to develop new proposals in dialogue that are shown through student works.

The evolution of the heritage concept means the identification of emerging heritage in the latest twentieth century within the professional practice: anthropological, industrial, modern, etc (A. Milovanović, A. Zorić, T. Popović, J. Ristić Trajković, A. Nikezić). In the way of opening up the concept, it is also possible to find the recovery of abandoned heritage, as historical settlements. This is the case of rural heritage, that has been recognized as such for some year nows. In fact, ICOMOS has published the *Principles concerning rural landscapes as heritage* (2017) to reinforce the importance of this context to balance, among other things, the growth of urban areas and the loss of identity and quality of life that is occurring in them.

This constant amplification of the conceptualization and therefore, what is considered heritage does not reflect on the institutional recognition. UNESCO identified this problem and initiated, in the context of the global strategy for a balance and representative World Heritage List (WHL), a pro-active approach with regard to the identification and documentation of less-represented heritage categories in the WHL, such as Modern Heritage (Architecture, Town Planning and Landscape Design of the Contemporary Age). Having the European cultural heritage as the case study, this problem of underrepresentation of Modern Heritage within UNESCO WHL still persists, raising awareness of the need for commitment, but also of the opportunity that modern heritage represents nowadays within the field of Heritage Studies (M. Loren-Méndez, R. F. Alonso-Jiménez, D. Pinzón-Ayala).

Chapter 1
The Outstanding Industrial Heritage of Thessaloniki. Options for a Sustainable Reuse

Maria Dousi

Abstract The issue of the protection, promotion and management of industrial heritage occupies today a dominant position in the debate on urban regeneration and the development of contemporary cities. In all the countries of the advanced world, with the change in production systems, the phenomenon of "de-industrialisation" has gradually appeared since 1950. As a result of this was the abandonment and desolation of large industrial units, which were located in the centres of cities. One of the consequent problems of this phenomenon is the way of managing and utilising the remaining stock of abandoned industrial facilities, which from being a source of productive activities have become "dead" and degraded sites. The challenge today, at the international level, is how these spaces will retain their character, be architecturally valuable, and become reusable nuclei for the future sustainable development of cities.

Introduction

Industrial heritage, as defined in the Statutes of the International Committee for the Conservation of Industrial Heritage (T.I.C.C.I.H.), and in particular in The Nizhny Tagil Charter For The Industrial Heritage (2003), consists of sites, structures, industrial and industrial equipment, means of transport, products, installations of all kinds, as well as their scientific documentation through written and oral testimonies related to them. According to the historian of Industrial Archaeology Jacques Pinard, the concept of "Industrial Heritage" includes a set of structures and objects (buildings, structures, machines, equipment, written or oral records, etc.), all of which deserve some attention because they contribute to the knowledge of the history of the productive activities of a country or a population (Pinard 1985).

The issue of the protection, promotion and management of industrial heritage occupies today a dominant position in the debate on urban regeneration and the

M. Dousi (✉)
Aristotle University, Thessaloniki, Greece
e-mail: mdoussi@arch.auth.gr

M. Loren-Méndez et al. (eds.), *Sustainable Heritage Management - Design: Notions, Methods and Techniques*, https://doi.org/10.1007/978-3-031-81970-4_1

development of contemporary cities. In all the countries of the advanced world, with the change in production systems, the phenomenon of "de-industrialisation" has gradually appeared since 1950. As a result of this was the abandonment and desolation of large industrial units, which were located in the centres of cities. One of the consequent problems of this phenomenon is the way of managing and utilising the remaining stock of abandoned industrial facilities, which from being a source of productive activities have become "dead" and degraded sites. The challenge today, at the international level, is how these spaces will retain their character, be architecturally valuable, and become reusable nuclei for the future sustainable development of cities.

This paper will attempt to examine the example of the city of Thessaloniki, which during the second half of the nineteenth to the first half of the twentieth century was a highly industrialised city, and in its urban fabric, there are nowadays very important abandoned large historical industrial complexes, which are awaiting restoration and sustainable reuse and their reintegration into the fabric of the city of today. The exploration will be based on the presentation of 3 very important industrial monuments through respective proposals for sustainable reuse in the context of Diploma Theses. These three historic industrial complexes are the Karolos Fix Brewery in the west and the Allatini Flour Mills and Allatini Ceramic Factory in the east of the city (Fig. 1.1).

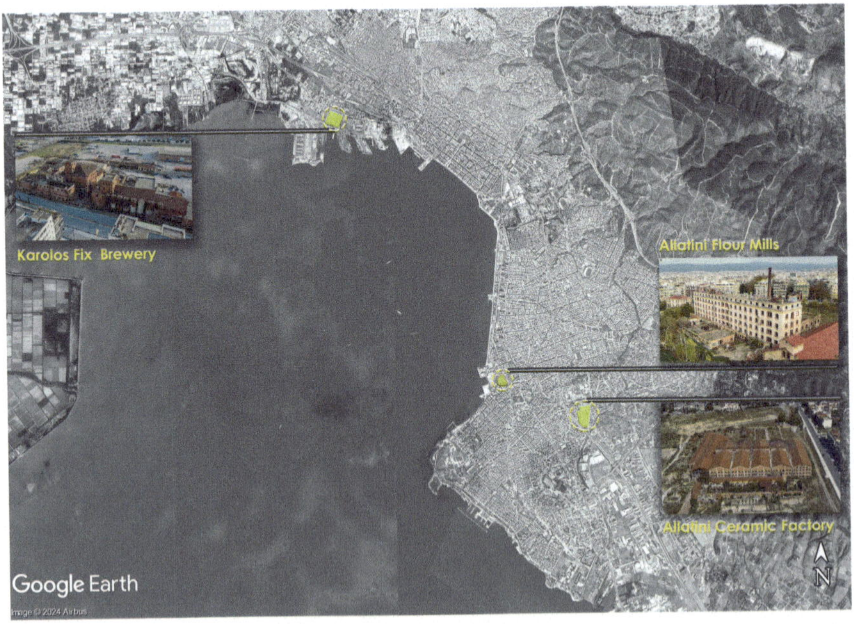

Fig. 1.1 Map of Thessaloniki, showing the three historic industrial complexes. To the west of the city the fix Brewery and to the east the Allatini Flour Mills and the Allatini Ceramic Factory. Background google earth, photos by the author

The Industrial Heritage of Thessaloniki—A Brief Historical Report

From the second half of the nineteenth century, when the first industrial facilities were established in Thessaloniki, the city was the most important administrative, strategic, commercial, and economic centre of south-eastern Europe after Constantinople. While other Greek cities in Macedonia (Edessa, Naoussa, Veroia) based their industrial development on water power, the industrial development of Thessaloniki was based on steam power (Kolonas et al. 1987). These first industrial units were characterised by the introduction of new construction methods and building materials which were already widely used in Europe (Dousi 2022, pp. 14a-14f).

Today's North Greek region, with Thessaloniki as the main focus from the second half of the nineteenth century until 1912, presents an important peculiarity due to the fact that, during the period under consideration, it belonged to the then modernising Ottoman Empire. In this context, important and large-scale industrial complexes were built, which still exist today and are listed monuments of our architectural heritage. The first factories were concentrated in the western zone of the city between the port and railway facilities. The factories in the eastern part of the city are built near streams or the sea for practical reasons. Typical examples of industrial complexes of the period are the industrial giants, the Allatini Flour Mills and the Fix Brewery.

At the beginning of the twentieth century and after its wide application in Belgium and France, a new material, reinforced concrete, was adopted in Thessaloniki. Thus, in 1910, the first frame with reinforced concrete columns and beams was constructed. The annexation of the city to the Greek state in 1912, the First World War, the devastating fire of 1917 and the arrival of refugees from Asia Minor after 1922 affected the industrial development. During the inter-war period, Thessaloniki was the industrial capital of Northern Greece. The technological innovations of the twentieth century, which were related to energy technologies and the production process, influenced the organisation of the sites. The use of electricity as the main source of energy allowed the free arrangement of industrial equipment and abolished the restrictions imposed by steam power. A typical example of an industrial complex of this period is the Allatini Ceramics (Kolonas et al. 1989).

The Industrial Heritage of Thessaloniki—A Sustainable Perspective

Through the reflection that has dominated the last decades at the international level, we have much to learn (Stratton 2000). Different views have prevailed, often conflicting, different practices have been tried out and already the results today provide us with important experience. However, it is characteristic that all views converge on a common ground. Today, internationally, state bodies, local authorities, social groups, owners and investors all agree, each from their own point of view, that abandoned industrial sites should be rescued and used in a sustainable way. In Greece, all these issues have not yet been well settled, but it is common knowledge that we are at a very critical point for the future of those parts of the industrial heritage that remain.

Thessaloniki, especially after the second half of the nineteenth century, until the first decades of the twentieth century, developed into a metropolis of industrial development in Greece. Numerous and diverse industrial activities, pioneering for European standards, were established in the city, which gave a huge boost to the economic development of the region and provided employment to thousands of people. They characterised and still characterise the fabric of the city because of their large size and imposing architecture. They are an important asset which today, in the majority of cases, remains unused. These important historical industrial complexes stand abandoned and degraded in the face of an embarrassed and indifferent society. It is fortunate that most of them have been designated as listed buildings and because of this they have escaped mass demolition (Dousi 2009, pp. 292–300).

The issue of preserving our industrial heritage is extremely important, as industries were inextricably linked to the productive activities and the development of each place, while at the same, time industrial facilities were and are landmarks and components of the image of cities. In other words, they shaped to a certain extent the social, economic, urban, architectural and technological history of cities. The urgent question is how industrial buildings and complexes, which are currently privately owned, because those owned by public bodies, we hope that they will have the conditions to be saved, will be able to be preserved, to keep their historical and architectural character and to be exploited through the prism of a balanced and sustainable development.

Today, two contradictory views prevail on this issue in Greece, which in broad terms are as follows:

- The view of those who support the absolute protection and preservation of industrial heritage, such as agencies and protection bodies, but who in many cases do not recognise the economic interests of the owners.
- The view of those who seek to exploit the prestige of industrial buildings and achieve maximum economic exploitation, or in the worst cases seek demolition, such as owners and construction companies.

The establishment and operation of the industrial facilities was intertwined with the city of Thessaloniki, where they were developed and its inhabitants. Therefore, the issue of their management and exploitation does not only concern their current owners, who rightly have a say in the matter, but the whole city. They are in most cases private properties, so in all likelihood any attempt at exploitation is potentially aimed at economic exploitation. Because of this fact, it seems rather unfeasible to use the facilities exclusively for museum purposes.

It is a fact that their exclusive museum use can ensure the absolute protection of structures and machinery and function didactically for the history of the city's industrial production, but due to the large size of the installations, there is a risk of creating "dead" spaces for the urban fabric, especially in the case of large industrial complexes. The solution, in our opinion, appears to be the result of a compromise, which will synthesise the two conflicting views towards a common goal. The important industrial heritage complexes should be protected and revitalised for reuse.

The revitalisation of industrial sites should be sought through the restoration of historic sites, the most important of which, together with their equipment, could accommodate museum uses and others could be reused for other compatible uses and activities combined with modern architectural interventions.

In the last decades of the twentieth century, most of the industries, with the expansion of the city of Thessaloniki, were located within the urban fabric, while at the same time, they completed their life cycle and the old shells were abandoned. In recent years, there has been an interest in the restoration and integration of new uses in the abandoned old industrial shells. The survey, analysis and evaluation of these historic industrial buildings is now a historical necessity in order to preserve and highlight these last examples of the city's industrial heritage (Dousi 2006, pp. 11–14).

Three Exemplary Proposals for Compatible Sustainable Re-Use

At the School of Architecture of the Aristotle University of Thessaloniki, a systematic effort is being made to raise awareness and educate students on issues of adaptive reuse of the city's industrial facilities. A brief presentation of reuse proposals at the level of diploma theses for three of the most important historical industrial complexes of Thessaloniki follows.

- Industrial complex of the brewery—Karolos Fix—Adaptive reuse in Museum uses and development areas of contemporary microbrewery

The complex of the brewery Karolos Fix is one of the most important preserved industrial heritage monuments in Greece. The complex started operating in 1893. In 1926, the majority of the company's shares were transferred to the Athenian company "Karolos Fix", which already owned three other large breweries in Athens. During the inter-war period, additions and alterations were made to the brewery's building complex and its mechanical equipment was modernised. The "Karolos Fix" company expanded the brewery after the war, with a juice production unit. The complex began to further expand its facilities over the years. One of the typical examples of these newer buildings is the bottling plant, which was designed by the architect Takis Zenetos and built in 1959 (Dousi 2005, p.16). The complex is a unique example of industrial heritage not only for the Greek but also for the European standards. The operation of the factory was discontinued in the early 1980s. In 1994, the Ministry of Culture and Tourism designated most of the historical complex as a historical monument along with its technological equipment. Since then the outstanding historic

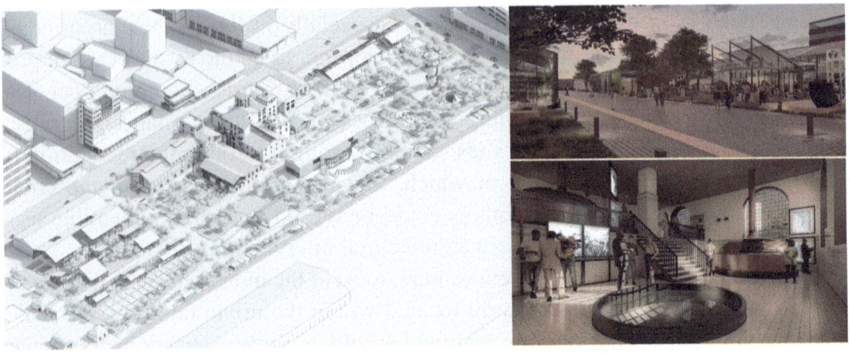

Fig. 1.2 Reuse of the historical industrial complex of Fix in Thessaloniki. General axonometric view of the project, where the general configurations of the site and photo-realistic illustrations are also visible. Diploma Design Thesis by Thalia Pelagini Kalfa and Sophia Mermiga Angeli. (School of Architecture, Aristotle University of Thessaloniki, February 2024, supervisor Maria Dousi)

industrial complex, which is privately owned, has been awaiting a serious restoration plan and rendering new compatible uses.

In the following, a proposal for a compatible and sustainable reuse of the complex is briefly presented through the Diploma thesis Diploma thesis by Thalia Pelagini Kalfa and Sophia Mermiga Angeli, School of Architecture of the Aristotle University of Thessaloniki, (February 2024), entitled: Zythos Narratives. Reuse of the historical brewery "Karolos FIX" in Thessaloniki (Fig. 1.2).

This Diploma design Thesis is a proposal to activate the historic industrial complex of the brewery "Fix" in Thessaloniki through an approach that focuses, on the one hand, on the historical value of the site and its past use and, on the other hand, on the multidimensional experience of the modern user. Taking into account the contemporary needs of the area, the neighbouring uses, the future activation of the surrounding areas, the evaluation of the complex and the production process, as well as the examination of contemporary trends in the brewery, a programmatic framework of interventions is created, aiming at a balance between the historical past and contemporary needs.

The vision is therefore based on three programmatic axes that will link the value of the industrial complex to the past, present and future. These are to preserve the tangible and intangible heritage (past), to activate the site and create a new landmark for the city (present) and to contribute to and promote new trends in brewing. The proposal is structured around three narratives. The narratives have a common design axis that varies according to the type of users. These narratives compose a grid of experiences that contribute to the creation of a complex dedicated to beer.

Regarding the preservation and continuity of historical memory, this is achieved through the interpretation of the former use, i.e. the creation of a self-referential museum, as well as the repetition of the preservation and interpretation of the former use, i.e. the continuity of the intangible heritage, the creation of microbrewery spaces. To activate the site and create a landmark, a large part of the complex is given back to the public by creating public open space, landscaping and parkland, as well as public uses that take over the complex. The integration of the complex into the present is achieved by creating spaces for secondary functions, which arise as needs from the evolution of the brewery into the present. Opportunities are created for the development of these functions, i.e. spaces for experimentation, workshops, office and design spaces, since a huge part of the brewery today is marketing and designing the brand identity of the product. In conjunction with the dissemination of brewing culture, the previously dormant historical space can be transformed into a living complex dedicated to beer.

- Industrial complex of Allatini Flour Mills–Adaptive reuse in a residential complex

In the site where the listed industrial complex of the Allatini Mills is located today, the first steam-powered flour mill of Thessaloniki, owned by a French company, was built in 1854. In 1882, the mill became the property of the Allatini Company. In 1898, the first steam mill burned down and in its place a new larger one was built according to the plans of the important Italian architect Vitaliano Poselli (1900) (Kolonas et al. 1980; Dousi and Nomikos 2000, pp. 267–278).

At the beginning of the twentieth century, the complex was one of the largest in the south-eastern Mediterranean. At the same time as the main building of the Mills, in 1900, additional buildings were built until the end of the 1920s, when the complex acquired its complete form. The complex then continued to modernise, passing into the ownership of new owners and expanding until 1950, when a devastating fire caused enormous damage to the most important buildings. During the 1960s and 1970s, the complex operates with continuous extensions and modernisation interventions, which reflect the aesthetics and technology of the time, without any concerns about the compatibility of the interventions with the historical structures. Allatini Mills operated continuously until the mid-1990s, completing 140 years of continuous operation, which is an achievement in the history of Greek industry. Today, after their closure, the removal of their mechanical equipment, and the removal of surrounding non-listed buildings, a serious plan for its sustainable restoration and reuse is still awaiting. (Dousi and Nomikos 2000).

In the following, a proposal for a compatible and sustainable reuse of the complex is briefly presented through the design thesis of Chrysa Karanikola and Silia Liatsi, School of Architecture, Aristotle University of Thessaloniki, July 2023, entitled: Mills Allatini | From the industrial past…to contemporary housing (Fig. 1.3).

Evaluating the existing situation of the industrial complex of the Allatini Mills and the direct consequences it brings to the wider neighbourhood, its chronic abandonment and the large area it occupies, it is proposed to revive, redesign and reintegrate it into the fabric of the city. The project proposes a series of strategic interventions to fully activate the complex and its surroundings, respecting the listed buildings and the needs of the local residents, while ensuring the timelessness and sustainability of the interventions.

Starting from the surrounding area management, it is proposed to divide it into individual functional units accessible to all, without strict boundaries. The creation of squares, rest areas, open-air exhibition spaces, leisure areas, large green areas and

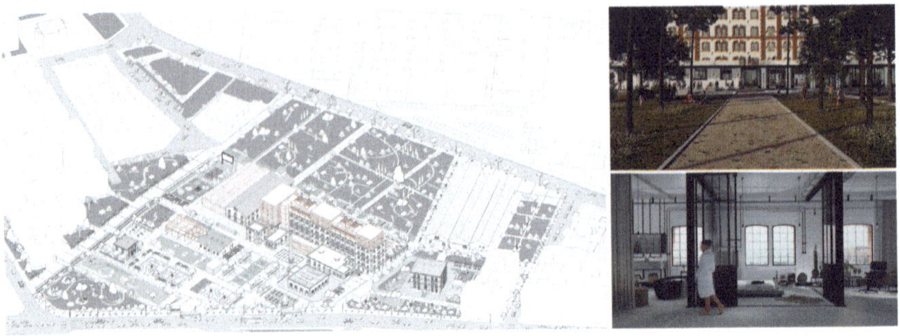

Fig. 1.3 Reuse of the Allatini Flour Mills industrial complex. General axonometric view of the complex (the redesigned parts of the complex are indicated in colour). General layout of the site, and photo-realistic drawings showing the "atmosphere", the materials and the configuration of the exterior and interior of the loft-type apartments, is also shown. Diploma Design Thesis of Chrysa Karanikola and Silia Liatsi. (School of Architecture, Aristotle University of Thessaloniki, July 2023, supervised by Maria Dousi)

numerous routes that encourage the use of bicycles are intended to fully activate the Allattini Mills, thus transforming an urban void of today into an attraction of tomorrow. The central flour mill is proposed to be used as a mill, while the other buildings in the complex are given complementary residential uses, such as small cooperative spaces, art studios, restaurants, commercial, leisure and cultural facilities and a number of open green spaces.

The former flour mill recovers its original appearance and is transformed into a complex of 33 loft-type residences on all floors, while the ground floor is used for communal purposes. It is also proposed to remove the top floor of the building and reconstruct it in line with the original layout. The recovery of the original structure (integration of the form) is carried out in accordance with the principles of the Venice Charter. It is a lightweight metal structure clad with perforated metal panels, which is placed in a recess in the roof of the listed building. The use of contemporary materials and the slight recess from the outline of the listed part reinforces the distinction between new and old, while integrating the new structure into the overall image of the building without altering its historical identity and integrity.

- Allatini Ceramic Factory | Adaptive reuse as a multifunctional centre for culture and education

The Allatini Ceramics Factory was founded in 1880 in the area of Pylaia. Apart from bricks and Marseilles-type tiles, the factory produced tiles, floor tiles and pipes. The quality of its products was excellent, competing with those of factories abroad. Even today, products bearing the "Fratelli Allatini" mark, clay blocks and tiles, found in historic buildings, are still in very good condition.

Since 1898, the company was part of the "Industrial and Commercial Company of Thessaloniki SA", which also owned the Allatini Flour Mills complex. In 1935, the original factory was destroyed by fire, and in 1939, new facilities were built.

From 1964 and for a period of thirty years, the factory operated as an independent company under the name "Allatini Mills". With the outbreak of the Second World War, it became impossible to start operations and the civil war followed. Thus the factory started production after 1955 with important additions being made in 1970. From the mid-1990s, the Company's production facilities under new ownership and management were transferred outside Thessaloniki and the complex was gradually abandoned. This huge industrial complex is also awaiting a scenario of sustainable reuse. (Dousi 2009, p 297).

In the following, a proposal for a compatible and sustainable reuse of the complex is briefly presented, through the Diploma design thesis by Apostolos Angelousis, School of Architecture, Aristotle University of Thessaloniki, September 2023, entitled: "Restoration and adaptive reuse of the Allatini Ceramics Factory in Thessaloniki (Fig. 1.4).

The old ceramic factory, which occupies a central position in the planning of the area, is located southeast of the new proposed park. A number of decisions are made for the development of the complex regarding the preservation and modification of the existing structures while respecting the history of the industrial building complex. A ceramic factory, compared to any other industrial complex, is itself a building-machine, enclosing within it gigantic and special structures that were essential to its production process. The oldest part of the original factory is chosen to function as a self-referential museum. In this part, the use of the existing industrial structures is a crucial element for the preservation of the historical and industrial heritage. Based on the analysis of the building stock, it has been decided to preserve certain industrial structures, combining their past character with contemporary needs. On the ground floor of the museum, one of the kilns used for firing the ceramic products is preserved, offering visitors an iconic glimpse into the production process. On the floor of the museum, one of the ceramic product drying facilities is preserved, highlighting the

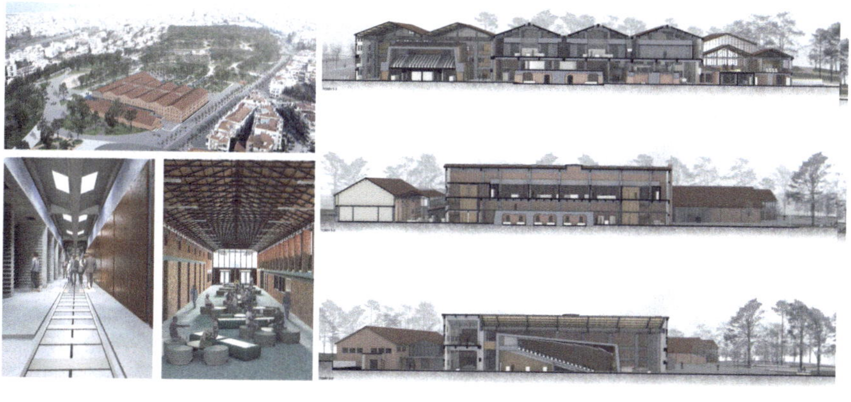

Fig. 1.4 Reuse of the industrial complex of Allatini Ceramic Factory. Photo-realistic illustration of the study area and the Allatini Ceramic Factory, views of the interiors and section drawings. Diploma Design Thesis of Apostolos Angelousis. (School of Architecture, Aristotle University of Thessaloniki, September 2023, supervisor Maria Dousi)

link with the past and offering a hands-on experience of industrial technology. The entrance to the museum and its ancillary uses, such as additional exhibition spaces, administrative spaces, public gathering spaces, multifunctional spaces and catering spaces, are chosen to be located in the part that communicate with the old wet processing building which was used to function as drying rooms. In the central part of the complex, the integration of educational and cultural uses was chosen as a central option for the integration of the complex in the contemporary context of the area. A part of the second furnace (Hoffmann furnace) was chosen to be kept in situ and adapted to accommodate exhibition uses on the ground floor and mezzanine. This decision highlights the historical significance of the site, while preserving the architectural character of the building. On the floor where the drying structures are preserved, the necessary modifications are being made to convert the former industrial facilities into educational spaces. Classrooms and presentation rooms will be created, enhancing the educational role of the site. At the same time, elements of the pre-existing industrial equipment are preserved, adding a rich historical experience to the educational environment. With these modifications, the building retains its authenticity while adapting to new functional requirements. In the later eastern part of the building complex, an amphitheatre of supra-local character is integrated within the historic shell, which is expected to become an important pole for the city, but will also highlight the historic industrial complex itself. In order to create this space, major interventions are being made to the building structure.

The combination of educational and cultural uses highlights the complex as a dynamic multidimensional cultural nucleus, integrating it into the city's multi-layered and diverse character and offering educational opportunities that combine the past with the present and give the historic industrial complex a dynamic sustainable perspective.

Epilogue

The reuse of industrial heritage is one of the most topical issues both from the point of view of balanced sustainable development of modern cities and from the point of view of contemporary architectural creativity on a global scale (Powell 1999). Among the quite rich international experience, we can mention the reuse of the Fiat factory in the Greater Turin area (by R. Piano), the Nestle factory in Noisier, France (by Reichen & Robert) and the gas tanks in Vienna (by J. Nouvel, Coop Himmelblau, M. Wehdorn, W. Holzbauer). From these examples from international experience it is clear that through private initiative and with the assistance of contemporary architectural creativity, it is possible to save historic industrial complexes and to make them dynamic nuclei for the balanced and sustainable development of modern cities. It is perhaps the last opportunity to promote and exploit this important heritage in Greece, before it is completely lost, or perhaps even worse, abused. We must promote and manage it in modern terms, with confidence, but also with respect for its culture and history. We are trying to raise awareness and activate citizens and institutions, but also to educate architects so that we can all contribute to this effort.

It is very encouraging that students of Architecture choose to study the industrial heritage of Thessaloniki and propose, through their diploma theses, visionary and feasible proposals for its compatible reuse and its active reintegration into the modern life of the city.

References

Dousi, Maria. 2005. The work of Takis Zenetos in the preserved industrial complex of fix in Thessaloniki. *Technografima* 278: 16.

Dousi, Maria. 2006. Thessaloniki's industrial Heritage. *Technografima* 307: 11–14.

Dousi, Maria, Nomikos, Michalis. 2000. Architectural analysis—documentation as a determining framework for the restoration interventions of the preserved industrial complex of Allatlni. In *1st National Conference: 'Mild Interventions and the Protection of Historic Structures*, 267–278. Thessaloniki: E.N.M.IYP.PO &.T.E.E-T.K.M.

Dousi, Maria, 2009. Industrial heritage management—the Thessaloniki experience. In The Industrial Heritage of the Dodecanese, 292–300. Rhodes.

Dousi, Maria. 2022. The iron structural systems of historic industrial buildings, the case of Thessaloniki. In *The Castles of Industry. Restoration, Reconstruction, Utilization*, 14a-14f. Thessaloniki: IANOS.

Kolonas, Vasilis, Papamattheaki, Lena. 1980. *The architect Vitaliano Poseli. His oeuvre in the 19th century*. Thessaloniki, Thessaloniki: Paratiritis.

Kolonas, Vasilis, Traganou Deligianni, Olga. 1987. *Principles of industry in Thessaloniki 1870–1912*. Thessaloniki: Hellenic Bank for Industrial Development.

Kolonas, Vasilis, Traganou Deligianni, Olga. 1989. *Thessaloniki 1912–1940, Industry and city*. Thessaloniki: Hellenic Bank for Industrial Development.

Pinard, Jacques. 1985. *L'archéologie industrielle*. Paris: Presses universitaires de France.

Powell, Keneth. 1999. *Architecture Reborn. The Conversion and Reconstruction of Old Buildings*. London: Laurence King Publishing.

Stratton, Michael. 2000. *Lndustrial Buildings, Conservation and Regeneration*. London: Taylor & Francis.

The International Committee for the Conservation of the Industrial Heritage (TICCIH). TICCIH operating rules and agreements. https://ticcih.org/about/about-ticcih/ Accessed 26 June 2024

The Nizhny Tagil Charter For The Industrial Heritage / July, 2003. https://ticcih.org/wp-content/uploads/2013/04/NTagilCharter.pdf. Accessed 26 June 2024

Chapter 2
Modernist Rural Landscape as Heritage: Challenging Value-Based and Multiscale Approach within Design Studio

Aleksandra Milovanović, Ana Zorić, Tamara Popović, Jelena Ristić Trajković, and Ana Nikezić

Keywords Value matrix · Design indicators · Education · Landscape as heritage

Introduction

Architectural education and practice are in an era of intense urban transformation in which two-folded research challenge is posed: (1) the construct of cultural heritage becomes one of the leading backbones of the problem-based approaches and analytical design strategies for problem-solving through design process, and (2) the construct of built heritage manifests an increasingly complex diversification framework within which the spectrum of different types of heritage manifests itself in relation to scale, contextual factors and ideological implications. Accordingly, the question of 'RE' design strategies, inheritance and cultural heritage plays an important role in the academic discussion on sustainability and the collective built heritage that represents the cultural product of the contemporary era. Following these contextual features within architectural schools, *EAAE Charter on Architectural Research* (EAAE 2018) highlights architecture both as a discipline and as a field of knowledge where connection with other disciplines and education is essential for the continual advancement of the discipline and understanding of inherited values of space, as well as 'heritage of the future'. Moreover, *UNESCO/UIA Charter for Architectural Education* (UNESCO/UIA 2017) recognizes that architectural heritage education is essential to (1) understanding sustainability, social context and sense of place, and (2) transforming the professional architectural mentality; and frames 'understanding

A. Milovanović (✉) · A. Zorić · T. Popović · J. Ristić Trajković · A. Nikezić
Faculty of Architecture, University of Belgrade, Belgrade, Serbia
e-mail: alekmil@arh.bg.ac.rs

M. Loren-Méndez et al. (eds.), *Sustainable Heritage Management - Design: Notions, Methods and Techniques*, https://doi.org/10.1007/978-3-031-81970-4_2

of heritage issues in the built environment' as objective of architectural education within the obligatory knowledge of cultural and artistic studies.

In the sense of contemporary urban paradigm, *New European Bauhaus (*NEB) foregrounds the two value-based framework dominantly supported by social dimension (European Commission 2020b): (1) line of respect—which stimulate the awareness of cultural heritage and nature, as well as respect for the environment, and (2) line of justice—which stimulate uniting values such as equality in education, collaboration and co-creation. Both of NEB frameworks indicate the need for the value-based and -oriented research approaches which could acknowledge the diverse and mutable cultural dimensions of built heritage that extend beyond aesthetic values and materiality. The notion of 'heritage value' is also addressed in Leeuwarden Declaration (ACE 2018) which recognizes the possibility for heritage sites to find new, mixed or extended uses and further contribute to increase of their social, environmental and economic value, as well as enhancement of cultural features. This research thesis correlates to the concept of *Baukultur* (European Ministers of Culture 2018) which encompasses the built environment from a multiscale perspective in order to understand the wider context of heritage entities and to fully consider their integration into urban environment and natural landscape (ACE 2018).

The chapter specific objective is to establish a value-based and multiscale matrix for heritage anticipation through (1) the development of the design matrix for evaluation which corresponds to the value-based architectural programming model, and (2) introducing modernist rural landscape as heritage to demonstrate anticipation of design values on multiscale manner. Accordingly, two research questions arise: (1) what are the key design values related to modernist rural landscape rehabilitation, and (2) how conceptual models operate in line with the different spatial levels and associated design values?

Landscape as Heritage: Need for Holistic Approaches

Following Leeuwarden Declaration (ACE 2018) heritage sites could 'create new dynamics and thereby contribute to the regeneration of the urban fabric' if a reflexive dialogue between past, present and future is activated, as well as if multiscale and territorial approaches are considered. This thesis implies the discussion on heritage construction on the level of a wider spatial system which integrates different urban patterns in hierarchical order such as landscape. Within this scope, 'value-based' principles and 'cultural landscape' initiated the establishment of a new paradigm for cultural heritage. According to Kryder-Reid (2020), 'heritage landscape' brings together two concepts both influenced by cultural construction: (1) landscape as a 'verb' rather than as a 'noun', and (2) landscape as a product of the shared values and beliefs. This twofold definition indicates the dynamic nature of landscape outside the physical and material aspects, as well as ideological construction of landscape introducing the notion of collective and cultural memory in heritage studies.

Considering the landscape's dynamic nature, reflected in the continuous population growth and the corresponding land use change; the protection and sustainability of environmental values stand out as one of the fundamental issues in the landscape design and planning process. Numerous authors indicate the importance of sustainable approaches to landscape design and planning (Haines-Young 2000, Antrop 2007, Wu 2008, Musacchio 2009, Milovanovic et al. 2020), but a set of comprehensive and operationally defined guidelines for landscape conservation has yet to be developed in an upcoming research periodical. The guidelines thus defined should include the landscape's physical, ecological, socio-economic, cultural and political components, with a distinct presence at all levels in time and space. The current research decade singles out sustainability as a leading topic in the academic dialogue, but there are still no comprehensive and operationally defined guidelines for sustainable landscape design and planning. Step forward to address this research question is realized by Musacchio (2009) who established the scientific basis for the design of landscape sustainability based on the six Es of landscape sustainability (environment, economic, equity, aesthetics, experience and ethics) in a form of conceptual framework for translational landscape research and practice of designed landscapes.

Methods and Materials

Having in mind that research operates within the platform of the disciplinary framework of architecture and, in particular, the field of architectural design, the general methodological setup combines research by design in the first phase and qualitative data analysis as an argumentative systematization in the second phase. Accordingly, the first phase is focused on exploring models for reprogramming of landscape, while the second is focused on explaining models and speculating on design values and indicators in line with the notion of landscape sustainability.

First Phase—Exploring Models within the Design Studio

This part of the research was conducted in the context of a studio-based environment in which different concepts were explored within a spatial framework of modernist rural landscape, further explained within the Case Study section. The research was challenged within the Study module—Design studio on 'heritage reprogramming' at the University of Belgrade—Faculty of Architecture involving 14 students (1st year of Master Academic Studies in Architecture) from February to May 2022.

The modernist rural landscape was defined as a case study having in mind the connection between modernist architectural doctrine and value-based architectural programming approach. Modernist heritage enabled comparative research of design values at different spatial levels—from the single spatial unit to the community and regional level, both indoors and outdoors (Ristić Trajković et al. 2021). Additionally, such a choice of research/design area implied research in the local context as an essential aspect of a sustainable approach to heritage rehabilitation. The modernist rural landscape (MRL) and related concepts (modern ruralization, modernist agricultural landscape, traditional agricultural landscape) are at the forefront for a global study on its role in the urban development framework, as well as for consideration of MRL potential for urban–rural revitalization and its benefits for heritage conservation. The value of these concepts in current studies is identified in accordance with the conceptual framework of cultural heritage and cultural landscape, as well as historical urban landscape (HUL), specifically heritage landscape and shared heritage (Bereskin 2019). In ideological terms, MRL is the result of (Djokic et al. 2020): (1) multiple contextual impact (including social, economic, demographic and technological aspects) which resulted in intensive transformation and collectivization of agricultural landscapes and landscape patterns in the socialist era over the Europe, and (2) the impact of the modern movement on the rural landscape, which began to unfold as early as in the mid-nineteenth century through experiments in utopian socialism and radical state reforms.

The spatial framework for design is situated in Third Belgrade, agrarian city located on the Danube's left riverbank, characterized by a specific landscape configuration and an urban morphology in line with the development of the PKB Agricultural Combine. It was predominantly intended for agriculture with a polycentric system of seven small-scale industrial settlements. The observatory framework covers an area of 37,000 ha–2110 ha of built environment and 34,890 ha of PKB's agricultural area.

The exploratory phase was tested through three different steps and in three different scales:

- Typo-morphological exploration that tests the relation between metropolitan area and MRL;
- Design brief—setting up design perspectives that test the relation between everyday life and architecture;
- Conceptual modelling that tests various spatial articulation and ambient sequence.

This research phase resulted in six conceptual models which differ according to design perspective, desired design outcome, as well as general notions considered as a research problem or potential for landscape rehabilitation through design (Table 2.1).

Second Phase—Explaining Models Through Two-Track Analysis

This part of the research was conducted as a qualitative data analysis based on 6 previously developed conceptual models. This research phase is comparative by

Table 2.1 Conceptual models

Conceptual model	Design perspective	Desired design outcome	Key words/notions
1	Circular economy	Redeveloping of local and regional identity through green blue infrastructure	–Natural landscape and resources –Heritage in harmony with nature –Reinterpretation of traditional industry (Industry 5.0)
2	Urban commons	Developing common spaces for shared functions through designing for adaptability	–Active social engagement –Creating a new culture –Prefabrication and modularity
3	Collective memory	Distinctiveness and sense of place through cultural and collective memory	–Place of cultural expression –Presentation of heritage urban –Exploration and cultural practice
4	Urban mobility	Establishing functional diversity through community-based approaches	–System of social interactions –Flexibility and transformability –Adaptive reuse
5	Leisure time	Promoting sustainable lifestyles through nature-based solutions	–Sequences of space –Living and leisure –Tourist potentials of heritage
6	Urban recreation	Stimulating environmental responsivity through cultural routes and itineraries	–Interaction of citizens with nature –Natural environments –Urban large-scale agriculture

nature as it recognizes, combines, systematizes and synthetizes individual proposals creating a unique and universal agenda for the rehabilitation of MRL.

The explanatory phase was performed through three steps:

- Development of Heritage Value Matrix (HVM) through recognition of key values related to the notion of landscape sustainability;
- Identification of role models and their analysis by pinpointing six design proposals and qualitative data analysis based on the recognized values;
- Performing of value-based and multiscale analysis in line with the proposed conceptual design framework.

Developing Heritage Value Matrix (HVM) for the Design of Landscape Sustainability

The HVM is recognized as a growing analytical tool in the study of architectural and urban heritage, which finds its methodological roots within the value-based approach of architectural programming (Hersherberg 2015; Milovanovic 2022). At the general level, HVM is a cross-cutting system in which two axes intersect: (1) problem-based approach in order to identify the scope that needs to be critically engaged through design process (problem-solving), and (2) values that decode the design indicators. The primary material for the development of the matrix includes two sources: (1) prominent umbrella frameworks related to the contemporary urban development paradigm established by European Commission, and (2) the Statements for Teaching through Design for Sustainability of the Built Environment and Heritage Awareness developed within the Erasmus + project HERSUS (IO3).

The first group of sources engages three prominent documents conceptualised as an integral part of the implementation strategy of the United Nations' 2030 Agenda (UN General Assembly 2015) and the sustainable development goals including (1) European Green Deal promoting commitment to tackling climate and environmental-related challenges (European Commission 2020a), (2) Renovation Wave for Europe which is an essential instrument for Europe to become climate neutral by 2050 (European Commission 2020c), and (3) New European Bauhaus expressing the EU's ambition of creating 'beautiful, sustainable, and inclusive places, products and ways of living' (European Commission 2020b). The study analysed the content of Communications from the Commission to the European Parliament, the Council, the European Economic and Social Committee and the Committee of the Regions—for each of the three mentioned umbrella frameworks. Special reference is focused on identifying basic design strategies, principles and indicators for achieving the goals of each framework.

The second group of sources is related to the HERSUS IO3 (Djokic et al. 2021), perceived as Vademecum on heritage and sustainability and reflected in a dual perspective on establishing statements on: (1) relevant Notions, Heritage Types, Design Approaches, Design Actions and Tools, and (2) their importance in education domain. Having in mind that HVM tends to systematize the scope of design indicators, IO3 content analysis was focused on following groups of terms (Djordjevic et al. 2022): (1) Notions (9 terms) as creations of discursive practices of heritage and sustainability, (2) Design Approaches (17 terms) as the analytical and problem-based approaches in the design process applied to treat and preserve a particular category, and (3) Design Actions (15 terms) as the processes of performing purposeful activities with the precise aim and values that are set. For each term, a general definition was analysed, as well as a pedagogical framework.

The first step in the development of the HVM for two-track qualitative analysis involved (1) content analysis of both groups of sources in order to derive the leading design indicators related to the heritage reprogramming and (2) clustering design indicators according to the value-based architectural programming methodology. The

HVM is structured in line with the six groups of design values derived from the scientific basis for the design of landscape sustainability (Musacchio 2009) —conceptual framework consists of the six Es of landscape sustainability—environment, economic, equity, aesthetics, experience and ethics.

Two-Track Qualitative Analysis: Value-Based and Multiscale Approach

Design proposals data sets that illustrate conceptual models for heritage reprogramming represent a central framework for qualitative assessment. In order to establish a comprehensive analysis of the proposed conceptual models, it was conducted in two main cross-cutting tracks—value track and multiscale track.

The value track is performed according to the framework of HVM. For each of the six conceptual models, a critical analysis of design values was conducted. It was performed through identifying their relevance/irrelevance for the perspective development or identifying a particular value as a concept generator for the design process.

The multiscale track is based on the relevant spatial levels for specific typo-morphological framework of engaged case study: single spatial unit / activity setting, building, community, settlement and (5) region/landscape. For each of the 6 proposed conceptual models, a critical analysis of spatial levels was conducted by identifying their influence/neutrality or identifying a particular spatial focus for concept development.

Findings and Discussion

This section will be structured in three parts. First, the heritage landscape value matrix will be elaborated in line with the 6 Es of landscape sustainability. Second, conceptual models will be discussed in a comparative manner through two-track analysis. Third, conceptual models for rehabilitation of MRL will be elaborated.

Heritage Landscape Value Matrix–Elaboration

Heritage Landscape Value Matrix is developed following the scientific basis for the design of landscape sustainability and analysed sources (EU umbrella frameworks and HERSUS IO3) (Fig. 2.1).

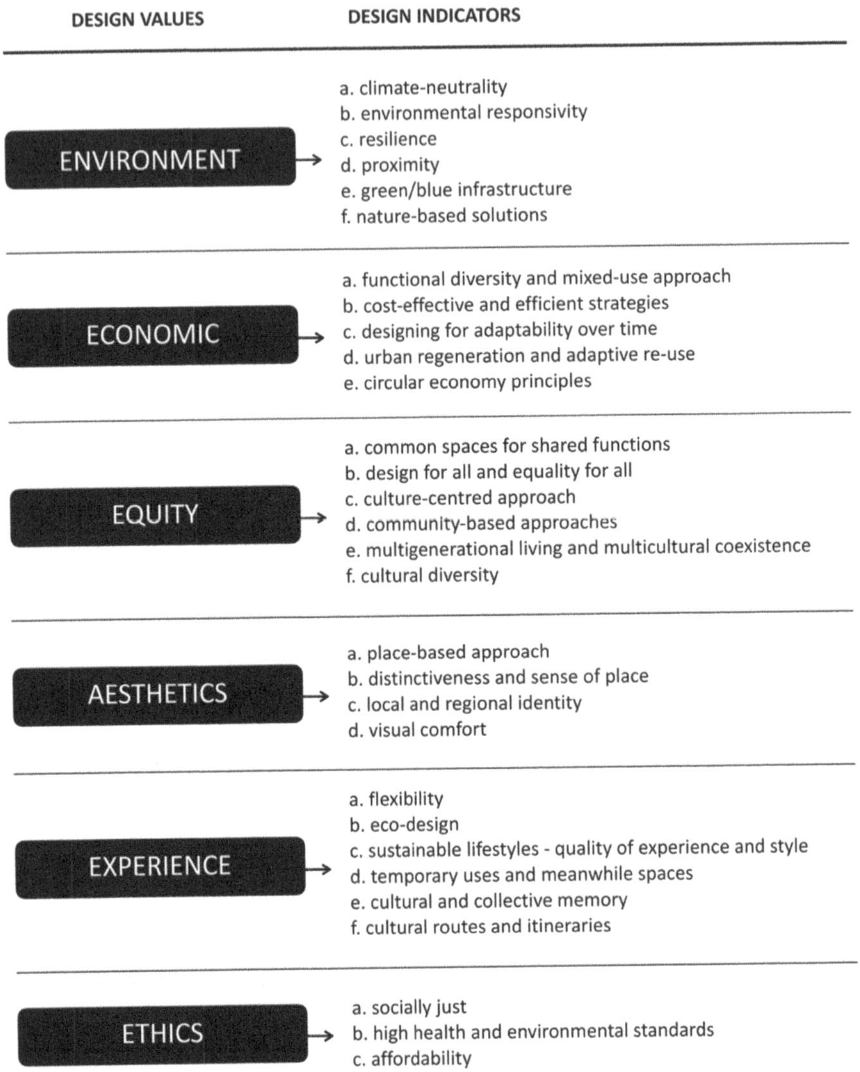

Fig. 2.1 HVM developed in line with the scientific basis for the design of landscape sustainability and analysed sources

Within the group of environmental design values, 6 design indicators were identified as follows: (1) notion—resilience and proximity, (2) design approaches—environmentally responsive design (indicator of environmental responsivity), climate sensitive design (indicator of climate neutrality), and green blue infrastructure as approach and (3) design actions—nature-based solutions (NBS). Within the analysis of both groups of sources, the highest level of concordance of environmental values was recognized, which is the result of the paradigms of the European Green Agreement, which are based on environmental values. Design approaches derived within this group of values include several common characteristics/ design principles: (1) adapting to the local climate and geomorphology, (2) utilizing energy and related environmental resources, (3) understanding of specific climate conditions and its impacts on design and (4) affirmation of natural and semi-natural entities designed to deliver a wide range of ecosystem services. However, these approaches differ according to the spatial scale of application—the focus of environmentally responsive design and climate sensitive design is focused predominantly on the object level, and construction detailing, while the approach aimed at creating green blue infrastructure focuses on the urban and regional planning scale and landscape scale. Finally, NBS are recognized and involved within the matrix as the concepts that bring nature into cities and those that are derived from nature.

Within the group of economic design values, 5 design indicators were identified as follows: (1) design actions—cost effective strategies, adaptive reuse and circular economy principles, and (2) design approaches—design for functional diversity, and design for adaptivity. The analysed sources show that the mentioned economic values are in close correlation with environmental indicators as a response to climate challenges and energy efficiency. On these bases, cost effective indicators, adaptive reuse and circular economy principles are perceived within the scope of the whole life cycle design that looks at environmental sustainability in a holistic manner.

Within the equity design values group, 6 design indicators were identified as follows: (1) notion—cultural diversity, and multigenerational living, and (2) design approaches—design for all, design for shared functions, culture-centred approach and community-based approach. The recognized indicators are predominantly derived from the framework of the NEB, whose vision is built around the core values of sustainability, aesthetics and inclusion. On these grounds, a culture led approach is affirmed as an important strategy for the achievement of the regenerative environment, and for creating a sense of shared responsibility within the community.

Within the group of aesthetic design values, 4 design indicators were identified as follows: (1) notion—distinctiveness and sense of place, and local and regional identity (cultural identity), and (2) design approaches—place-based approach, and visual comfort. This group of values, similar to equity values, has its basis in the determinants of the NEB initiative and the Baukultur concept, which affirms the perception of the total visual milieu. These indicators contribute to the establishment of a reflective dialogue between heritage and contemporary architecture.

Within the experiential design values group, 6 design indicators were identified: (1) notion—sustainable lifestyles, and cultural and collective memory; (2) design approaches—design for flexibility, and eco design; and (3) design actions—temporary uses and meanwhile spaces, and cultural routes and itineraries. This group of values contains the heterogeneous nature of indicators, which correlate with other groups of values and indicators, which confirms that architectural experience and experience in the perception and understanding of heritage is built on multiple benefits of cultural, social, environmental, and economic aspects.

Within the group of ethics design values, 3 indicators were identified as follows: (1) design actions—design for social justice and design for affordability, and (2) design approaches—consideration of high health and environmental standards. This group of values is the least developed and represents a direct normative repercussion of environmental and equity values. The indicator of social justice is correlated with the approaches designed for all, and equality for all, as well as the notion of multigenerational living and cultural diversity, while health and environmental standards and affordability represent the articulation of environmental indicators into economic value.

Explaining Models Through Two-Track Analysis

Value-Based Analysis Track

Observing the global distribution of indicators in all concepts, the value-based analysis track showed that conceptual design proposals are mainly based on an aesthetic approach, but that there are also dominant other indicators. The research shows that based on a qualitative analysis of the design indicators distribution according to whether they are particular concept generators, relevant or irrelevant, the centre of gravity of the solution is to the greatest extent on aesthetic values, which is completely natural for architectural design, because it is at the core of the formation of architectural quality and visibility, and that all other groups of values and their associated indicators indicated the necessity to integrally observe different spatial aspects and their effects on achieving landscape sustainability.

Observing design indicators distribution by their value group (Table 2.2), it was recognized that environmental values are dominantly based on the identification of natural resources and their designation as heritage. It was recognized that the MRL contains exceptional natural features conditioned by agriculture as an industry that shapes that landscape. Hence, the economic group of values tends to operationalize natural resources and enable circular flows that contribute to the adaptability of space, and affirm economic principles. Additionally, it was recognized that the design indicator of functional diversity within the economic values has a dominantly indirect role in the design process, meaning that it arises as the design process's final outcome, and not as a primary goal. The group of equity values is dominantly based on the sociological scale of space to integrate culture-centred approach and community-based approaches. In the context of the MRL, the influence of agricultural production and collectivism, i.e. the collective movement, on the creation of specific urban patterns is particularly specific. The experience design values group is manifested to the greatest extent in accordance with the ideological layers of the MRL, within which the issue of cultural and collective memory is affirmed, as well as contemporary reinterpretation of intangible heritage values, which can be translated into new sustainable lifestyles. In this sense, these design indicators within experience design values are considered through a directed approach scenario that tries to establish development phases or some usage scenarios—cultural routes, itineraries. The group of ethics values in this case is manifested as a direct repercussion of other groups of values in accordance with the design thesis to establish adequate standards of space and equality from a social perspective.

Table 2.2 Identification of relevant design values/indicators for conceptual models

Design values	Design indicators	1	2	3	4	5	6
1. Environment	a	+	+	x	x	+	+
	b	•	+	+	+	•	•
	c	•	+	x	x	+	•
	d	x	•	•	•	•	•
	e	•	+	x	x	•	•
	f	•	•	x	x	•	•
2. Economic	a	x	+	+	•	+	+
	b	+	+	x	+	x	+
	c	+	•	x	•	+	•
	d	x	+	•	•	x	x
	e	•	+	x	+	x	•
3. Equity	a	x	•	•	•	•	•
	b	x	•	+	+	+	X
	c	x	•	•	•	+	+
	d	+	•	•	•	•	+
	e	x	•	x	•	+	x
	f	x	+	+	+	+	x
4. Aesthetics	a	•	•	•	•	•	•
	b	•	•	•	•	•	•
	c	•	•	•	•	•	•
	d	+	+	•	•	•	+
5. Experience	a	•	+	+	•	+	•
	b	•	+	x	x	•	+
	c	+	+	•	•	•	•
	d	x	x	+	x	•	+
	e	+	•	•	+	•	•
	f	•	+	•	•	•	•
6. Ethics	a	x	+	+	•	+	•
	b	•	+	+	x	•	•
	c	x	+	x	•	+	+

Index: x—Irrelevant; + —Relevant; •—Particular concept generator

Table 2.3 Identification of relevant spatial levels for conceptual models

Spatial level	Conceptual model					
	1	2	3	4	5	6
Activity setting	x	x	+	x	+	+
Building	x	+	+	+	x	x
Block	x	+	•	+	+	x
Community	+	•	+	•	+	+
Region/landscape	•	+	x	+	•	•

Index: x—neutral; + —influential; •—particular spatial focus

Multiscale Analysis Track

Insight into the multiscale track (Table 2.3)—conducted by identifying the influence/neutrality of particular scale for design development or by identifying a particular spatial focus for concept development—confirms the multiscale approach as a leading prospectus for achieving a reflexive dialogue between heritage and landscape. Findings point out that heritage entities should be understood in their broader context to argumentatively establish a value framework in line with the critical urban environment. From the value-based analysis, it is recognized that the specific framework of conceptual models in line with the spatial levels has completely different value indices, identified conceptual models mainly contain an identical relationship of particular spatial focus with other value factors. The exception is the distribution of the indices for model 2 and model 4, which are nevertheless different in terms of values. This indicates the necessity of parallel consideration of values and their distribution related to different spatial levels within the design process. Particular spatial focus is primarily assigned to region/landscape level followed by community level.

Conceptual Models for Rehabilitation of Modernist Rural Landscape

Particular design values are recognized as concept generators in the case that they have at least 50% of associated design indicators to that group. Accordingly, design approach lines are defined following the hierarchy of influence of a certain design values group on the development of a conceptual solution.

The first conceptual model is based on the design perspective of circular economy, which in the case of MRL is manifested on a region/landscape scale. This approach starts from the identification of natural resources (including hidden natural resources) by placing them at the service of redeveloping local and regional identity. The concept of circular economy in this case represents an indirect economic design indicator and is generated within the scope of environmental design values including environmental responsivity, resilience, green blue infrastructure and NBS. It is a large-scale architecture that represents a system solution that is built in parallel at the level of the whole (landscape) and at the level of individual elements or typologies with a special environmental role. This design perspective in a long-term timeframe has a role in improving ethics design values through the establishment of high health and environmental standards for users on the community level. The conceptual model 1 represents a demonstration of the reinterpretation of the traditional MRL industry, which indicates the potential of Industry 5.0 for its rehabilitation.

The second conceptual model is based on the design perspective of urban commons, which manifests a multiscale approach within the MRL framework—from the level of landscape towards the building. This approach has a double focus: (1) urban commons is considered from the aspect of adaptive reuse of an inactive industrial complex in public property and (2) urban commons is examined in the context of creating common spaces for shared functions. The notion of urban commons is generated within the scope of equity design values through engaging design for all, culture-centred and community-based approaches. It is a middle scale intervention based on the principles of adaptive conversion and designing for adaptability. In a long-term perspective, the design perspective based on urban commons supports cost effective economic strategies, and it encourages cultural and collective memory significant for the community level. Conceptual model 2 represents a demonstration of industrial heritage reuse in MRL and the affirmation of community principles.

The third conceptual model is based on the idea of collective memory as an ideological implication of MRL with a focus on the community level as social scale, and the level of the industrial complex as architectural scale. This approach starts from the identification and interpretation of cultural values as well as personal and collective memory related to the industrial activity. The idea of collective memory has a role in (1) creating a place of cultural expression with the intention of transmitting historical records and (2) presenting the MRL heritage with the intention of stimulating the creation of new cultural practices. This approach is dominantly based on aesthetic design values through a place-based approach and stimulation of the sense of place, identity and visual comfort followed by the indirect influence of experiential design values related to the cultural and collective memory. It is a middle scale intervention based on the principles of urban rehabilitation and the creation of physical extensions within inherited spaces of industrial heritage. In a long-term perspective, this design perspective contributes to the creation of cultural itineraries as a record of historical heritage and a record of the creation of a new future heritage.

The fourth conceptual model is based on the urban mobility paradigm that manifests itself on a larger scale—challenging the relationship of MRL and the wider metropolitan area of the city. This approach starts from the recognition of the underdeveloped infrastructure network within the MRL and the insufficient connectivity of urban patterns, and accordingly insufficiently established social interaction. The issue of urban mobility has a double perspective aimed to: (1) improve the indicator of urban proximity within the environmental group of design values, and (2) encourage the formation of shared spaces for shared functions and community enhancement.

At its core, this approach is based on the principles of adaptability and reuse with a focus on achieving maximum flexibility of space, which would provide a framework for the implementation of various scenarios and programmes. In the long-term perspective, this design perspective contributes to the affirmation of sustainable lifestyles through establishing an architectural framework as a system of social interactions. The conceptual model 4 is a demonstration of community-based approaches in design that focus on ethics, design values of social justice (creating multigenerational living, and space of cultural diversity) and affordability (principles of flexibility and transformability).

The fifth conceptual model is based on the consideration of leisure time viewed in two extreme spatial levels—at the landscape level as a manifestation of cultural features of space, and activity setting as a framework of user behaviour. This approach starts from observing the open public space and trying to form a system of routes and programme sequences. Accordingly, this notion represents a double value design foundation that includes: (1) consideration of the formation of temporary uses and meanwhile spaces based on eco design principles, which contribute to the affirmation of experiential values, and (2) the establishment of an infrastructural system of open public spaces through NBS, which contributes to the affirmation of environmental values. It is an environmentally sensitive architecture that represents a systemic solution created through the design of sequences of spaces and micro-typologies that encourage climate neutrality. In a broader perspective, this design perspective has a role in improving the tourist potentials of heritage, considering that it enables the articulation of infrastructure into a cultural route. The conceptual model 5 demonstrates how sustainable lifestyles are promoted through environmentally friendly architecture.

The sixth conceptual model is based on the thematic framework of urban recreation, which introduces the issue of interaction between citizens and nature at the sociological level of the community and its corresponding spatial level. This approach is based on the identification of agriculture as a generator of MRL and the search for modern formats of urban agriculture in the city. In this sense, the theme of urban recreation transforms the issue of urban agriculture from a dominantly environmental and economic category to an experiential and ethics category. This means that agriculture of resources and total natural environment are engaged in the service of creating cultural routes and itineraries. It is an urban large-scale agritecture that introduces modern production systems and NBS for the revival of the MRL. This design perspective in the long term has a role in stimulating environmental responsibility and signifies the ideological values of the MRL whose generator is agriculture. The conceptual model 6 represents an urban prototype established on a large scale of agricultural landscape that represents both a tourist sensation and an extended daily life of residents.

Concluding Remarks

The concluding remarks were developed according to the initial research questions outlined at the beginning of this chapter: (1) what are the key design values related to MRL rehabilitation, and (2) how conceptual models operate in line with the different spatial levels and associated design values?

The answer to the first question was found in the application of value-based analysis that allowed the identification of the conditionality between the groups of design values and indicators. The conducted research indicated that landscape heritage, and particularly MRL, initiate design indicators in all value groups relevant to the conceptual framework of landscape sustainability. However, the centre of gravity of all conceptual models is in the group of aesthetic values in accordance with the place-based approach, while other groups of values are manifested in the service of creating specific design programmes that contribute to the redistribution or reprogramming of patterns at different spatial levels.

The answer to the second question was found in the application of multiscale analysis that decodes the operational framework of each conceptual model in line with the different spatial levels and associated design values. When considering the index distribution in relation to the spatial levels, it was recognized that it can be defined three approach lines in accordance with scale direction within design process: (1) bottom-up approach—design from activity setting towards landscape level, (2) middle scale approach—design from community level towards two directions—towards activity setting and towards landscape level, and (3) top-down approach—design from landscape level towards single spatial units.

In synthesis, the chapter demonstrated three main research pillars: (1) the importance for the development of the theoretical framework based on soft systematic paradigm that intertwines 'landscape as text' and value-based approaches for the evaluation of heritage; (2) the potential of design studio learning environment for the implementation of complex analytical methodologies for the development of conceptual models for landscape rehabilitation; and (3) the research potential of heritage value matrix for understanding the construct of landscape heritage from multiscale and value-based perspective.

References

ACE Architects' Council of Europe. 2018. *Leeuwarden Declaration—Adaptive Re-Use of the Built Heritage: Preserving and Enhancing the Values of Our Built Heritage for Future Generations*. ACE: Brussels, Belgium.

Antrop, Marc. 2007. Reflecting upon 25 years of landscape ecology. *Landscape Ecology* 22: 1441–1443.

Bereskin, Emily. 2019. Modern rural landscapes in contemporary heritage imaginaries: The case of Germany's Southern Oderbruch. In SHS Web of Conferences, vol. 63, p. 11002. EDP Sciences.

Djokić, Vladan, Aleksandra Milovanović, and Jelena Ristić Trajković. 2020. The textuality of the modernist rural landscape: Belgrade agricultural combine (PKB) as a driver of the urban development of third Belgrade. *Land* 9: 452.

Djokić, V., Nikezić, A., Loren-Méndez, M., Sakantamis, K., Philokyprou, M., Sorbo, E., Péral-López, J. 2021. *Statements: Teaching through Design for Sustainability of the Bult Environment and Heritage Awareness*. Belgrade: University of Belgrade—Faculty of Architecture. Available at: https://hersus.org/statements-for-teaching-through-design-for-sustainability-of-the-built-environment-and-heritage-awareness/. Accessed on 11 February 2023.

Đorđević, A., A. Milovanović, M.P. Milojević, A. Zorić, M. Pešić, J. Ristić Trajković, A. Nikezić, and V. Djokić. 2022. Developing methodological framework for addressing sustainability and heritage in architectural higher education—insights from HERSUS project. *Sustainability* 14 (8): 4597.

European Commission. 2020a. Communication from the Commission to the European Parliament, the Council, the European Economic and Social Committee and the Committee of the Regions, The European Green Deal, COM/2019/640 Final; European Commission: Brussels, Belgium.

European Commission. 2020b. Communication from the Commission to the European Parliament, the Council, the European Economic and Social Committee and the Committee of the Regions, New European Bauhaus, Beautiful, Sustainable, Together, COM/2021/573 Final; European Commission: Brussels, Belgium.

European Commission. 2020c. Communication From the Commission to the European Parliament, the Council, the European Economic and Social Committee and the Committee of the Regions, A Renovation Wave for Europe—Greening our Buildings, Creating Jobs, Improving Lives, COM/ 2020/662 Final; European Commission: Brussels, Belgium.

EAAE European Association for Architectural Education. 2018. *Principles and Practices of Architectural Education: A Position Paper of the EAAE Education Academy*; EAAE: Porto, Portugal.

European Ministers of Culture. 2018. Davos Declaration: Towards a HIGH-quality 'Baukultur' for Europe; Federal Department of Home Affairs, Federal Office of Culture: Davos, Switzerland.

Haines-Young, Roy. 2000. Sustainable development and sustainable landscapes: Defining a new paradigm for landscape ecology. *Fennia* 178: 7–14.

Hershberger, Robert. 2015. *Architectural programming and predesign manager*. Routledge.

Kryder-Reid, Elizabeth. 2020. "Heritage Landscapes". In Encyclopedia of Global Archaeology, pp. 4947–4952. Cham: Springer International Publishing.

Milovanović, Aleksandra, Danijela Milovanović Rodić, and Marija Maruna. 2020. Eighty-year review of the evolution of landscape ecology: From a spatial planning perspective. *Landscape Ecology* 35: 2141–2161.

Milovanović, Aleksandra. 2022. *The Role of the Methodology of Architectural Programming in the Design Process of Space Morphology: Example of Third Belgrade*. Doctoral Disseration. University of Belgrade—Faculty of Architecture.

Musacchio, Laura R. 2009. The scientific basis for the design of landscape sustainability: A conceptual framework for translational landscape research and practice of designed landscapes and the six ES of landscape sustainability. *Landscape Ecology* 24: 993–1013.

Trajković, Ristić, Aleksandra Milovanović Jelena, and Ana Nikezić. 2021. Reprogramming modernist heritage: Enhancing social wellbeing by value-based programming approach in architectural design. *Sustainability* 13: 11111.

UN General Assembly. 2015. Transforming Our World: The 2030 Agenda for Sustainable Development; United Nations: New York, NY, 927 USA.

UNESCO-UIA Validation Council for Architectural Education. 2017. *Charter UNESCO/UIA for Architectural Education*; International Union of Architects: Paris, France.

Wu, Jianguo Jingle. 2008. Making the case for landscape ecology an effective approach to urban sustainability. *Landscape Journal* 27: 41–50.

Chapter 3
Modern Heritage in the Context of UNESCO. Recognition, Geographical Representation and Criteria of European Cultural Heritage in the World Heritage List: France, Germany, Italy and Spain

Mar Loren-Méndez, Roberto F. Alonso-Jiménez, and Daniel Pinzón-Ayala

Abstract The present chapter focuses on Modern Heritage (MH) of the built environment in the context of European cultural heritage: the fact that almost 50% of the World Cultural Heritage within the UNESCO World Heritage List (WHL) is located in Europe—449 declared properties out of 900 worldwide—it makes this continent a proper case study for MH. Being the cultural heritage the most represented in the WHL, the analysis can help understand current and future tendencies, as well as unbalances in the inclusion of MH in the WHL. The study will work on the first four countries of the list with more cultural heritage, all of them Europeans: France, Germany, Italy, and Spain. Firstly, the present research offers an identification and geographical location of MH. It continues with an analysis and reflection of the most represented criteria when applied to MH. Finally, it zooms in the predominant programs and uses commonly valued within MH. It aims to unveil the problem of underrepresentation of MH within WHL, raising awareness among researchers, students, and faculty members of architecture of the need for commitment, but also of the opportunity that Modern Heritage represents.

Keywords Heritage types and categories · Industrial heritage · Global Strategy · Contemporary architecture · Urban heritage

---

M. Loren-Méndez (✉) · R. F. Alonso-Jiménez · D. Pinzón-Ayala
Universidad de Sevilla, Seville, Spain
e-mail: marloren@us.es

M. Loren-Méndez et al. (eds.), *Sustainable Heritage Management - Design: Notions, Methods and Techniques*, https://doi.org/10.1007/978-3-031-81970-4_3

Introduction

The World Heritage List. Its Emergence, Disproportions and Challenges

The search and relevance for heritage and cultural diversity has been in the foundation of the 1972 World Heritage Convention (WHC). It is the World Heritage List (WHL), the main tool to recognize and integrate heritage worldwide. In 1978, the Operational Guidelines for its Implementation (OGI) established the development of the WHL—and the WHL in danger—as the first two out of four functions of the World Heritage Committee. Between 1987 and 1993, UNESCO decided to analyze the list, the heritage types included and their geographical location:

> This study revealed a strong imbalance between the different types of properties, as well as in their geographical distribution, with a high concentration in some areas to the detriment of others. Of the 410 properties inscribed, 304 were cultural properties and only 90 natural and 16 mixed, the vast majority being located in developed regions of the world, particularly in Europe (Ministerio de Cultura y Deportes 2023).

At that moment, the WHL already revealed that the cultural heritage was much more represented than natural or mixed heritage, ascending to 74%. Regarding geographic unbalance, this cultural heritage was mainly concentrated in Europe.

The beginning of the nineties constituted a breakthrough for the WHC and its List: the World Heritage Centre was created by UNESCO responding to the great success of the WHL, and at the same time, a new Executive Committee of the General Assembly of the International Council of Monuments and Sites (ICOMOS) was elected, whose president and general secretary were not Europeans resulting in a "more open approach and the insistence that the contributions of the various regions of the world be taken more into account" (Luxen 2019, p. 195). Nevertheless, UNESCO headquarters were located in Paris since its creation and has remained there ever since, while the United Nations headquarters was moved out of Europe to New York after the Second World War (Fernández and Silva 2018, p. 148).

It is in this context where it emerges the Global Strategy in 1994, which aim is a balanced, representative and credible WHL. The Nara Conference on Authenticity (UNESCO 1994) was key for putting the value of cultural and heritage diversity in the center of the heritage reflection of the WHC, in the context of a more flexible approach to the concept of authenticity in order to open up the WHL. It is also worth mentioning that this conference took place outside Europe.

> The diversity of cultures and heritage in our world is an irreplaceable source of spiritual and intellectual richness for all humankind. The protection and enhancement of cultural and heritage diversity in our world should be actively promoted as an essential aspect of human development (UNESCO 1994, pp. 1-2).

50 years after the WHC and almost 30 after Nara, both the cultural heritage disproportion and the European predominance persist in the WHL: in 2023 cultural heritage accounts for more than 77% of the list. It should be noted that this study

was developed prior to the 45th session of the World Heritage Committee and the new additions to the list in 2023. As a note of caution, it is confirmed that the data presented in this study covers the WHL up to 2022.

UNESCO Modern Heritage Program: Problematic and Opportunities

Together with this geographical balance, UNESCO also initiated a pro-active approach with regard to the identification and documentation of less-represented heritage categories in the WHL. One of these categories is Modern Heritage (MH); in 2001 the World Heritage Centre, together with ICOMOS and DoCoMOMO International started the Modern Heritage Program (MHP)—MH is defined by Francesco Bandarin within the Program as the built heritage of the Modern Era—with the goal of identifying and listing Modern Heritage (Loren-Méndez 2021).

In the context of such program, it was pointed out the vulnerability of MH, due to weak legal protection and low appreciation among the general public and institutions (Van Oers and Haraguchi 2003). As Francesco Bandarin pointed out: "there should be a pro-active approach to identify and document those less-represented categories of heritage, such as Modern Heritage" (Van Oers and Haraguchi 2003, p. 4).

Despite the good intentions of the program, we have to wait until 2017 for the next UNESCO monographic publication, in which the director of the World Heritage Centre, Mechtild Rössler, acknowledges that this heritage type is still underrepresented and points to it as a field of opportunity (Aprile and Doubleday 2017).

The present study aims to unveil that this problem of underrepresentation of MH within UNESCO WHL still persists in 2023, in the 20th Anniversary of the MHP. It also contributes to raising awareness among researchers, students and faculty members of architecture of the need for commitment, but also of the opportunity that modern heritage represents.

In order to do so, the present chapter focuses on MH in the context of European cultural heritage: the fact that almost 50% of the World Cultural Heritage within the WHL is located in Europe—449 declared properties out of 900 worldwide—it makes this continent a proper case study for MH. Being the cultural heritage the most represented in the WHL, the analysis can help understand the present and future tendencies, and unbalancing issues in the inclusion of MH in the WHL.

The study will work on the first four countries of the List with more cultural heritage, all of them Europeans: France, Germany, Italy, and Spain. Firstly, the present research offers an identification and geographical location of MH. It continues with an analysis and reflection of the most represented criteria when applied to MH. Finally, it zooms in the predominant programs and uses commonly valued within MH.

Presence and Absence of Modern Heritage: A Geographical Perspective

In the first place, the study identifies the number of MH properties of each country. As shown in the following Table 3.1, these are the four countries with more declared Cultural heritage in the WHL, all belonging to Europe and, therefore, with a significant weight in terms of MH properties. The Table 3.2 offers a summary of modern heritage properties of these four countries: Germany, France, Italy and Spain.

In these four countries, transboundary heritages have been documented which, in some cases, are present in some of them at the same time. This causes a clear distortion in the numerical calculations, as it even affects countries outside Europe. For this reason, it has been decided not to include them in this study.

A representative example of this data analysis limitations regarding the transboundary heritages, is The Great Spa Towns of Europe. This transnational serial property comprises eleven spa towns, located in seven European countries: Baden bei Wien (Austria); Spa (Belgium); Františkovy Lázně; Karlovy Vary; Mariánské Lázně (Czechia); Vichy (France); Bad Ems; Baden-Baden; Bad Kissingen (Germany); Montecatini Terme (Italy); and City of Bath (United Kingdom). That international European spa culture developed from the early eighteenth century to the 1930s, and it exemplifies the democratization of the modern urban programs, responding to the social accessibility to hygiene and leisure.

One of the most obvious limitations establishing the selection criteria is due to the complexity of sorting out in certain items the weight and relevance of the contribution of the nineteenth and twentieth centuries. Some of them belong entirely to the nineteenth or/and twentieth century. However, others have been built throughout different periods of time; in these cases, the Evaluation and Declaration documents have been reviewed in order to confirm their relevance during the nineteenth or/and twentieth centuries. In them, we can highlight issues linked to the modern attitude when intervening in historical legacies, contributing not only to their conservation and continuity over time, but also to incorporating new interpretations. In this sense, we can point out the German property of Wartburg Castle:

> Wartburg Castle blends superbly into its forest surroundings and is in many ways "the ideal castle". Although it has retained some original sections from the feudal period, the form it acquired during the nineteenth century reconstitution gives a good idea of what this fortress might have been at the height of its military and seigneurial power. It was during his exile at Wartburg Castle that Martin Luther translated the New Testament into German (World Heritage Convention 1999a).

Table 3.1 Cultural heritage properties in the WHL in the world top 4

State name	Italy	Spain	Germany	France
MH national	**3**	**4**	**9**	**8**
MH transboundary	2	1	4	2
National	**48**	**41**	**40**	**39**
Transboundary	5	2	8	4

Table 3.2 Summary of modern heritage properties in Germany, France, Italy and Spain

Property denomination	Date	Criteria	State name
From the Great Saltworks of Salins-les-Bains to the Royal Saltworks of Arc-et-Senans, the Production of Open-pan Salt	1982	(i)(ii)(iv)	France
Works of Antoni Gaudí	1984	(i)(ii)(iv)	Spain
Strasbourg, Grande-Île and Neustadt	1988	(ii)(iv)	France
Paris, Banks of the Seine	1991	(i)(ii)(iv)	France
Völklingen Ironworks	1994	(ii)(iv)	Germany
Crespi d'Adda	1995	(iv)(v)	Italy
Bauhaus and its Sites in Weimar, Dessau and Bernau	1996	(ii)(iv)(vi)	Germany
Palau de la Música Catalana and Hospital de Sant Pau, Barcelona	1997	(i)(ii)(iv)	Spain
Historic Fortified City of Carcassonne	1997	(ii)(iv)	France
Museumsinsel (Museum Island), Berlin	1999	(ii)(iv)	Germany
Wartburg Castle	1999	(iii)(vi)	Germany
Zollverein Coal Mine Industrial Complex in Essen	2001	(ii)(iii)	Germany
Le Havre, the City Rebuilt by Auguste Perret	2005	(ii)(iv)	France
Vizcaya Bridge	2006	(i)(ii)	Spain
Berlin Modernism Housing Estates	2008	(ii)(iv)	Germany
Fagus Factory in Alfeld	2011	(ii)(iv)	Germany
Nord-Pas de Calais Mining Basin	2012	(ii)(iv)(vi)	France
Champagne Hillsides, Houses and Cellars	2015	(iii)(iv)(vi)	France
Speicherstadt and Kontorhaus District with Chilehaus	2015	(iv)	Germany
Ivrea, industrial city of the twentieth century	2018	(iv)	Italy
Le Colline del Prosecco di Conegliano e Valdobbiadene	2019	(v)	Italy
Nice, Winter Resort Town of the Riviera	2021	(ii)	France
Mathildenhöhe Darmstadt	2021	(ii)(iv)	Germany
Paseo del Prado and Buen Retiro, a landscape of Arts and Sciences	2021	(ii)(iv)(vi)	Spain

As can be seen in Fig. 3.1, although the four countries have relatively similar figures in terms of declared cultural heritage, there is an imbalance when we look at declared MH. Thus, the positions are reversed, with Germany and France coming first in the ranking, leaving Spain and Italy, whose combined declarations do not even reach those of France, at the bottom of the list.

The evolution of the declarations in these four countries alone shows that those affecting MH have remained constant over the last forty years, with no significant alterations—acceleration or deceleration—detected (Fig. 3.2). Although the declarations affecting cultural heritage have experienced periods of upturn (1996/1997), as well as downturns (2001/2003), in no case have they had a direct consequence on

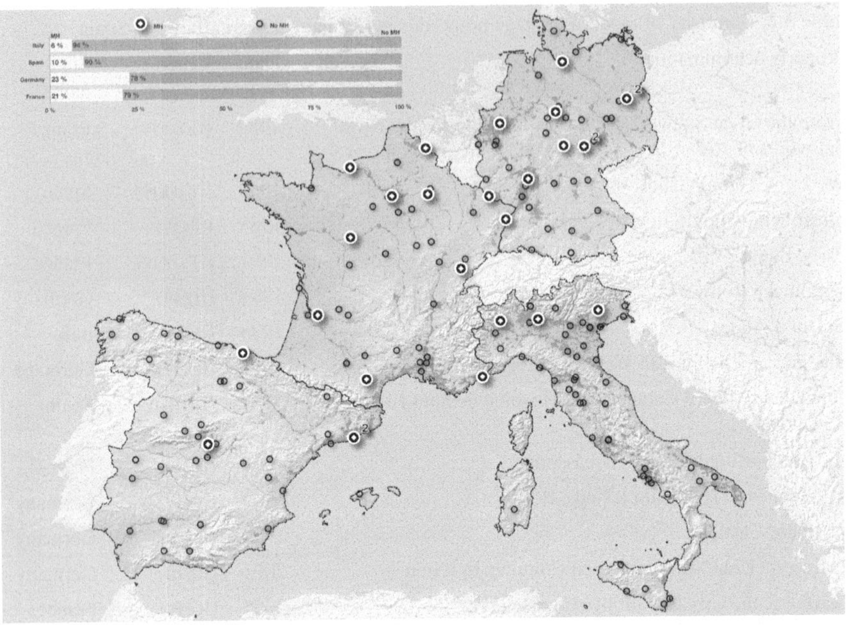

Fig. 3.1 Map with location of cultural properties in the World Heritage List of Italy, Spain, Germany, and France, highlighting the Modern Heritage. Diagram with proportion of Modern Heritage per state. *Source* Produced by authors based on UNESCO World Heritage List information

MH declarations, which very rarely exceed two declarations in the same year. The separation of the lines representing CH and MH declarations in these four countries has increased over the years, with no signs of significant changes in this situation in the short term.

Criteria of Modern Heritage in World Heritage List

Within the diversity of scales, situations and uses of the properties included on the World Heritage List, the justification and evaluation of all of them is based on the six cultural heritage criteria established by UNESCO. This framework determined by the international organization allows for an approach based on comparable issues in the different properties. In order to detect the main aspects identified with the valuation of MH properties, we proceed to analyze the inclusion and representativeness of the different criteria in those properties considered as MH within the four countries that form part of the study (Fig. 3.3).

Among the different criteria, it is evident that criteria (ii) and (iv) are the most represented within these properties. These two criteria are linked to the exchange capacity and explicitly mentioned three scales: architecture, urban planning and

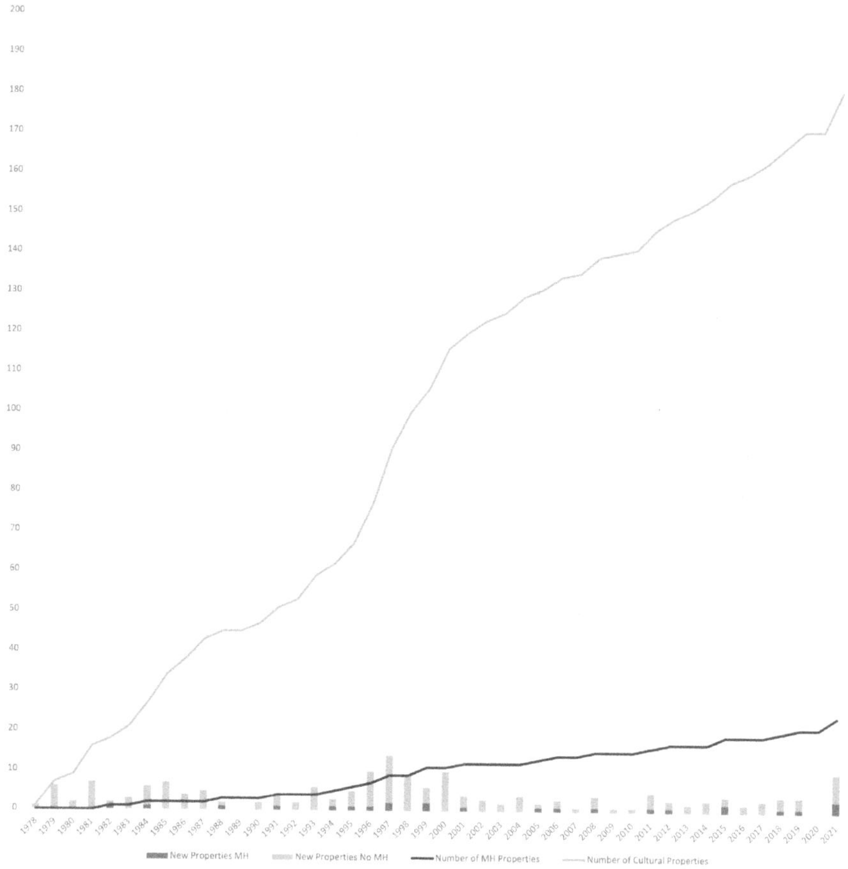

Fig. 3.2 Evolution of cultural heritage included on the World Heritage List in France, Germany, Italy and Spain. Inclusion of cultural modern heritage properties and comparison with cultural properties in these countries. *Source* Produced by authors based on UNESCO World Heritage List information

technology as essential parts, with a specific reference in both criteria to their technological contribution.

Criterion (ii): "to exhibit an important interchange of human values, over a span of time or within a cultural area of the world, on developments in architecture or technology, monumental arts, town-planning or landscape design" (World Heritage Committee 2021).

Criterion (iv): "to be an outstanding example of a type of building, architectural or technological ensemble or landscape which illustrates (a) significant stage(s) in human history" (World Heritage Committee 2021).

While all six cultural criteria are represented within the sample of the 26 properties considered MH, those properties linked to at least one of the criteria (ii) or (iv) account for more than 91% (22/24). Two groups can be identified: criteria (ii) and

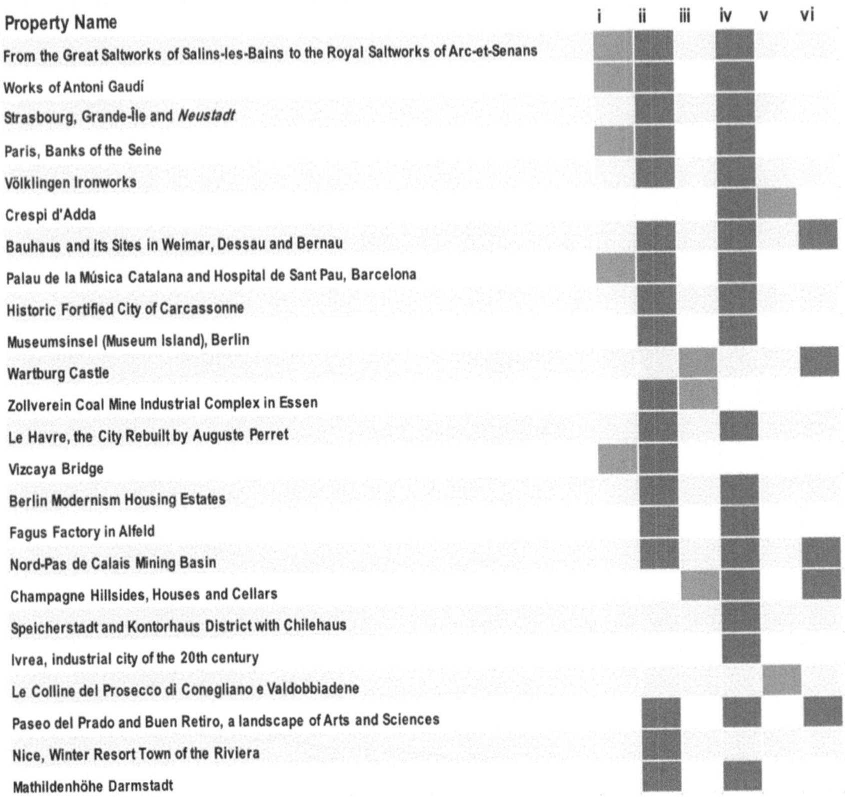

Fig. 3.3 Diagram of criteria under which the inscription of properties is justified. *Source* Produced by authors based on UNESCO World Heritage List information

(iv) as predominant ones, and criteria (i), (iii), (v) and (vi) as less-represented criteria with 5, 3, 2 and 5 properties linked, respectively.

Regarding the impact of criteria (ii) and (iv), 20 and 21 properties, respectively, the number of properties that present this pair of criteria jointly (17) stands out, which represents 65% of the total. Of these 17 properties, in 11 of them this pair is the only two criteria justified in the declaration. An example of this situation can be the justification of the criteria of Strasbourg, Grande-Île and Neustadt:

> Criterion (ii): French and Germanic influences have shaped the Grande-Île and Neustadt. They have enabled the emergence of a unique expression coming from the two cultures, which is especially conveyed in the fields of architecture and urbanism. […]. The Neustadt, a modern city forged by Haussmannian influences, and a model of urbanism, also embodies the theories of Camillo Sitte
>
> Criterion (iv): […] From 1871 onwards, the face of the town was profoundly modified by the construction of an ambitious urbanistic project, leading to the emergence of a modern, functional city, emblematic of the technical advances and hygienist policies that were emerging at the turn of the 19th and 20th centuries (World Heritage Committee 2017).

Studying the criteria under which the inscription of each property is proposed, we can observe that recurrent issues appear: improve social condition, development of activities prolonged in time, typological evolution, technological innovation, urban values, reconstruction; all of these linked to different fields that stand out for their evolution in the nineteenth and twentieth centuries. Among the justifications, the consideration as outstanding examples and the synthesis, combination and exchange between generations and cultures, stand out as main topics mostly linked to architecture and urban design.

From the study developed, a connection of MH properties with its materiality, whose main components are architecture, technology, urbanism and landscape can be glimpsed. These issues are particularly represented in relation to the evolution of the different sectors of society and especially to industrial development. With the outbreak of the second Industrial Revolution and the French Revolution, the complexity of a reality characterized by change meant the diversification of the images and uses that make up modernity. The application and appearance of new ideas, materials and morphologies, ends up shaping a material support and an identity linked to what we can now consider modern heritage.

Programs and Uses of Modern Heritage in World Heritage List

The predominant programs analysis of MH in the WHL is also key to understand the uses identified and commonly valued within MH.

In the first place, the modern intervention to cultural heritage is already valued and considered of Outstanding Universal Value within certain declarations. Together with the Wartburg Castle referred above, the Historic Fortified City of Carcassonne, an outstanding example of a medieval fortified city, "…is also of exceptional importance because of the lengthy restoration campaign undertaken by Viollet-le-Duc, one of the founders of the modern science of conservation." (World Heritage Convention 1997). These Declarations, while valuing the modern approach to heritage, are still focused on medieval architecture and city.

Industrial heritage is indeed the most predominant one, with 50% out of the 24 MH—Germany 5 out of 9 (55%); France 3 out of 8 (37,5%); Italy 3 out of 3 (100%); Spain 1 out of 4 (25%)—identified in these four countries. It comprises different uses and scales, which range from elements such a bridge; to urban ports or industrial towns; and also territorial extractive complexes or rural landscapes.

It is very symptomatic that in Spain the Vizcaya Bridge constitutes the only property we can identify with industrial heritage (Fig. 3.4), and there is no National MH directly valuing factories, industrial towns or productive landscapes, either agricultural or extractive. In contrast, the first MH declaration in Spain of 1984 includes seven buildings by Antoni Gaudí in Barcelona, Spain (Fig. 3.5). Although such declaration is valuing the technological innovation of his work and its subsequent influence, it still predominates the genius of the author, and above all the artistic and creative expression of progress. This declaration finally protects unique and unrepeatable works, which are referred to as "monuments" in the Declaration, and not so much as examples as technological, social and political processes and changes, within the definition of Modernization by Marshall Berman (1982) (Conti 2017, p. 8).

> Seven properties built by the architect Antoni Gaudí (1852–1926) in or near Barcelona testify to Gaudí's exceptional creative contribution to the development of architecture and building technology in the late 19th and early 20th centuries. These monuments represent an eclectic, as well as a very personal, style which was given free reign in the design of gardens, sculpture and all decorative arts, as well as architecture (World Heritage Convention 1984).

It is symptomatic that the declaration includes an industrial settlement such as the Colonia Güell, of which it protects its crypt as a Gaudian work, but nothing else is said regarding the industrial program. The 1997 Declaration of two other works of Catalan Modernism—the Palau de la Música Catalana and the Hospital de Sant Pau by Domenech i Montaner—recoccurs in such artistic expression and unveils a national aspiration.

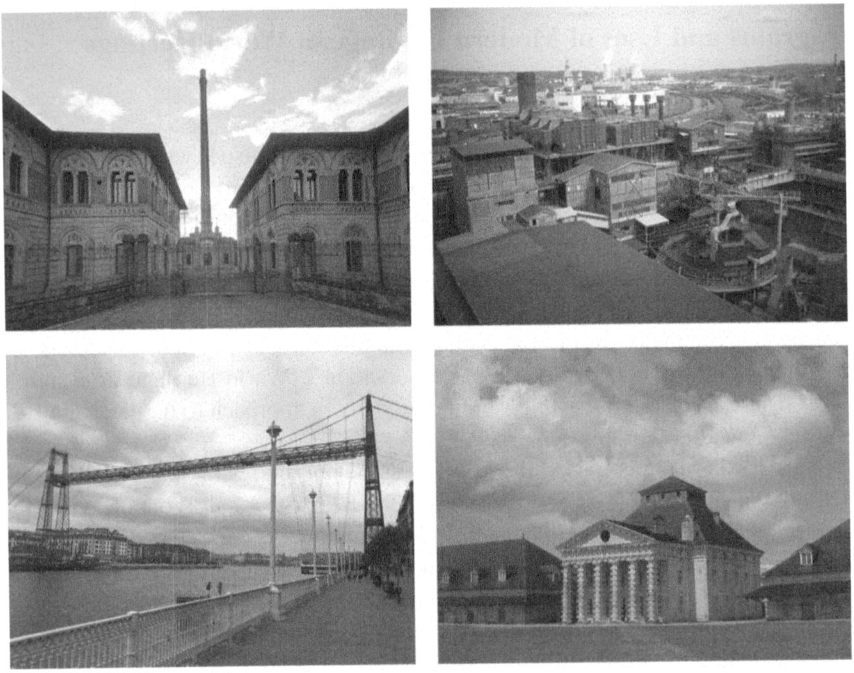

Fig. 3.4 Up-Left: Crespi d'Adda. Xiquinhosilva. 2015, https://flic.kr/p/KGxgAr. Accessed on 25th November 2022; Up-Right: Völklingen ironworks. Martin Abegglen. 2013, https://flic.kr/p/euGrE2. Accessed on 5th March 2023; Bottom-Left: Vizcaya Bridge. Daniel Pinzón-Ayala. 2017. Botton-Right: From the Great Saltworks of Salins-les-Bains to the Royal Saltworks of Arc-et-Senans, the Production of Open-pan Salt. 2007. saragoldsmith, https://flic.kr/p/2cWKbq

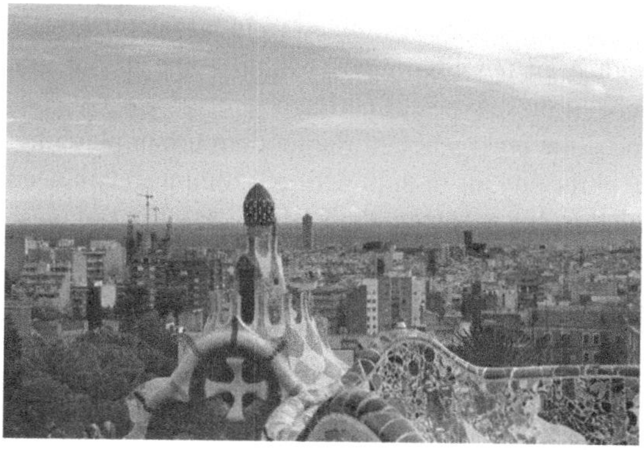

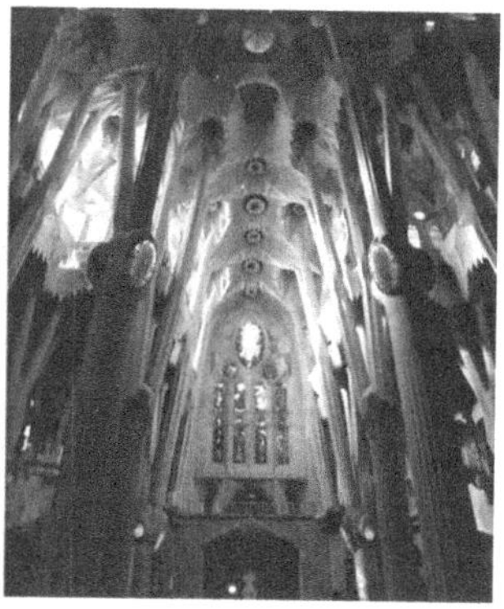

Fig. 3.5 Works of Antoni Gaudi, Left: Viewpoint of Park Güell; Right: Inside of Sagrada Familia. *Source* Roberto Alonso-Jiménez (2018)

In Italy, the three Modern Heritage Properties can be considered industrial heritage, concentrating this time in valuing industrial towns such as Crespi d'Adda (Fig. 3.4) and Ivrea, this last one valuing specifically architecture and urban planning of the 1930s and 1960s; and the rural landscape of Le Colline del Prosecco di Conegliano e Valdobbiadene. France includes two extractive/productive landscapes, such as the Great Saltworks of Salins-les-Bains to the Royal Saltworks of Arc-et-Senans, the Production of Open-pan Salt and Nord-Pas de Calais Mining Basin, and one rural iconic landscape such as Champagne Hillsides, Houses and Cellars. Finally, Germany comprises extractive landscapes such as Zollverein Coal Mine Industrial Complex in Essen; factory complexes such as Völklingen Ironworks (Fig. 3.4); Fagus Factory in Alfeld, this latter the only industrial productive heritage which is specifically valuing its architecture linked to the Modern Movement. Together with it, the Speicherstadt and Kontorhaus District with Chilehaus are declaring modern port warehouses, offices and port-related businesses.

This already institutionalized and assumed value of industrial heritage, linked to the last two centuries of our history, to our MH, was exceptional in the twentieth century: only one industrial heritage property out of the twelve properties in these four countries were declared in the 1980s and two in the 1990s. The early declaration of the Royal Saltworks in France in 1982 can be explained for its iconic dimension within the Enlightenment, being a neoclassical assembly. It is in fact considered "… the first major achievement of industrial architecture, reflecting the ideal of progress of the Enlightenment. The vast, semicircular complex was designed to permit a rational and hierarchical organization of work…" (World Heritage Convention 1982).

In the beginning of the nineties the German declaration of the Völklingen Ironworks began an explicit identification of this country with MH, and specifically with the industrial heritage. The Crespi d'Adda industrial town Italian declaration in 1995 can also be explained by the neo gothic approach to this industrial town. Declarations of rural productive landscapes, modern architectures linked to industry, infrastructural elements and even urban industrial neighborhoods, would come in our era, within the context of the emergence of these types of heritage.

The city as a space of Modernity, the emergence of the modern urban planning, with its rational conceptualization of the city, its monofunctional zoning and planned extensions are also present in the valuing on MH from the WHL declarations of these countries. "Improved qualities in life in cities by facilitating movement and providing green spaces and infrastructure was a key concern of positions linked to Modernity" (Conti 2017, p. 14).

The declarations include both the experimental early Modernist built environment, such as the German modern architecture, urban planning and landscape design of Darmstadt Artists' colony in Mathildenhöhe (World Heritage Convention 2021a), and the Postwar approach to the modern reconstruction of the cities, of which Le Havre by Auguste Perret in France is considered "an outstanding post-war example of urban planning and architecture based on the unity of methodology and the use of prefabrication, the systematic utilization of a modular grid, and the innovative exploitation of the potential of concrete" (World Heritage Convention 2005).

The city also defines the new programs, within the social urban aspirations of Modernity. In 1999, the Museumsinsel (Museum Island) of Berlin constitutes a social phenomenon. Specifically, the democratization of art: "The five museums on the Museumsinsel in Berlin, built between 1824 and 1930, are the realization of a visionary project and show the evolution of approaches to museum design over the course of the twentieth century" (World Heritage Convention 1999b). The Paseo del Prado and Buen Retiro, in Madrid, has also recently been incorporated to the list in 2021, as another iconic example of planned urban development, consolidating the modern social accessibility to culture (World Heritage Convention 2021b). Nice has been declared as the Winter Resort Town of the Riviera, considering the emergence and consolidation of this emergent tourist program, which started to shape the cities and their architectures (World Heritage Convention 2021c).

Housing lies indeed in the core of the social aspirations of modern architecture and urban planning. However, there is only one MH declaration: Berlin Modernism Housing Estates, which was declared in 2008:

> The property consists of six housing estates that testify to innovative housing policies from 1910 to 1933, especially during the Weimar Republic, when the city of Berlin was particularly progressive socially, politically and culturally. The property is an outstanding example of the building reform movement that contributed to improving housing and living conditions for people with low incomes through novel approaches to town-planning, architecture and garden design. The estates also provide exceptional examples of new urban and architectural typologies, featuring fresh design solutions, as well as technical and aesthetic innovations (World Heritage Convention 2008).

The Declaration of Bauhaus and its Sites in Weimar, Dessau and Bernau specifically valued Bauhaus architecture as "Classical Modernism", within a radical reform of architecture and design (World Heritage Convention 1996). This means that the assessment of the Modern Movement in Europe is still pretty much confined to its

German origin: the universal design and social aspirations still lacking for future declarations, geographically and temporally diversified.

Conclusions

After the research carried out, the disproportion of the WHL persists: firstly, the unbalance between Cultural Heritage and Natural and Mixed heritage far from going down, has gone up since the study of 1994. Secondly, European Cultural Heritage is still the most represented in the WHL. Finally, and taking France, Germany, Italy and Spain as representative case studies, MH is a small part of the cultural heritage declarations, constituting less than 19%. Taking into account that Humanity has built more in the last two centuries than in the rest of its long history this phenomenon is worth noticing.

If we take a closer look at the evolution of declarations affecting cultural heritage, they have experienced periods of upturn, with exponential increases—such as 2000—as well as falls, but in any case, these logics have a direct consequence on the evolution of MH. Efforts to promote MH do not seem to have borne fruit.

The unequal commitment to the MH shows very clearly which were the objectives set by each country when it came to reflecting itself in WHL. These intentions are clearly linked to issues such as national identity or the trademark image they wish to project to the rest of the world:

> The pressure of the States Parties to incorporate new assets is increasing, given that, more than a recognition of their values, they increase the attraction and competitiveness of the territories; both by accrediting them as outstanding for both cultural and, to a lesser extent, natural tourism, and by providing added value to their products (Fernández and Silva 2020, pp. 106-107).

Obviously, Spain and Italy have made a greater effort to obtain the recognition of historical examples, the Spanish case being even more evident as it has entered the twentieth century in a very timid manner and with architectures that are not the most representative of the international modern trends, neither in terms of architectural language nor in terms of use, but rather singular and unique properties clearly marked by authorship. The two Catalan properties are clear samples of it: Antoni Gaudí's works and Palau de la Música Catalana with Hospital de Sant Pau, both in Barcelona. On the other hand, Germany shows a clear commitment to highlighting cases linked to the industrial phenomenon (Zollverein Coal Mine Industrial Complex in Essen) or to social housing as a major contribution of contemporary architecture in the last century (Berlin Modernism Housing Estates).

It is understandable that the valuation of the modern scientific approach to intervention in historical heritage is already present within the MH Declarations, for it is implicitly valuating historical heritage of other periods.

We can confirm that the consideration and appreciation of industrial heritage has been institutionalized at the international level in the UNESCO context. It has thus overcome a restrictive approach based solely on its ethnological value, and its valuation as cultural heritage has already been assumed. While in Germany as we outlined above, the industrial heritage is very present, diverse and assumed as an essential part of its country values, in Spain this industrial heritage is absent. In France and Italy it is moderately integrated, but rarely identified with modern architectures, with the exception of the Industrial city of Ivrea, in Italy. The city is already identified with MH, both as an urban planning ensemble, and the space for modern programs, such as culture, housing and leisure, based on a social vocation and accessibility, being still very scarce and exceptional. Some programs are totally absent, such as cemeteries, hospitals and sanatorium, for naming some of them.

Finally, Modern Heritage can still be considered an emergent heritage, and as such, its assessment is still lacking depth and complexity in valuing this type of heritage: although incorporating issues based on social, geopolitical and economic aspects, it is still more linked to the formal or morphological aspects.

References

Aprile, Helen, and Doubleday, Gina. 2017. *World Heritage Review* 85.

Berman, Marshall. 1982. *All that is Solid Melts into Air: The Experience of Modernity*. New York: Simon & Schuster.

Conti, Alfredo. 2017. Modern Heritage and the World Heritage Convention. *World Heritage Review* 85: 8–14.

Fernández Salinas, Víctor, and Silva Pérez, Rocio. 2020. La Unesco y la Lista del Patrimonio Mundial. Una lectura territorial. In *El patrimonio cultural en la provincia de Ciego de Ávila (Cuba): análisis y propuesta de ida y vuelta*, ed. Antonio Ortega Ruiz, 73–112. Sevilla: Universidad Internacional de Andalucía.

Fernández Salinas, Víctor, and Silva Pérez, Rocio. 2018. La lista del patrimonio mundial: misión y disfunción-The World Heritage List: Mission and Dysfunction. *Revista Otarq: Otras Arqueologías* 3: 147–167.

Loren-Méndez. 2021. Heritage Types: Modern Heritage. In Statements for Sustainability and Heritage Awareness, eds. Vladan Djokic, Ana Nikezic, Mar Loren-Méndez, Konstantinos Sakamantis, Maria Philokyprou, Emanuela Sorbo, and José Peral, 89-96. Belgrado: University of Belgrade.

Loren-Méndez, Mar, Alonso-Jiménez, Roberto F, and Pinzón-Ayala, Daniel. 2023. Recognition of the universal value of less-represented heritage categories: The case of modern heritage of our built environment. In *Internacional handbook on research and design for the sustainability heritage*, eds. Mar Loren-Méndez, Marta Garcia-Casasola, Vladan Djokic, Ana Nikezic, Konstantinos Sakamantis, Maria Philokyprou, and Emanuela Sorbo, 85–93. Belgrade, Serbia: University of Belgrade.

Luxen, Jean-Louis. 2019. The Nara document: Its achievement and its limits (Trans: Magar, V.). *Conversaciones*. 8:194–208.

Van Oers, Ron, and Haraguchi, Sachiko eds. 2003. *Identification and Documentation of Modern Heritage*. Paris: Unesco World Heritage Centre.

Ministerio de Cultura y Deporte. 2023. Lista de Patrimonio Mundial de Unesco. https://www.culturaydeporte.gob.es/cultura/areas/patrimonio/mc/patrimoniomundial/unesco-patrimoniomundial/lista-de-patrimonio-mundial.html. Accessed 06 February 2023.

UNESCO. 1994. Nara Document on Authenticity. In Nara Conference on Authenticity—Conférence de Nara sur l'authenticité, Japan 1994, Proceedings, ed. Knut Einar Larsen, xxi-xxv. Trondheim: Tapir Publishers.

World Heritage Committee. 2017. Decision 41 COM 8B.32. In *41st Session of World Heritage Committee.* Krakow: UNESCO. https://whc.unesco.org/en/decisions/6904. Accessed on 09 March 2023.

World Heritage Committee. 2021. WHC.21/01. *Operational Guidelines for the Implementation of the World Heritage Convention.*

World Heritage Convention. 1982. From the Great Saltworks of Salins-les-Bains to the Royal Saltworks of Arc-et-Senans, the Production of Open-pan Salt. https://whc.unesco.org/en/list/203. Accessed on 06 Feb 2023.

World Heritage Convention. 1984. Works of Antoni Gaudí. https://whc.unesco.org/en/list/320. Accessed on 06 Feb 2023.

World Heritage Convention. 1996. Bauhaus and its Sites in Weimar, Dessau and Bernau. https://whc.unesco.org/en/list/729. Accessed on 06 Feb 2023.

World Heritage Convention. 1997. Historic Fortified City of Carcassonne. https://whc.unesco.org/en/list/345. Accessed on 06 Feb 2023.

World Heritage Convention. 1999a. Wartburg Castle. https://whc.unesco.org/en/list/897/. Accessed on 06 Feb 2023.

World Heritage Convention. 1999b. Museumsinsel (Museum Island), Berlin. https://whc.unesco.org/en/list/896/. Accessed on 06 Feb 2023.

World Heritage Convention. 2005. Le Havre, the City Rebuilt by Auguste Perret. https://whc.unesco.org/en/list/1181. Accessed on 06 Feb 2023.

World Heritage Convention. 2008. Berlin Modernism Housing Estates. https://whc.unesco.org/en/list/1239. Accessed on 06 Feb 2023.

World Heritage Convention. 2021a. Mathildenhöhe Darmstadt. https://whc.unesco.org/en/list/1614. Accessed on 06 Feb 2023.

World Heritage Convention. 2021b. Paseo del Prado y el Buen Retiro, paisaje de las artes y las ciencias. https://whc.unesco.org/en/list/1618. Accessed on 06 Feb 2023.

World Heritage Convention. 2021c. Nice, Winter Resort Town of the Riviera. https://whc.unesco.org/en/list/1635. Accessed on 06 Feb 2023.

Part II
Methodologies for Sustainable Heritage Management

One of the principles that characterizes conservation management is the multidisciplinary and holistic perspective, also called inclusive in this case, to reinforce the importance of citizen participation and to highlight the transfer of knowledge as the main objective of heritage understood from its significance.

The case study presented, the Ocean Swimming Pool on Leça da Palmeira beach, designed by the architect Álvaro Siza and built between 1961 and 1966, is a reference for the approach to modern heritage management in a building listed as a National Monument in Portugal (2011). It is one of the most relevant projects in the recent history of architecture. This is a proposal framed in the *Keeping It Modern* Grant, awarded by the Getty Foundation in 2020. A rigorous and complex management plan, presents the main steps of the technical process without losing the connection with the citizens (T. Cunha Ferreira, P. Murilo Freitas, H. Mendonça) . A sustainable methodology that will open up a new way of managing modern heritage, has an important challenge for the twenty-first century: to comply with the UN Sustainable Development Goals (SDGs) (United Nations 2015).

Heritage and sustainability form a binomial from which to care for the environment, where the figure of the Historic Urban Landscape (HUL) acquires particular relevance. A tool that needs to examine *how the critical history and theory of sustainability intersect with the theory and practice of architectural heritage conservation* (P. Pyla, D. Zacharia). A relationship between knowledge and proposal that needs to be promoted from the learning methodologies applied in schools of architecture. Students will work with different tools that attempt to measure the positive and negative environmental impact of heritage conservation.

The holistic approach and the preservation of the environment also imply the introduction of bioclimatic design in the proposals. In this case, through the architectural competitions with examples in Greece, the contribution gives answers to professionals who need to be able to manage both requirements: heritage conservation and sustainable solutions (K. Sakantamis, A. Chatzidimitriou).

From a research perspective, the updating of methodologies must close the gap between urban planning and heritage interventions, where the key is the recognition of values and the consideration of authenticity (M. García-Casasola, J. Rey-Pérez).

Methodologies for sustainable heritage management must base decisions on the identification of values and the impact of the intervention on them. This approach is not well formalized. As stated in the Faro Convention (2005): *the Value of Cultural Heritage for Society*, the main objective of conservation is to work with heritage communities to share knowledge and perceptions, to preserve and add cultural values, to construct meaning.

Chapter 4
Holistic and Inclusive Approach in the Conservation Management Planning of Modern Heritage

Teresa Cunha Ferreira, Pedro Murilo Freitas, and Hugo Mendonça

Keywords Conservation management plan · Holistic methods · Modern architectural heritage · Community-based strategies

Introduction

Conservation Management Plans are a multi-staged process that allows and promotes active preservation of buildings and places (Kerr 2013). Anchored in the Burra Charter (ICOMOS Australia 2013), they provide a cyclical framework in which the conservation of heritage is carried out by involving communities' values to foster integrated management. It is often argued that the closest the relation of the various rightsholders in assessing significance and developing policies for achieving common agreements on how to maintain it, the more efficient conservation is achieved (Mason 2002). However, in the present-day context of high complexity, holistic approaches foster more resilient and adaptive conservation planning. They must be strongly related to monitoring as a strategy to assess impacts and increase resilience toward rapid changes (Court et al. 2022).

Literature review on high-impact journals shows that most cited papers associating holistic methodologies are often linked with the historic urban landscape (HUL) preservation field, where the set of interdisciplinary themes is very wide and needs overview strategies to embrace the complexity and scale of a cultural significant settlement (Manal et al. 2019). However, 'policy making', as a process of creating, formulating, adopting, implementing, and evaluating actions for a common public

T. Cunha Ferreira (✉) · P. M. Freitas · H. Mendonça
CEAU, Faculty of Architecture, University of Porto, Porto, Portugal
e-mail: tferreira@arq.up.pt

M. Loren-Méndez et al. (eds.), *Sustainable Heritage Management - Design: Notions, Methods and Techniques*, https://doi.org/10.1007/978-3-031-81970-4_4

goal (Knill and Tosun 2008), tends to be overlooked as an activity that can also be applied to significant constructions, presenting intricated challenges.

In buildings that developed recent heritage status, such as modern architecture, questions might arise regarding planning conservation management: values might not be clear enough to justify careful maintenance; stakeholders might not be aware of current risks and their responsibilities to comply with updated conservation standards; and there might be no adequate understanding of the variety of interdisciplinary topics to manage and how to achieve future expected conditions. These questions have been at the core of modern architectural heritage debate since their cultural significance recognition made intervention practices evolve from developing experimental (and frequently aesthetical) renovations of derelict masterpieces toward the establishment of programs and recommendations aimed to foster good practices (ICOMOS ISC20C 2017, 2021).

So, as these problems are also being seen more and more interlocked, meaning that there are still untackled experiences to face in preserving modern buildings to expand these practices (Torrent 2022), which are the methods and strategies for a holistic approach to the management and conservation of modern heritage?

To address that research question, this paper presents the management and conservation policies for the Ocean Swimming Pool, designed by the architect Álvaro Siza and built from 1961 to 1966. The building is a landmark of twentieth century architecture and has a strong relation with local communities, being listed as National Monument (in 2011) and in the Tentative List for World Heritage (in 2017). However, due to its location on the Leça da Palmeira beach, it suffers from coastal erosion that is also being aggravated by climate change. The most notable deficiency in the building's use was related to the plumbing system, as many of the pipes had either been destroyed by corrosion or were clogged with sand, compromising the operating conditions of the Ocean Swimming Pool. The exposed concrete structures presented moderate damage, namely vertical cracks, distributed cracking, and spalling.

The Ocean Swimming Pool Conservation Management Plan was developed under the *Keeping It Modern* Grant, awarded by the Getty Foundation in 2020. It enabled inspection and diagnosis to be carried out through a series of in situ non-destructive or minor destructive tests and laboratory tests which informed the localized repair of exposed concrete in the context of a significant conservation campaign promoted by the Municipal Council of Matosinhos between 2018 and 2021. This commission was entrusted to Álvaro Siza, who sought to preserve the integrity and authenticity of the preexistence, maximizing the preservation of original materials, including their texture and patina. His intervention also envisaged the extension of the bathing complex to the north, enhancing the site's future viability and functionality.

The focus of this article is to demonstrate, through a real case study, how a holistic approach was implemented for the conservation and management of a modern building, providing several policies with contributions from different disciplines.

Methodology

Following the aforementioned *Keeping It Modern* Grant, the methodological framework was provided according to (Table 4.1).

First, to develop any plan for heritage sites or buildings, since they fulfill an important cultural and social function, it is important to assess which characteristics qualify them as significant for a given community. According to many recommendations, cultural significance is understood as the set of qualities, both tangible and intangible, which make a place of value to society (Kerr 2013).

Under this perspective, an integrated strategy must be developed, especially to deal with the complexity raised in the Conservation Management Planning process. The building assessment methodology developed for the Ocean Swimming Pool was based on two main axes: (i) 'documentary evidence' and (ii) 'physical evidence', as presented below (Table 4.2).

'Documentary evidence' included the research on different kinds of sources from eight archives (written, graphic, photographic, and cartographic materials, among others), published material and oral sources.

Using the building fabric as the most reliable source of information, the analysis of the 'physical evidence' was also of major importance in the process, supported by a geometric survey (using laser scanner and photogrammetry methods), and building condition assessment using non-destructive and minor destructive testing. Inspection and diagnosis included a condition survey, evaluation of the out-of-plane deformations of the walls, and different types of testing to determine the mechanical and chemical properties of different materials (not only but especially concrete) used in the building.

Table 4.1 Ocean swimming pool conservation management planning process

Stage	Theme	Procedure
Stage 1	Understanding the place	Gather documentary and physical evidence;
Stage 2	Assessing significance	Identify attributes and heritage values; Develop a statement of significance;
Stage 3	Building condition assessment and vulnerabilities	Assess physical condition; Identify vulnerabilities and risks;
Stage 4	Management and conservation policies	Develop overarching policies covering governance and interpretation; Develop site policies on planning constraints, risk assessment, climate change adaptation, and landscape preservation; Develop detailed conservation and management policies;
Stage 5	Monitoring and implementation	Establish implementation priorities and phasing; Monitor implementation process;

Table 4.2 Ocean swimming pool's building assessment methodology

Evidence	Type	Procedure
Documentary evidence	Archival sources	Textual, drawings, photographic, cartographic
	Published material	Books, book chapters, journal articles, academic works;
	Oral sources	Álvaro Siza, collaborators, engineers, contractors, municipal council, site managers, operational staff, experts, users, and visitors;
Physical evidence	Surveys	Laser scanner, photogrammetry, condition survey, wall out-of-plane deformations evaluation
	Non-destructive and minor destructive testing	Concrete carbonation testing, concrete and steel Rebars mechanical properties testing, chloride content testing, concrete testing with Schmidt hammer, sonic testing, color analysis based on spectrophotometry, ground penetrating radar testing;

Community-Based Strategies

In addition to building assessments performed by specialists, community participation and stakeholder integration is a key factor for accessing significance and drafting conservation policies. Since communities are the ones that sustain and continuously produce values in relation to different attributes, integrating them in the process is among the best practices for heritage conservation. This implies shifting from top-down, reactive, and regulatory approaches toward bottom-up, proactive, and participating approaches. (Rossetti et al. 2022).

The variety of issues raised by communities in the assessment of cultural significance of the Ocean Swimming Pool demonstrated the potential of citizens' engagement to build more inclusive and democratic planning policies. Moreover, the broad concept of 'community' adopted goes beyond users and visitors to encompass the perspective of academics, staff, concessionaires, and the Municipal Council. The latter are important to provide a perspective on the building's daily management and use also by including site managers. Their experience is fundamental to gathering information about existing maintenance policies.

Table 4.3 presents the correlation between stakeholders and participation strategies adopted for assessing cultural significance and vulnerabilities.

The semi-structured interviews with site managers were critical sources for understanding the perspective of the daily management actors. The interview protocol and the survey design applied to users and visitors were methodologically influenced by the experiences of referenced case studies (Fayad and Buckley 2019), particularly the 'IMAGINE Ballarat' project and its 'love, change, imagine' questionnaire method that enables to tie values and attributes recognition to conservation policymaking.

Regarding data analysis, the research followed the taxonomy developed by Tarrafa Silva and Pereira Roders (2012). The systematic nature of this content analysis

Table 4.3 Participation strategies adopted

Actors	Participation strategies
Institutional (listing justification)	Document analysis
Experts (architects and critics)	Interviews 'Sharing memories' webinar
Site managers (municipal council, Matosinhos sport and building staff)	Interviews Surveys (Love, Change, Imagine)
Users and visitors	Surveys (Love, Change, Imagine)
Virtual community	Social media coding
Students	Word cloud activity
Children	Drawing activity

methodology enabled the draft of a statement of significance, toward holistic and inclusive conservation policies (ICOMOS 2020).

Conservation and Management Policies

Conservation and Management policies are a fundamental tool to endorse a careful management of change. Hence, in conservation planning, they act as strategic guidelines that activate a cycle involving monitoring and future evaluation procedures (Knill and Tosun 2008). Two issues of concern, however, must be considered: if these policies convey properly the assessed demands (in heritage sites or buildings, usually expressed in the aforementioned statement of significance), and if they were conceived interlaced into a management system formed by all rights-holders of a heritage site with power enough to formulate, adopt and implement conservative measures.

The Ocean Swimming Pool's Conservation Management Plan was developed through a set of policies ranging broad subjects, from planning to maintenance. Following a statement of significance developed previously, they were framed to:

- Respect and reinforce the primacy of Álvaro Siza's design principles in understanding significance and guiding all management and change.
- Retain and respect the Ocean Swimming Pool's cultural significance.
- Ensure adaptations and new uses are compatible with the significant values of the place while preventing uses that could result in its progressive degradation.
- Retain the authenticity, integrity, character, and quality of the building and its various elements, providing guidance for the care and conservation of significant elements and fabric.
- Retain the integrity of the original structural systems, materials, and finishes, while allowing for necessary upgrading.

- Draw attention to the need for coordination of planning, continuity of conservation, and other expert advice, toward cyclical maintenance.

Any future adaptations or new uses must be compatible with the significant values of the place and aligned with Álvaro Siza's design principles which should guide all management decisions and change.

Thus, policies were organized under three main intervention areas, numbered and highlighted in separate boxes: (A) *Overarching Policies*, related to Álvaro Siza's design principles, stakeholders' action definition, Sustainable Development Goals, interpretation, and archives; and two groups of *Operational Policies*: (B) *Site Policies*, incorporating strategies for the safeguarding and spatial planning of the site, risk management and climate change adaptation; and (C) *Conservation and Management Policies*, providing guidance for material conservation, detailed actions of periodic maintenance, and operational management and use (Table 4.4).

Overarching Policies

Besides the first integration into a conceptual model of policy formulation, the definition of general policies is essential for regulating all decision-making, i.e., to prevent the implementation of ad-hoc or circumstantial decisions that do not consider the conservation of the building's values and significance. Therefore, in the Ocean Swimming Pool, not only any future changes or interventions should be preceded by a detailed investigation, as they should be analyzed by an Advisory Body gathering stakeholders and experts and ensuring respect for Álvaro Siza's design principles. They are meant to be the primary source of guidance and inspiration for any proposed changes to the place as well as its ongoing management, conservation, and use.

Stakeholders

The stakeholders of the Ocean Swimming Pool include the General Directorate for Cultural Heritage, responsible for heritage safeguarding on a national level, the Municipal Council of Matosinhos (which owns the Ocean Swimming Pool), Matosinhos Sport (a municipally owned corporation charged with the management of sports facilities), and Casa da Arquitectura (responsible for organizing guided tours) on a local level, Álvaro Siza and the different users.

It is very important that the site's management roles and responsibilities are clearly defined, assigned, and understood by all stakeholders. The most important stakeholders for the safeguarding of the Ocean Swimming Pool are the ones with the most power and interest. However, it is important to find ways to engage with the stakeholders with less power or interest to enhance involvement and thus achieve more success in preserving and managing the site.

Table 4.4 Ocean swimming pool policy structure

Class		Policy	Description
A	Overarching policies	A.1	Advisory body
		A.2	Álvaro Siza's design principles
		A.3	Investigation and knowledge
		A.4	Engagement of stakeholders
		A.5	Administrative control
		A.6	Sustainable development goals
		A.7	Interpretation plan
		A.8	Communication strategy
		A.9	Community engagement
		A.10	Archives and collections
B	Operational policies: Site policies	B.1	Spatial planning
		B.2	Cultural heritage safeguarding
		B.3	Buffer zone preservation
		B.4	Coastal protection regulation
		B.5	Risk management plan
		B.6	Green procurement
		B.7	Climate change adaptation
		B.8	Landscape management plan
C	Operational policies: Conservation and maintenance policies	C.1	Significance and tolerance for change
		C.2	Maintenance plan
		C.3	Inspection routines
		C.4	Archival database
		C.5	Building information management system BIM
		C.6	Accessibility plan
		C.7	Capacity and entrance control
		C.8	Housekeeping routines
		C.9	Security system
		C.10	Lifesaving and rescue
		C.11	Fire prevention plan
		C.12	Risk awareness educational plan
		C.13	Use regulation
		C.14	Health emergency plan

Sustainable Development Goals

As an important challenge for the twenty-first century, the Ocean Swimming Pool Conservation Management Plan seeks to comply with the UN's Sustainable Development Goals (SDGs) (United Nations 2015), contributing with a set of measures that will maintain the good practices the site has been applying so far. It is also the utmost intent to adopt new SDGs that will further complement the building role in the world's inclusivity, education, shared knowledge, and sustainability as a recognized cultural heritage.

Interpretation, Communication, and Community Engagement

To convey the cultural significance and the heritage values of the Ocean Swimming Pool to the public, policies that foster interaction with visitors and users are a fundamental part of the Conservation Management Plan so that they can establish responsible use and natural care for the building. As expressed by the ICOMOS Charter for the Interpretation and Presentation of Cultural Heritage Sites, 'interpretation' refers to 'the full range of potential activities intended to heighten public awareness and enhance understanding of cultural heritage sites' (ICOMOS ICIP 2008).

On the Ocean Swimming Pool, this can represent a series of active strategies that provide users with information about the site and keep them actively involved in acquiring more knowledge about the building's heritage values (Fig. 4.1). It is also expected that interpretation allows people to engage and create new narratives of the building through experience and facilitate cultural significance recognition while maintaining public awareness about the conservation strategies proposed in the Conservation Management Plan.

The interpretation of the Ocean Swimming Pools should be made through effective communication by the Municipal Council of Matosinhos, Matosinhos Sport, and Casa da Arquitectura. It would also be important to provide context about the significance of the Ocean Swimming Pool in the framework of buildings designed

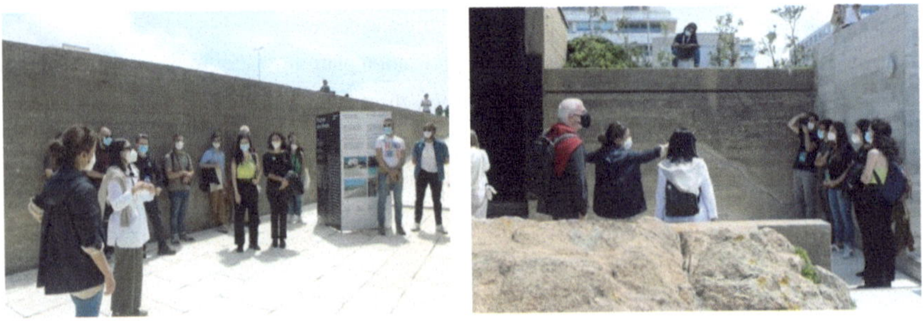

Fig. 4.1 Guided tours to the ocean swimming pool during Porto open house. Photo: Frederico Barbosa, 2021

by the architect Álvaro Siza, including those that are included on the World Heritage Tentative List (UNESCO World Heritage Center 2017). This knowledge must be communicated through easy, direct, and concise activities with the public. Hence, the Interpretation Plan should be based on comprehensive research to identify significant themes and interpretative opportunities that should be adapted according to the different audiences for the interpretative activities, providing a framework for managing visitors. The Plan should be regularly reviewed to incorporate lessons taken from the results of previous activities.

Operational Policies

Site Policies

Site policies frame the management and conservation of the Ocean Swimming Pool within its territorial, landscape, and spatial/urban planning context, as to define policies and guidelines for the management, safeguarding, and risk assessment of the site.

Planning and Safeguarding. Being listed as a National Monument, the Ocean Swimming Pool conservation management must comply with the Fundamental Law of the Portuguese Cultural Heritage (Law No. 107/2001), which forbids any interventions, inside or outside listed properties or any change of use likely to affect it, without prior consent from the cultural heritage administration. The listing process determines the establishment of a Buffer Zone, defined by Ordinance No. 608/2012, to ensure the landscape and visual integration of the property, including the surrounding areas that are relevant to the context. Licenses for construction works that alter the topography, alignments, heights and the distribution of volumes and roofs or the exterior finishes of buildings cannot be granted without prior assent from the cultural heritage administration. Due to its location, it is also subject to the Caminha-Espinho Coastal Zone Program, which determines that new constructions, soil sealing actions, changes to the coastline, and activities that result in the direct destruction of relevant marine ecosystems or rock outcrops are forbidden along the Coastal Protection Strip in which the property is set.

Risk Management and Climate Change Adaptation. It is crucial to acknowledge and understand the impacts that natural and anthropic hazards, in particular those whose effects are intensified by climate change, have on the Ocean Swimming Pool, to ensure that they are prevented, mitigated, or, when that is not possible, that adaptation measures are implemented. In the context of the Conservation Management Plan, these hazards were identified (Fig. 4.2), alongside scenarios and vulnerabilities ranked according to their likelihood and severity to assess their risk level. It is essential to adapt to the present and future impacts they will have on the property to assure preparedness when such events occur through a flexible Risk Management

Plan that considers multiple scenarios and emergency/post-disaster lines of action that would rapidly and efficiently restore the situation back to normal.

Landscape Management. The Ocean Swimming Pool's landscape integration is a crucial element of its cultural significance, as it was one of the main principles behind Álvaro Siza's design approach. Therefore, the visual relationships between the property and the surrounding landscape must be retained (Fig. 4.3), ensuring regular maintenance, cutting and pruning of the vegetation, beach cleaning, and strict control of furniture and signage placement.

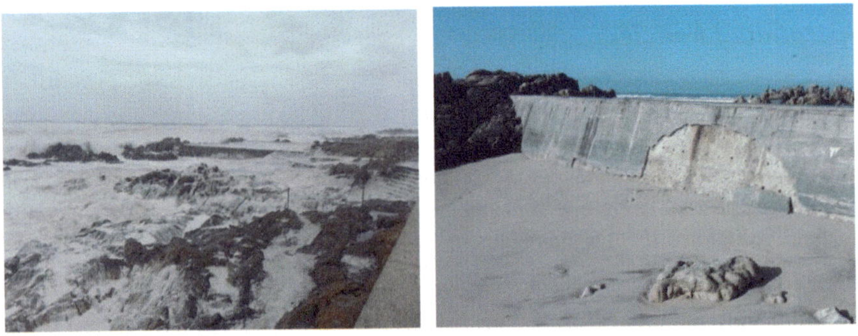

Fig. 4.2 Ocean Swimming Pool during storm surge and concrete detachment as a result of wave impact. Photo: Civil Protection Department of the Municipal Council of Matosinhos, 2017; 2018

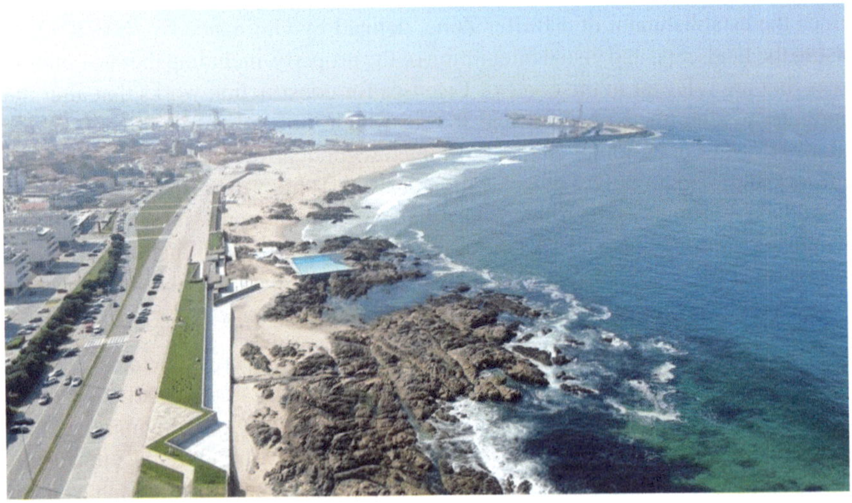

Fig. 4.3 Visual continuity toward the south and the port of Leixões. Photo: Pixel, 2021

Conservation and Maintenance Policies

Conservation and Maintenance Policies aim to provide non-exhaustive guidance for future conservation and maintenance actions (both preventive and reactive) responding to threats and vulnerabilities, daily wear and tear, or interventions related to functional upgrades or adaptations responding to new requirements.

Tolerance for Change. The tolerance for change levels of the most significant areas and components of the Ocean Swimming Pool were identified to provide a framework for managing change while retaining the site's cultural significance. Future changes or conservation interventions must not interfere with the components and spaces of higher significance, hence the importance of clearly establishing each area's tolerance for change to inform management decisions and conservation strategies adopted in future interventions. The followed methodology is aligned with the tolerance for change framework at heritage places (Burke 2017) and supported by the assessment of cultural significance alongside archival and documentary research.

Inspection and Maintenance. Based on the effects the identified hazards can have on the Ocean Swimming Pool elements' condition, it is possible to generate management and maintenance actions to guarantee its good condition and physical integrity. Therefore, it is possible to ensure preventive conservation that minimizes physical changes through specialized advice in conservation decisions and actions. Thus, specific lines of action, involving inspection, cleaning, proactive, and repair actions, were developed for each constructive element to maintain its integrity and at the same time preserve the site's significance (Fig. 4.4). In the context of a Maintenance Plan, it is important to consider an estimate of annual costs for each maintenance action and to develop cyclical maintenance and treatment manuals to establish priorities and short and long-term interventions, thus maintaining a proactive conservation approach.

Operational Management. Having in mind the Ocean Swimming Pool is a fully functioning public facility that has been open uninterruptedly since its opening in 1965, policies regarding operational management must be developed to ensure the

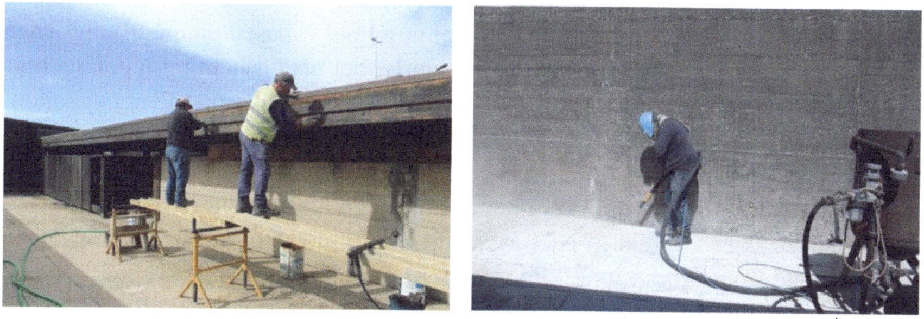

Fig. 4.4 Ocean swimming pool during conservation actions. Photo: Frederico Barbosa, 2020

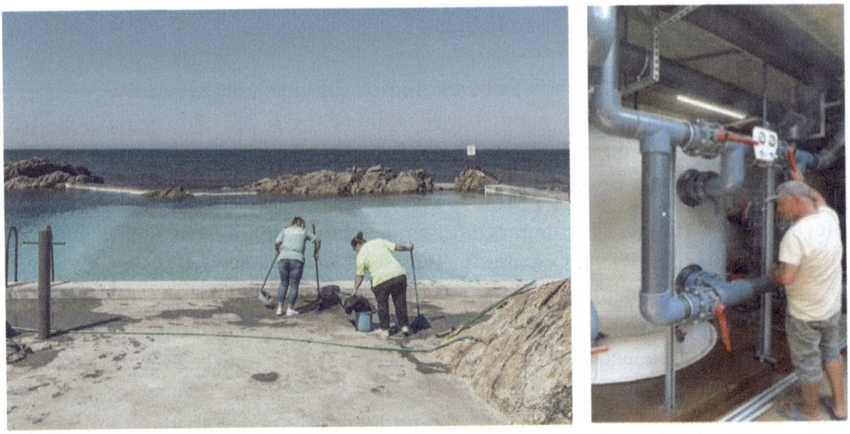

Fig. 4.5 Housekeeping routines. Photo: left, Inês D'Orey, 2021; right, Frederico Barbosa, 2021

correct functioning of the building while minimizing impacts of visitation and intensive use. To accomplish this goal, a broad set of policies were defined, ranging from occupancy control to prevent damage from exceeding the designated capacity to housekeeping routines permanently updated to the best technical standards and the implementation of a security system that prevents misuse during operating hours and acts of vandalism during closing hours (Fig. 4.5).

Monitoring, Implementation, and Phasing

After policy formulation, adequate monitoring inside a well-functioning management system allows the development of a sustainable intervention culture. As mentioned before, this allows the stakeholders to act based on preventive measures instead of reactive ones (Cunha Ferreira 2018). Under a policy cycle, this is also a fundamental process to develop experiences, qualify and quantify problems, and ensure revisions through informed specialized reports.

Thus, the monitoring of the Ocean Swimming Pool is based on qualifying when and how corrective measures are taken regularly, but also accepting input such as spontaneous reporting, or even suggesting new forms of articulation and organization. This system facilitates multidisciplinary decision-making to achieve conservation objectives. Besides, information surveyed by different experts is necessary, as well as field observation by researchers or specialists and assessment of legislation and its possible changes. When impact assessments are not required by national and supra-national frameworks for all interventions affecting the property, employing a Heritage Impact Assessment or 'stand-alone' impact assessment (UNESCO 2022), may widen the reach of the management system.

The implementation of policies should also respect feasibility, emergency, and available resources and funds, leading to phasing strategies. It is important to give emphasis to a 5-year strategy for actions that can be immediately applied (such as legislation principles and management system development) toward long-term actions that might need specific adaptation and coordination.

Table 4.5 presents a model of a phasing strategy in which the Ocean Swimming Pool's policies should be implemented for a period up to 60 months.

Conclusions

Conservation of heritage sites is considered a complex activity. It demands from architects, urban planners, engineers, historians, and other actors the combination of a sensitive perception of existing features and the ability to read it as a valuable resource for humanity. Also, the increasing context of vulnerability related with climate change, conflicts and other factors, demands resilient and adaptive strategies toward rapid changes scenarios.

On the other hand, traditional top-down approaches, even if well executed by heritage institutions, are being subjected to severe criticism for neglecting the idea that 'heritage', by definition, frames broader relations to society that wants to participate in democratic scenarios. Thus, either way or another, approaches that do not pin up updated twenty-first century heritage definitions nowadays, do not fully achieve the overarching objective of any Conservation Management Plan, i.e., to preserve heritage assets for future generations.

In this sense, holistic and inclusive approaches are important tools to achieve this goal. As a result of a careful array of assessments, it maintains the planner as a peer into a larger group of interdisciplinary stakeholders, constantly mediating different challenges. That is why participatory methodologies are crucial: they provide not only a variety of information but also, make the planner commit the whole process with the communities that will, in the future, adopt and implement the different actions collectively co-created. Embodying this, the Ocean Swimming Pool Conservation Management Plan proves that modern heritage does not differ from other assets in planning conservation. As affirmed by Ana Tostões (2022, p. 17), it settles a new 'operative relationship between the practice of preservation, the act of valuing the building's materiality, and deciding on a suitable preservation strategy'.

Limitations should be mentioned in developing a plan under a global health emergency in 2020 and 2021. However, it only demonstrated the importance of engaging stakeholders and monitoring planning in the current context of global vulnerabilities.

Finally, this work confirmed that holistic and inclusive approaches in conservation management planning are an important tool that helps to improve resiliency. However, it must be noted that it is only by their inclusiveness and flexibility that they enable effective operational actions of mitigation or adaptation toward rapid changes scenarios.

Table 4.5 Policy description with implementation strategies for the ocean swimming pool

	Policy	Description	Phasing (in months)					
			0	12	24	36	48	60
A	A.1	Advisory body	•					
	A.2.1	Álvaro Siza's design principles—program	•					
	A.2.2	Álvaro Siza's design principles—interventions	•					
	A.2.3	Álvaro Siza's design principles—urban policies		•				
	A.3	Investigation and knowledge	•					
	A.4	Engagement of stakeholders		•				
	A.5	Administrative control	•					
	A.6	Sustainable development goals		•				
	A.7	Interpretation plan			•			
	A.8.1	Communication strategy—online platform			•			
	A.8.2	Communication strategy—shared knowledge network				•		
	A.9.1	Community engagement—learning programs				•		
	A.9.2	Community engagement—broadening access					•	
	A.10.1	Archives and collections—archives					•	
	A.10.2	Archives and collections—collections						•
B	B.1	Spatial Planning	•					
	B.2	Cultural heritage safeguarding	•					
	B.3	Buffer zone preservation	•					
	B.4	Coastal protection regulation	•					
	B.5.1	Risk management plan—mitigation preparedness			•			
	B.5.2	Risk management plan—disaster recovery				•		
	B.6.1	Green procurement—environmental analysis			•			
	B.6.2	Green procurement—strategies				•		

(continued)

Table 4.5 (continued)

	Policy	Description	Phasing (in months)					
			0	12	24	36	48	60
	B.7.1	Climate change adaptation—adaptation measures			•			
	B.7.2	Climate change adaptation—permanent flooding				•		
	B.7.3	Climate change adaptation—staff training				•		
	B.7.4	Climate change adaptation—archival information						
	B.8.1	Landscape management plan—landscape integration		•				
	B.8.2	Landscape management plan—urban planning		•				
	B.8.3	Landscape management plan—visual relations		•				
	B.8.4	Landscape management plan—Landscaping			•			
	B.8.5	Landscape management plan—beach cleaning			•			
	B.8.6	Landscape management plan—placement of furniture			•			
	B.8.7	Landscape management plan—placement of signage			•			
C	C.1	Significance and tolerance for change	•					
	C.2.1	Maintenance plan—framework		•				
	C.2.2	Maintenance plan—costs		•				
	C.2.3	Maintenance plan—cyclical maintenance			•			
	C.3	Inspection routines		•				
	C.4	Archival database					•	
	C.5	Building information management system BIM						•
	C.6	Accessibility plan		•				
	C.7.1	Capacity and entrance control—occupancy		•				
	C.7.2	Capacity and entrance control—visitor information			•			

(continued)

Table 4.5 (continued)

	Policy	Description	Phasing (in months)					
			0	12	24	36	48	60
	C.8	Housekeeping routines		•				
	C.9	Security system		•				
	C.10	Lifesaving and rescue		•				
	C.11	Fire prevention plan		•				
	C.12	Risk awareness educational plan				•		
	C.13	Use regulation	•					
	C.14	Health emergency plan	•					

Acknowledgements This research was funded by the Getty Foundation, under the 'Keeping It Modern' Program (Grant ORG-202047064), as well as by the European Regional Development Fund (ERDF) through 820 COMPETE 2020—Operational Program for Competitiveness and Internationalization 821 (OPCI), and by national funds through FCT/MCTES (PIDDAC), under the scope of the projects POCI-01-0145-FEDER-822 007744, 2020.01980.CEECIND.

References

ICOMOS Australia. 2013. The Burra Charter: The Australia ICOMOS Charter for Places of Cultural Significance. Burnwood: ICOMOS Australia. https://australia.icomos.org/wp-content/uploads/The-Burra-Charter-2013-Adopted-31.10.2013.pdf

Burke, S. 2017. Tolerance for change: Introducing a concept and a challenge to ICOMOS members, in Conservation Turn—Return to Conservation: Tolerance for Change, Limits of Change. Firenze: Edizioni Polistampa. p. 77–94.

Court, S., Jo, E., Mackay, R., Murai, M., and Therivel, R. 2022. Guidance and toolkit for impact assessments in a World Heritage Context. Manual. UNESCO, ICCROM, ICOMOS and IUCN, Paris, France; Rome, Italy; Charenton-le-Pont, France; Gland, Switzerland. https://whc.unesco.org/en/guidance-toolkit-impact-assessments

Cunha Ferreira T. 2018. Bridging planned conservation and community empowerment: Portuguese case studies. *Journal of Cultural Heritage Management and Sustainable Development* 8 (2): 179–193. https://doi.org/10.1108/JCHMSD-05-2017-0029

Fayad, S., and K. Buckley. 2019. The transformational power of the HUL approach: Lessons from Ballarat, Australia, 2012–2017, in Pereira Roders, A.; Bandarin, F. (Eds.) *Reshaping Urban Conservation. Creativity, Heritage and the City*, vol. 2. Singapore: Springer.

ICOMOS ICIP. 2008. The ICOMOS Charter for the interpretation and Presentation of Cultural Heritage Sites. ICOMOS ICIP. http://www.enamecharter.org/principles_0.html

ICOMOS ISC20C. 2017. Approaches to the Conservation of Twentieth-Century Cultural Heritage Madrid—New Delhi Document. ICOMOS ISC20C. https://openarchive.icomos.org/id/eprint/2692

ICOMOS ISC20C. 2021. The Cádiz Document. InnovaConcrete Guidelines for Conservation of Concrete Heritage. ICOMOS ISC20C. https://openarchive.icomos.org/id/eprint/2578

ICOMOS. 2020. European Quality Principles for EU-funded Interventions with potential impact upon Cultural Heritage. Paris: ICOMOS. https://openarchive.icomos.org/id/eprint/2436/

Kerr, J.S. 2013. *Conservation Plan*, 7th ed. ICOMOS Australia.

Knill, C., and J. Tosun. 2008. Policy Making. Working Paper 01/2008. KOPS: University of Kontanz. https://kops.uni-konstanz.de/server/api/core/bitstreams/9f7f2258-ffe2-4c27-864c-4a30abfe8e8a/content

Manal, G., C. Houbart, and J. Teller. 2019. The Historic Urban Landscape approach to urban management: A systematic review. *International Journal of Heritage Studies* 25 (10): 999–1019. https://doi.org/10.1080/13527258.2018.1552615

Mason, R. 2002. Assessing Values in Conservation Planning: Methodological Issues and Choices In: de la Torre, M. (Eds.). Assessing the Values of Cultural Heritage. Los Angeles: The Getty Conservation Institute. https://www.getty.edu/conservation/publications_resources/pdf_publications/pdf/assessing.pdf

Rosetti, I., C.M. Ferreira Bertrand Cabral, A. Pereira Roders, M. Jacobs, and R. Albuquerque. 2022. Heritage and sustainability: Regulating participation. *Sustainability* 14 (3): 1674. https://doi.org/10.3390/su14031674

Tarrafa Silva, A., and A. Pereira Roders. 2012. Cultural Heritage Management and Heritage (Impact) Assessments, in Michell, K., Bowen, P. and Cattell, K. (Eds.) *Proceedings of the Joint CIB W070, W092 & TG72 International Conference on Facilities Management, Procurement Systems and Public Private Partnership.* Cape Town: Department of Construction Economics and Management. University of Cape Town. p. 375–382.

Torrent, H. 2022. What to do with modern tradition? In *Modern Heritage: Reuse, Renovation, Restoration*, ed. A. Tostões, 18–22. Basel/Lisbon: Birkhäuser/Docomomo International.

Tostões, A. 2022. Why preserving memory matters for building a wonderful world. In *Modern Heritage: Reuse, Renovation, Restoration*, ed. A. Tostões, 12–17. Basel/Lisbon: Birkhäuser/Docomomo International.

UNESCO. 2022. *Guidance and toolkit for impact assessments in a World Heritage Context.* Resource Manual. Paris: UNESCO. https://whc.unesco.org/en/guidance-toolkit-impact-assessments

UNESCO World Heritage Center. 2017. *Ensemble of Álvaro Siza's Architecture Works in Portugal.* Paris: UNESCO. https://whc.unesco.org/en/tentativelists/6224

United Nations. 2015. Transforming our world: the 2030 Agenda for Sustainable Development. A/RES/70/1. New York: UN. https://sdgs.un.org/2030agenda

Chapter 5
Architectural Heritage Conservation from the Perspective of Environmental History

Panayiota Pyla and Danae Zacharia

Keywords Architectural theory · Design pedagogy · Critical sustainability · Rethinking modernism

Introduction

This article examines how the critical history and theory of sustainability intersect with the theory and practice of architectural heritage conservation. It focuses on the outcomes of the course 'Architecture and a Critical History of Ecology' which interrogates how the emergence of environmental consciousness since the mid-twentieth century influenced architecture and its social, technological, and design experimentations. This historically minded course taught at the Department of Architecture at the University of Cyprus (Course code ARH 511) has been designed from scratch to provide an alternative perspective and also an important companion to more technically oriented courses on environmental technologies in architectural design. It enhances student's theoretical knowledge and criticality by cultivating an understanding of the socio-political implications of recent concepts around 'sustainability,' 'green' and 'climate' that shape contemporary architectural theory and practice. Placing both the vocabulary and rhetoric of the contemporary 'sustainability culture' (Parr 2009) under the microscope, lectures and critical readings examine issues such as how nature is constructed in public perception; the repercussions of viewing it as a 'resource' for human use; and how 'resource conservation' is also tied to resource consumption. The course also contextualizes design strategies that highlight 'environmental sensitivity', 'ecodevelopment', 'green architecture', and 'environmental

P. Pyla (✉) · D. Zacharia
University of Cyprus, Nicosia, Cyprus
e-mail: pyla.panayiota@ucy.ac.cy

M. Loren-Méndez et al. (eds.), *Sustainable Heritage Management - Design: Notions, Methods and Techniques*, https://doi.org/10.1007/978-3-031-81970-4_5

building assessments' in their specific historicities. Class readings and discussions focus on different types of design trends, organized around weekly guiding themes that stimulate critical reflection on 'green' strategies. (For details see Pyla 2020).

This chapter casts the spotlight on a key theme of the course, namely 'heritage conservation.' Students are introduced to this theme through lectures and critical readings that expose environmental dilemmas of conservation, and they are asked to further explore the theme through short assignments and longer research projects that sharpen students' critical tools and encourage connections between readings and design projects. Reflecting on student work that focused on conservation, this chapter aspires to systematically expose how the critical histories of ecology that are studied in the course also inform a critical understanding of the role of heritage and heritage conservation. In order to allow for consistency in the object of study here, the discussion below focuses on one type of student output, namely the collages they produced in response to assigned readings and relevant discussions. Unlike research papers which required systematic investigation of specific locally built projects, interviews, and mapping of sorts, the collages appealed to other types of abilities, that are perhaps closer to students' design and artistic skills. The assignment invited students to produce an image that dramatized an impression or concern they formed following a specific reading. The collages are therefore conceptual or abstracted responses, and not products of long-lasting research. They called on students to stage, or even exaggerate the relationships of heritage to environmental issues, in order to invite the class to engage in deeper reflection on the responsibilities and potentials of heritage practices. Using the work produced from this conceptual exercise, and highlighting the class discussion provoked by them, this chapter aims to showcase some of the complex questions on heritage tackled by the course: Is the conservation of buildings a means of achieving sustainable development, or does building conservation compete with the priorities of sustainability? Does the very practice of heritage conservation have environmental costs? And how does the issue of material authenticity, which is very important to architectural artifacts, become entangled with environmental priorities for zero emissions and energy conservation? Below we highlight specific environmental dilemmas of heritage preservation, indicated in the subtitles, that have emerged from the course.

Toxicities of Conserving

Perhaps the most contentious environmental issue faced by heritage professionals is the fate of toxic materials that are embedded during the construction of buildings. No matter how celebrated buildings are at the moment of their creation, their heritage value changes when they are proven, or even suspected, to be harmful to the environment. The reading of Amy Swift on 'inherited toxicity' (2011) captured precisely this predicament by exposing the 'ideological conflict' encapsulated in Frank Lloyd Wright's Price Tower—a building of undeniable significance as architectural and world heritage but at the same time a potential health hazard due to the toxicity of its distinctive copper elements. As Swift discusses, the wide use of copper in 20th-century architecture presents antagonisms between environmentalism and conservation in the management of these inherited embedded toxins. Concerns over environmental and human health led to testing carried out on the tower's copper eaves that eventually dismissed the possibility of toxic runoff migrating into groundwater, absolving both building and caretakers of bringing harm to humans and the environment. Swift acknowledges that due to limitations of the soil testing, the possibility that further testing might indicate an increased risk could not be eliminated. She preemptively defends the conservation of the copper elements, arguing that if interventions became necessary, they should allow the building to keep its aesthetic marker, while minimizing the risk to human or environmental health.

Swift's analysis inspired class student Orestis Pavlides to use Fallingwater, a better-known building by the same architect, to exaggerate concerns about embedded toxins in modern heritage buildings. (Fig. 5.1) Although the toxicity of Fallingwater is entirely imagined by Pavlides, his collage utilizes Wright's masterpiece to pick up on two understated findings of Swift's article that lie in the shadows of her main argument. The first is that the link between construction materials and their surroundings is not always explicit. As indicated in the above discussion, Swift acknowledged that testing on the Price Tower stopped short of investigating this link fully because it focused only on the most straightforward route of toxins migrating into groundwater; runoff falling directly into the soil, which was assumed to have the greatest level of toxins. Testing thereby overlooked alternative routes, such as gutters, that runoff could use to migrate to groundwater. The collage, by comparison, astutely emphasizes how toxins seep into the environment, as they can be seen literally flowing through the waterfall. Pavlides even borrowed the green color of patinated copper to represent the toxins, highlighting the irony about the dubious environmental impact of all things green.

Overall, the collage calls attention to the multiplicity of intertwinements of buildings and their environments, and the choice to assign an imaginary toxicity to Fallingwater, which is highly acclaimed for the way it blends with its natural surroundings, was particularly provocative. Fallingwater has also been characterized as a "contradiction in sustainable design" (Alter 2020) exactly because of its intimate relationship with nature and more specifically, the constant maintenance it requires on account of its proximity to the waterfall over which it is built. The student astutely chose to exaggerate that contradiction, highlighting yet another finding of class discussions that emerged from a critical reading of Swift's article, and further explored through our own (the instructors') retrospective reflection on the collage; that for Swift, and perhaps many in the architectural community in general, environmental concerns come second to heritage value. At the time of this collage's creation, neither of the two Wright buildings were World Heritage Sites,[1] but since then, Fallingwater was inscribed onto UNESCO's list of World Heritage Sites in 2019, and Price Tower

[1] Fallingwater is one of the buildings included in the World Heritage Site 'The 20th-Century Architecture of Frank Lloyd Wright.' (UNESCO 2023)

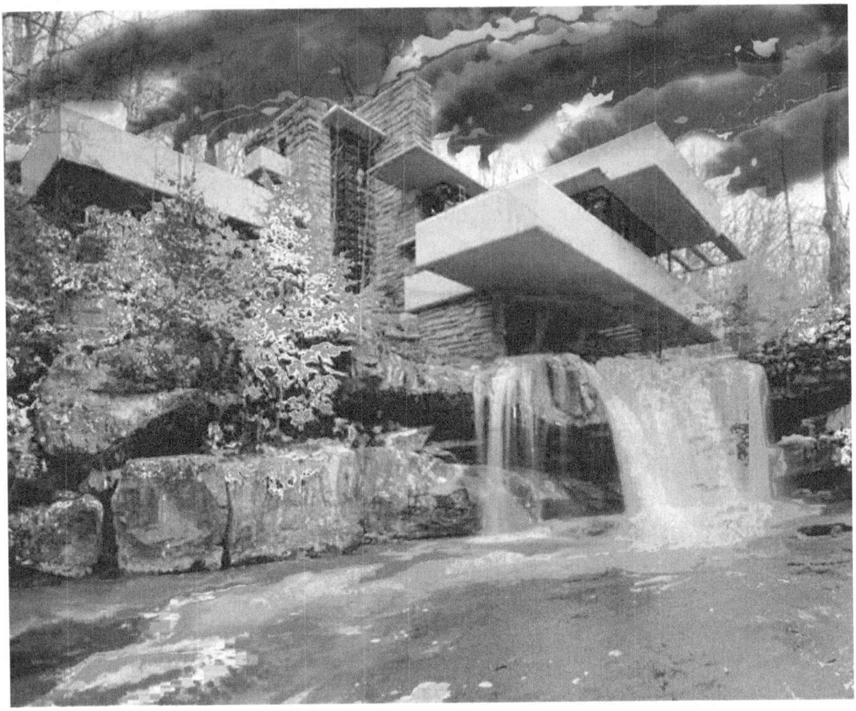

Fig. 5.1 Collage by student titled "Falling Toxic." *Source* ARH 511 Class Project by Orestis Pavlides, University of Cyprus, 2013

was nominated for inclusion in 2015 but has not made the list yet. In the hypothetical scenario where construction materials are found to be toxic, one wonders if the buildings' classification as World Heritage Sites would tip the scales in favor of conservation, no matter what the environmental cost. Even in the absence of toxic materials, students questioned whether the act of conserving Fallingwater as architectural heritage is in itself unsustainable, not just in terms of expense, but also the resources and energy used to maintain it. The provocations set forth by this collage therefore pose the following challenging questions: Are the clear and present dangers of toxic waste sometimes downplayed in favor of a vaguer notion of a modern building's alleged 'fit' with its surrounding environment? How would classification as a World Heritage Site affect the dynamics of heritage conservation and the priorities of environmentalism, especially in the present day when both modern heritage and the global climate are both under unprecedented threat?

Environmental Wastes of Material Conservation

To interrogate whether and how the conservation of buildings contributes to resource conservation, and supports environmental protection at large, the collage by Elena Rousou and Kleitos Kleitou examined the refurbishment of the Secretariat Building, one of the buildings of the United Nations headquarters complex in New York City. (Fig. 5.2) The renovation of the entire complex, named the Capital Master Plan (CMP), is a highly acclaimed model for conserving buildings in the name of sustainability. The CMP sought to renovate buildings of the complex rather than demolish them, so as to avoid the resource waste that would have resulted from the demolition of multiple large-scale buildings. A conscious effort was made to preserve the aesthetics of the Secretariat Building specifically, by replicating the existing glass façade of its iconic curtain wall with energy-saving glass. Recognizing this as a 'green' strategy, and influenced by Adrian Parr's critique of "greening" strategies and "greenwashing", the students chose to pose this question: What happened to the original glass of the Secretariat Building that was replaced as part of the renovation? To expose the environmental costs of discarded building materials the students created a stop-motion GIF that depicts the Secretariat Building prior to refurbishment (the first frame), during refurbishment (the middle frames), and after refurbishment (the fourth and final frame). The middle frames show the building foregrounded by a pile of shattered glass and debris, implying that this is waste produced by the refurbishment. A billboard atop the building reads 'soon to be green,' while in the final frame, this has been replaced with endorsements from LEED; a well-known green building rating system. The students intended to highlight the irony of awarding the Secretariat Building (which had indeed achieved the highest level of LEED certification), when the process, according to them, produced other forms of waste that burdened the environment.

Influenced by William McDonough's dictum 'Waste Equals Food,' the students appear to be making a statement that aligns with the argument advanced by the architect and his partner ecological chemist Michael Braungart: that we "should remove the word 'waste' from our vocabulary and start using the word product instead" (McDonough 2004). At the same time, their collage draws on Michael Zaretsky's critique of LEED in 'LEED After 10 Years' (2010) another key reading of the course, that exposes the blind spots of this particular green building rating system. Prompted by the student's more critical look into the celebrated refurbishment, the instructors researched the topic further, discovering that contrary to the students' initial assumption, the glass was not discarded but was recycled and used in the refurbishment. More specifically, the glass was ground into pozzolan (an environmentally-friendly cement replacement) and used to create pavers that were then used to tile the plaza at the foot of the building. This is said to reduce carbon emissions by replacing up to 30–50% of the cement in concrete. Zaretsky's exposition implores us to consider how the true fate of the glass, which scores credits in the LEED rating system for recycling and using regional materials, may produce other forms of waste. One such possibility is that the invention of innovative materials that repurpose discarded glass in order to

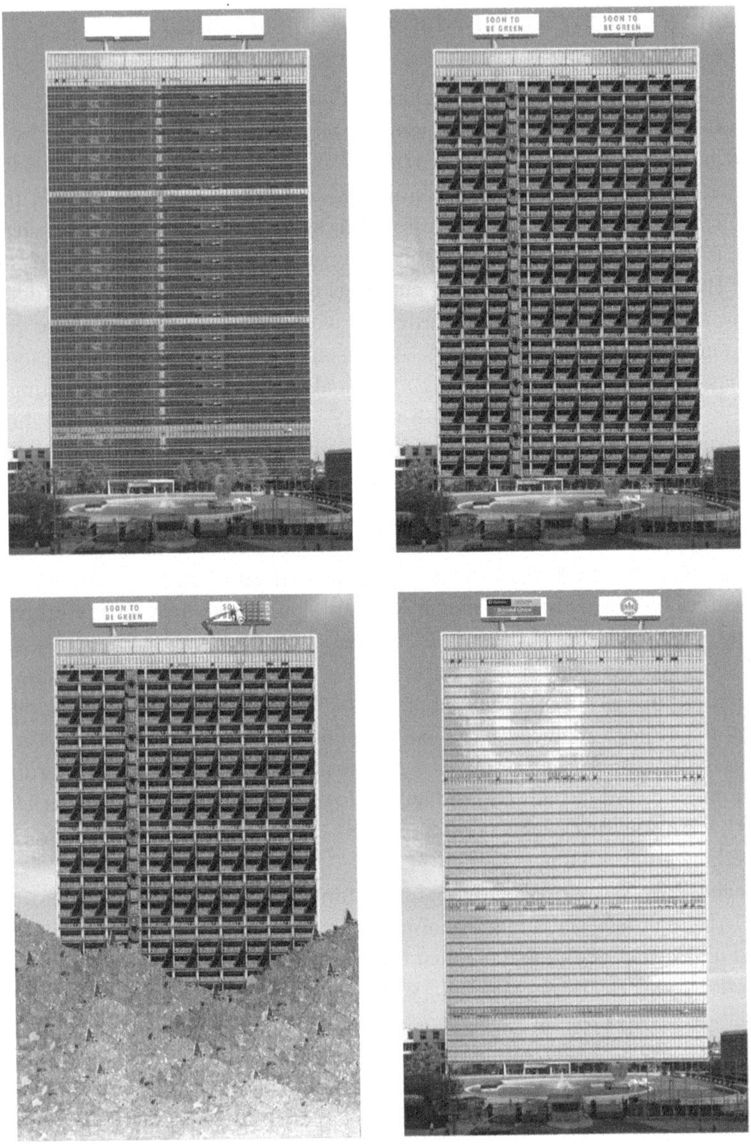

Fig. 5.2 A series of frames from a stopmotion GIF contemplating the relationships of material reuse and waste in the refurbishment of the Secretariat Building, UN headquarters. *Source* ARH 511 Class Project by Elena Rousou and Kleitos Kleitou, 2021

prevent it from ending up in landfill, actually spurs the production of more concrete-containing materials. How does heritage value play a part in this process? Does the notoriety of the building and the symbolic gesture to return the material from its iconic façade to the project, encourage the creation of even more paved areas? How justified is it to label this material as 'green'? This line of questioning highlights that paving a larger surface area than necessary with pavers containing pozzolan would be awarded by LEED, even if it contained more concrete than a smaller surface area of concrete pavers.

Discerning 'Natural' from not

Course readings collectively raised the issue that built artifacts are harming air, soil, and water quality through excessive energy and water consumption, carbon emissions, and toxic construction materials. This prompted many students to seek ways to convey the impact of buildings on the natural environment in their collages. Barbara Basile's provocative image, titled 'The Anti-Heritage of Venice,' (Fig. 5.3) highlighted the other side of this argument by casting the spotlight, not on new constructions, but on existing and much cherished built heritage. Basile's collage considers nature, rather than built artifacts, as the potential agent of harm, demonstrating how elements that we typically value as 'natural,' such as sea water and sea life, can have a detrimental effect on valuable built heritage. Her collage depicts the complex ecology of mould and fish that erodes the underside of the architectural treasures of Venice. Basile is not simply creating an artificial counterargument to the effect of 'if buildings harm animals, animals also harm buildings'; instead, she is exposing the complexity of environmental degradation, where both buildings and animals stand to lose. This is because if the seawater and humidity of Venice are eroding age-old building stones, it is not simply a 'natural process;' it is very much tied to industrial and agricultural waste, as well as other human activities such as unchecked tourism development. Basile's collage thus underlines the complicated entwinement of human and animal histories, as well as the interconnectedness of built and natural landscapes. Such interconnectedness was explored in readings such as 'Kiss Nature Goodbye: Marketing the Great Outdoors' (Beardsley 2000) and 'Smoky Mirrors and Unreflected Vampires: From Eco-revelation to Eco-relevance in Landscape Design' (France 2000), that demonstrate how our understanding of 'nature' changes through partnerships between humans and nature. This is particularly evident in the commodification of nature which has spurred the creation of simulations of nature as seen in the Beardsley text, as well as in eco-revelatory design as explored in the France text.

In the context of these readings, Basile's response captures the fluidity of the term 'natural,' leading us to question the tensions that arise through the coexistence of humans and nature and how this relationship 'threatens' architectural heritage. Two main threats to Venice's heritage—the rising sea level, which is responsible for the narrative that Venice is 'sinking,' and building decay which is accelerated by unnatural waves—are problems that can be linked directly to humans. On the

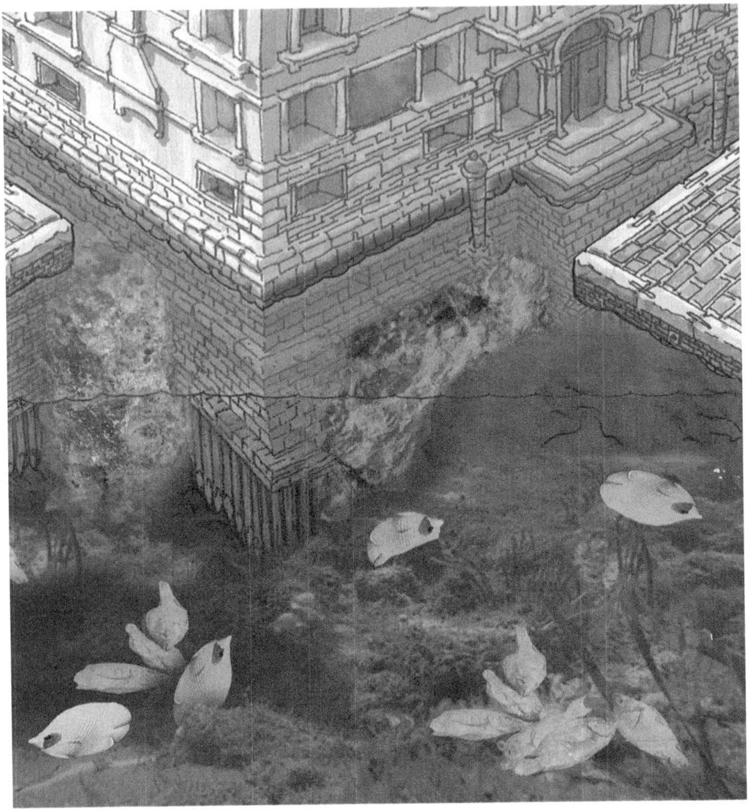

Fig. 5.3 Drawing pointing to 'role reversals' between nature and buildings. *Source* ARH 511 Class Project by Barbara Basile, 2013

one hand, climate change, but also industrial and agricultural waste deposited in the canals, causes the canals to become shallower over time. The erosion of the buildings, on the other hand, is caused by unnatural waves due to increased traffic in the canals which is especially high during peak tourism seasons. In addition to erosion, the presence of chemicals in the water (also a side effect of industrial waste) stimulates algae multiplication that accelerates building decay. The discussion provoked by this collage, not only reveals that nature can become the scapegoat for the decay of buildings, but that further antagonisms may develop between the two. Decaying buildings have become part of Venice's appeal to tourists, and since the economy of Venice relies heavily on tourism, conservation efforts will need to delicately navigate competing priorities with direct consequences on heritage.

Material Authenticity

An emergent and recurring issue faced by striving for sustainability in heritage conservation is considering how it impacts material authenticity. Whether it involves the management of toxic components, replacement of components with more energy-efficient ones, or restoration of components to their 'natural' form, the conservation of buildings with significant historical and architectural value is confronted with a challenge. As highlighted by Swift's text, a dilemma emerges when 'authentic' construction materials are proven to have a detrimental impact on the environment and human health. In the case of the Price Tower, for example, the green color of the patinated copper was important in that it was envisioned by the architect who likened the tower to a tree that had escaped from the overcrowded forest. (Swift 2011) The copper elements are therefore inextricably tied to the building's significance, which poses the dilemma of whether to sacrifice the authenticity of architecture for the sake of the environment, or as suggested by Swift, to merely minimize the risk using other interventions–even if this compromises its aesthetic authenticity anyway. Conversely, in the renovation of the Secretariat Building, the authenticity of the original glass façade was weighed against its energy losses, and this led to a decision to replace the original glass with 'state-of-the-art' double-glazed panels. This sacrificed material authenticity, but ensured aesthetic continuity, in order to gain energy efficiency. The multi-layered and ambiguous definitions of authenticity in architectural production, demonstrated by the above-mentioned examples, was explored further through the course by a student research project. The students concluded that, ultimately, such dilemmas arrive at a crucial moment where key actors are to decide which authenticity matters more. Their analysis also highlighted that deciding can become even more complex when the materials specified in an architect's design differ from the materials used in the final construction. By posing the question: 'Which of a building's many forms over time constitutes its most authentic self?' The students challenge the priority inferred by Swift's analysis of Price Tower, highlighting that the answer would invariably differ from the perspective of an architect, architectural historian, owner, or user. Their revelation therefore provides a more critical angle for answering the above question for Price Tower, Fallingwater, and the Secretariat Building.

The case of Venice highlights a further complication of determining material authenticity: that materials undergo changes over time and such changes may contribute to a building's aesthetic significance. Whether these are considered natural or not (as in the case of decaying buildings in Venice), the atmosphere and charm of Venice is deeply rooted in the decaying aesthetic of its buildings. This is equally true for unnatural changes such as the copper elements of Price Tower which were patinated on site using an acid bath. The perishable aspect of heritage, which derives from the durability of construction materials, prompts us to consider the lifespan of architectural artifacts. This was captured through another student collage, by Katerina Nikoletti, that likened a building to a milk carton (Fig. 5.4). At first glance, the collage implied the need to carefully consider the parameters for the longevity of a building, as indicated by the 'best before' date stated on the carton. At the same time, it made a reference to the building's afterlife through additional instructions to 'then recycle.' The subtext of the collage, inferred by the absurdity of the collage, is that buildings are more complex. This is because, unlike milk, a building is not 'homogenized,' it is composed of different parts that have different expiration dates (e.g. pipes live less than stones), but also because buildings are anchored in social, cultural, and historical contexts specific to them. In this way, the collage not only challenges us to think about the loss of heritage that results from 'expired' architecture, but also, the ways in which this intersects with ecological concerns. This is best explained through the example of the Secretariat Building, where the existing glass can be considered to have 'expired.' If the building were to be demolished on account of an expired component, this would prematurely end the life of countless other components that had not reached the end of their lifespan. If the said components are not recyclable, this creates waste. It is with this logic that heritage conservation, which gained prominence with the reversal of obsolescence in architectural thinking (Abramson 2016), was and is employed as a sustainable strategy. Expiration and obsolescence, often used interchangeably, are both anticipated with the passing of time and sometimes have common causes and effects. By referencing expiration in her collage, Nikoletti suggests that the durability of the building is not only predicted, but prescribed, as is the case with cartons of milk which are expected to state clearly when their content expires.

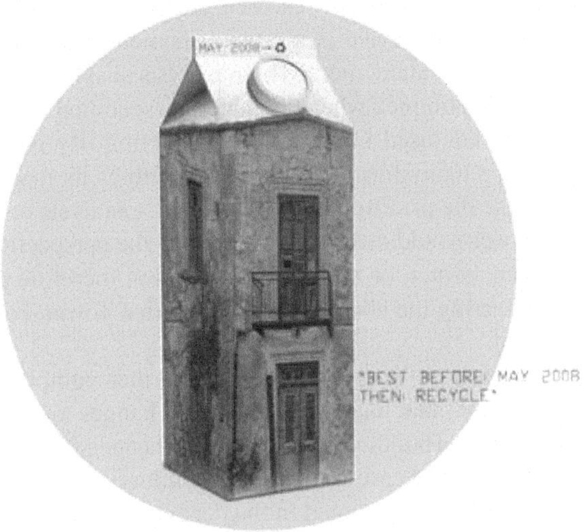

Fig. 5.4 Sketch contemplating if buildings have an expiration date. *Source* ARH 511 Class Project by Katerina Nikoletti, 2019

Conclusion

The course described in this chapter aims to challenge key student assumptions about the positive and negative environmental effects of heritage conservation. Using critical and theoretical tools from architectural and environmental history, the chapter, and the student work it showcases, pose key questions about heritage that are useful to any contemporary practice in heritage conservation. The student collages did not go as far as to suggest new or alternative methods of practice to counter the problems they highlighted, however, they have shed light on specific contentious realities encountered in the profession; that heritage can at times be toxic, that conservation may be wasteful if it requires excessive use of resources, and that the built environment cannot be separated from 'nature' and the 'natural,' as these are not categories outside the human, the technological and the social. These realizations can be crucial in informing principles and priorities for advancing green strategies, devising or revising environmental assessment rating systems, or developing new modes of thinking for bringing together ecological design and heritage conservation.

References

Abramson, Daniel M. 2016. *Obsolescence: An Architectural History*. Chicago: The University of Chicago Press.

Lloyd, Alter. 2020. Fallingwater: A Contradiction in Sustainable Design. Treehugger. https://www.treehugger.com/fallingwater-contradiction-sustainable-design-4859188

Beardsley, John. 2000. Kiss nature goodbye: Marketing the great outdoors. *Harvard Design Magazine Winter /spring* 2000: 60–67.

France, Robert. 2000. Smoky mirrors and unreflected vampires: From eco-revelation to eco-relevance in landscape design. *Harvard Design Magazine Winter /spring* 2000: 36–40.

William, McDonough. 2004. Design, Ecology, Ethics and the Making of Things a sermon given at the Cathedral of St. John the Divine, New York City (1993). In *The Sustainable Urban Development Reader*, ed. Stephen M. Wheeler and Timothy Beatley, 181–187. London and New York: Routledge.

Parr, Adrian. 2009. *Hijacking Sustainability*. Cambridge, MA: MIT Press.

Panayiota, Pyla. 2020. Architecture and a Critical History of Ecology, ARH 511: Student Projects, 2018–2020. Department of Architecture, University of Cyprus. https://296214fb-3882-4652-b49a-49af3c5216cd.filesusr.com/ugd/2b0b61_06c79347e93444e988535e11369095c6.pdf. Accessed 14 January 2023.

Swift, Amy. 2011. Inherited toxicity: An expanded concept of sustainability for preservation. *Docomomo Journal* 44: 58–67. https://doi.org/10.52200/44.A.KES8ABO1.

UNESCO World Heritage Centre. 2023. 1992–2023. World Heritage List. UNESCO. https://whc.unesco.org/en/list/

Michael, Zaretsky. 2010. LEED after Ten Years. In *New Directions in Sustainable Design*, ed. Adrian Parr and Michael Zaretsky, 191–201. London: Routledge.

Chapter 6
Objectives, Frameworks, Processes, and Outcomes: The Role of Bioclimatic Design Architectural Competitions, for Historic Open Public Spaces, in Greece

Konstantinos Sakantamis and Angeliki Chatzidimitriou

Keywords Environmental design · Sustainability · Urban design competitions · Historic sites · Cultural heritage preservation

Introduction

Escaping the architectural circles and entering the public domain, the issue of architectural competitions is once again relevant in Greece. Recent high-profile articles (Sakelaridou 2022; Papaioannou 2023) have raised issues of democracy/equal representation of Greek Architects in the context of competitions concerning major public works. In fact, Greece's latest legal framework (Ministry of Environment and Energy 2021) for the procurement of national or international architectural/urban design competitions dictates that their undertaking is mandatory in the case of public works that are of significance and prominence. Nevertheless, recent alterations to the above framework make it possible (in the case of private funding donations and/or devolved organizational patterns) for the state to hold such competitions by invitation, while delegating Greek architects to the role of the side-kick of renowned

A. Chatzidimitriou
Department of Architecture, School of Engineering, University of Patras, Patras, Greece

K. Sakantamis (✉) · A. Chatzidimitriou
School of Architecture, Faculty of Engineering, Aristotle University of Thessaloniki, Thessaloniki, Greece
e-mail: ksakanta@arch.auth.gr

School of Applied Arts and Sustainable Design, Hellenic Open University, Patra, Greece

M. Loren-Méndez et al. (eds.), *Sustainable Heritage Management - Design: Notions, Methods and Techniques*, https://doi.org/10.1007/978-3-031-81970-4_6

international architectural offices, as in the case of the competition for the extension to the Archaeological Museum, in Athens. The public discourse instigated by the particular competition is now further perplexed as the winning architects—David Chipperfield Architects (in cooperation with Alexandros N. Tombazis & Associates SA) can now also boast the Pritzker Prize for 2023. Nevertheless, the above clearly denotes that the Greek state upholds architectural competitions as a realistic best practice when dealing with major public places and indeed in the particular case of historic open public spaces. At the same time, the increasing attention given to environmental issues, in the light of ongoing climate change, has spurred a recent trend in the procurement of "bioclimatic design" competitions for such open spaces, which, in most cases, hold significant historic importance and built heritage.

The paper develops its critique to the current practice of such bioclimatic competitions in Greece through the study of three such competitions, procured for central locations, for the cities of Serres, Ioannina, and Heraklion. In the first two cases, the shortlisted designs are re-visited through a computer simulation, evaluated, and ranked according to environmental indicators, while the results are discussed in relation to the outcome of each competition. The methodology employed in these two cases has been used in several post-graduate research thesis (Kalaitzidou 2022, Moleiro Dale 2021, Gentsidis 2020, Atsalinou 2020, Tsigaki 2019) and has the added value of an educational perspective, facilitating the development of scientific environmental design evaluation skills through the critical appraisal of major public competition rankings/outcomes. In the case of Heraklion, for which one of the authors was a member of the judging panel of the competition, the paper reflects on the internal dynamics of a judging panel operating under the "bioclimatic" framework, drawing on official minutes and critiques to the process offered by members of the judging panel and the local authorities involved.

Methods for Assessing Architectural Competition Design Proposals

The assessment of environmental parameters in bioclimatic architectural competitions based on specific criteria and targets is necessary to substantiate and validate the quality of the proposals. The assessment criteria selection depends on the scope of the competition, the project site, the targeted uses, and the expected outcome, and may take into account quantitative and qualitative criteria. The measurable data for the environmental performance evaluation may include summer temperature reduction and UHI mitigation, comfort indices improvement, wind velocity, pollutants concentrations, building energy savings, etc. Qualitative criteria such as improvement of biodiversity, resilience enhancement, social connectivity and interaction, accessibility, sustainable mobility, and ventilation potential can also be documented by measurable indices such as built area ratio, green coverage, species variations, water elements, permeable surface area, uses distribution, pedestrian and cycling paths area, air quality indices, sky view factor, etc. Both measurable and qualitative criteria and assessment methods for microclimate development in open spaces have been proposed by earlier research projects such as RUROS (Nikolopoulou et al. 2004) and set as targets in funding programs such as the Bioclimatic upgrade of public open spaces programme, in Greece, (CRES 2011), and the Natural Capital Financing Facility within the EU framework (EIB n.d.) while further reporting indicators are being developed to allow monitoring socio-economic and environmental impact of urban sustainability interventions and to account for natural capital (European Commission, Directorate-General for Environment 2018, 2019).

The available tools for the environmental evaluation may include specialized simulation software for microclimate development and computational fluid dynamics tools, basic solar shading, and comfort indices calculation tools as well as simple data sheets, area measurements, and calculations. The evaluation process using such tools is based on the comparison between baseline conditions and proposed design conditions for each examined parameter and index. Improvement accomplished by the proposal is tested by either meeting specific target thresholds or by the extent of achieved change of the examined indices. In the case of architectural competitions assessment, monitoring tools and measured data can also be useful, but only to assist in baseline conditions evaluation and software validation and/or to be used as boundary conditions input for initializing simulations. Recent studies (Baeza et al. 2021) have proposed the use of a common tool for rapid analysis and decision support in urban design competitions to be used by both participants in the design process and juries in the proposal evaluation process.

Bioclimatic Design Competitions—Awards Review

Two post-graduate research studies (Tarasi 2019; Solidaki 2019) performed an evaluation of the awarded proposals from architectural competitions procured at Serres and Ioannina in Greece. The assessment of the proposals was made on the basis of their environmental performance and their effect on pedestrian comfort conditions, using microclimate simulations with analytic ENVI-met software. The critical microclimate parameters of air temperature, relative humidity, wind speed, surface temperature, and PMV indices were estimated through simulations for each proposal, and the results were compared to simulation results of the baseline conditions to calculate the potential improvement, on a typically warm summer day. The microclimate simulation results were also associated with urban form and materiality indicators such as ground surface material proportions, vegetation cover, and trees distribution for each case to further assess the impact of design parameters on the environmental conditions.

The Emporiou Square Competition in Serres

The architectural competition for the regeneration of Emporiou Square in Serres was announced in 2015 by the Technical Services Directorate of the Municipality of Serres, in collaboration with the Technical Chamber of Greece, Department of Central Macedonia (TEE/TKM). Located in the city center, the square has been a busy commercial hub since the last century, currently hosting multiple commercial activities, services, and urban transport, surrounded by midrise buildings of conventional construction and cultural monuments such as the neighboring ottoman baths. Twenty-eight entries were submitted to the competition, among which three equal first prizes and three honorable mentions were awarded by the jury.

Tarasi (2019) evaluated the three prize awarded proposals, on the basis of environmental performance and influence on pedestrian comfort conditions, using ENVI-met software for microclimate parameters simulation. The evaluation methods included:

- Comparison of the results of the initial state of the area with the results of each of the proposals,
- Assessment of the effect of the various forms of planting, materials, etc. in the microclimate parameters and pedestrian comfort, and
- Evaluation of the contribution of greenery to the effectiveness of the design of the square.

The key indicators of microclimate conditions, i.e. air temperature, wind speed, surface temperature, and pedestrian thermal comfort indices PMV, were estimated by simulations of each awarded proposal, and compared against the original conditions at the site (Fig. 6.1). The simulation results were also associated with design parameters such as the proportions of ground materials, the amount of vegetation cover, and the distribution scheme of trees, in the analyzed scenarios. The spatially averaged values of the simulation results for the entire area, derived from ten different points of the square, were used to assess the proposals' effects on air temperature, wind speed, surface temperature, and PMV indices at noon and evening of a warm day in July. Additionally, four key spots within the square were selected to compare the daily fluctuations of the same parameters between the proposals and the baseline conditions and to analytically assess the temporal effects of the variations of spatial design, materials selection, and vegetation distribution (Fig. 6.2).

The comparison revealed that Proposal 2 had the greatest area averaged improvement over the baseline conditions on a typical summer day with a reduction of air and surface temperatures up to 0.63 °C and 5.2 °C, respectively, corresponding to reduction of 2.2% and 15.8%, respectively, and a reduction of comfort indices up to 16.1%. The comparison of spot results also showed the impact of the amount and the arrangement of trees and of the ground surface shading on microclimate and pedestrian comfort. More specifically, results indicated that:

- Shading of the ground surface had the greatest positive effect on microclimate.
- The amount of trees proportionally reduced temperatures during the daytime, but diminished the cooling potential of the square in the evening.

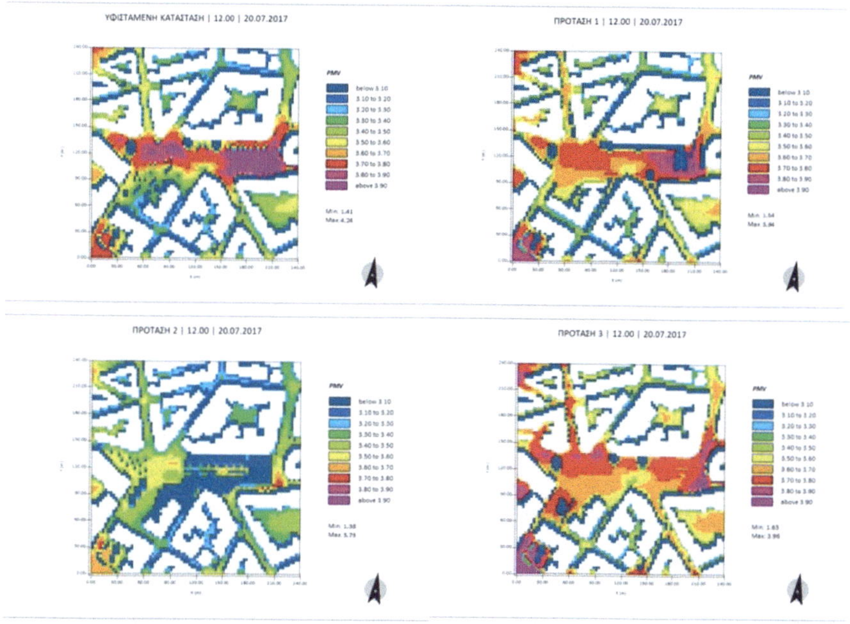

Fig. 6.1 PMV indices by simulation results for the baseline conditions of Emporiou square (top left), Proposal 1, Proposal 2, and Proposal 3 (Tarasi 2019)

- Tree arrangement played a critical role in microclimate improvement: linear tree distribution presented a better overall effect for the whole day than individual trees or clusters of trees.
- At unshaded areas, the ground reflectivity determined the results: high reflectivity improved ambient temperatures but deteriorated PMV indices.

The Pyrros Square Competition in Ioannina

The municipality of Ioannina, Greece, launched a panhellenic architectural competition for the redesign of Pyrros square, in 2017, calling for preliminary urban design proposals for a new identity for Ioannina's central square with respect to the historic context and bioclimatic, environmentally responsive approach. The site is a significant open space in the city, surrounded by public functions such as the Law Courts and the Prefecture building, as well as by a listed modernist building, the *Oasis* café by Aris Konstantinidis, built in 1971. The prizes announcement of 1st, 2nd, and 3rd awards and three honorable mentions, in 2018, was followed by an intense public discourse on the awarded proposals and triggered the research study (Solidaki 2019)

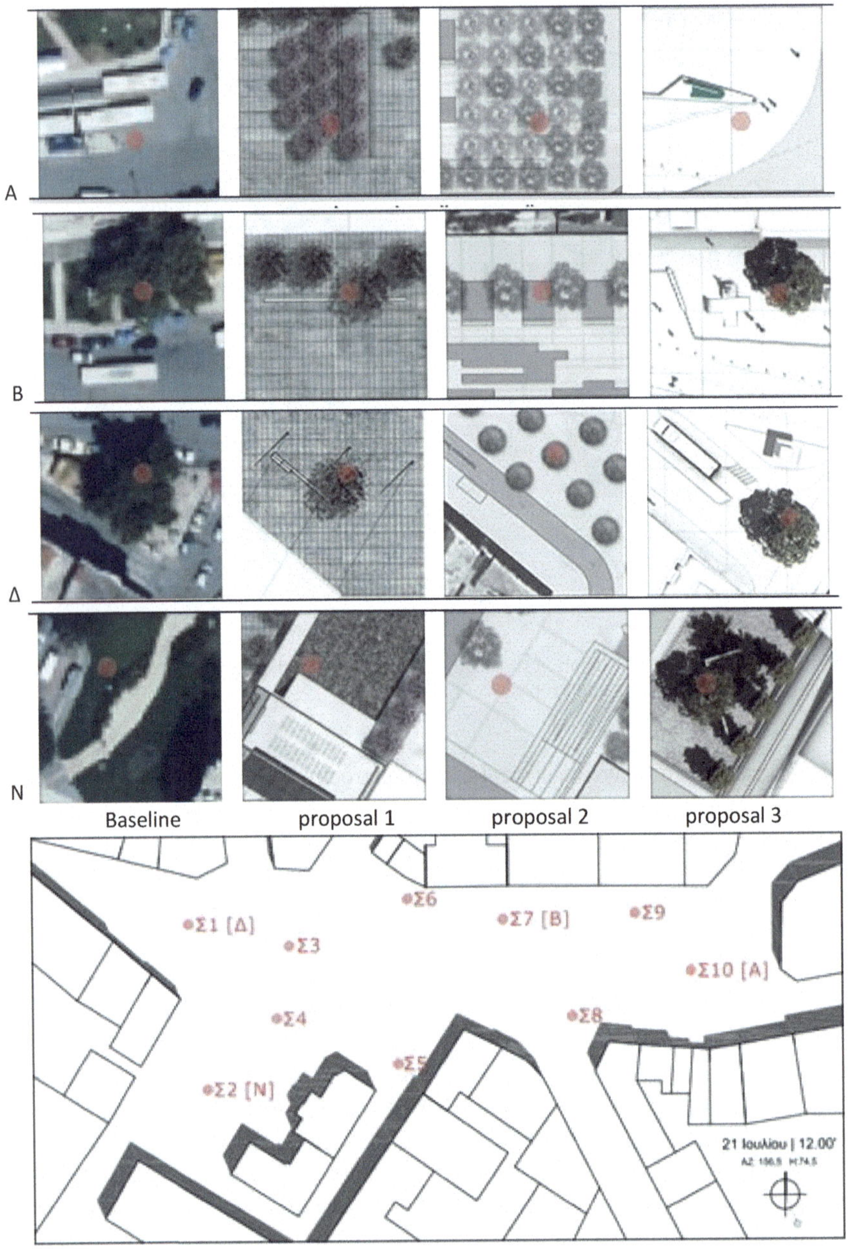

Fig. 6.2 Materials and vegetation at four key spots of Emporiou square (A, B, Δ, N) in the baseline conditions and the three awarded proposals (Tarasi 2019)

for the evaluation of their environmental performance considering that the competition brief did not set specific quantitable bioclimatic targets neither defined methods for environmental improvement documentation.

The existing conditions of the project site and the proposals of the first three awarded entries of the competition were also simulated with ENVI-met software (Figs. 6.3 and 6.4) and assessed according to microclimate parameters and pedestrian comfort (Fig. 6.5). Air temperature, relative humidity, wind velocity, and direction were the critical parameters evaluated in relation to the proposed design, ground surface materials, and green and tree coverage. The assessment results reveal differences in microclimate development among the proposals; however, the potential improvement by each proposal revealed by the simulations did not correspond directly to the awarded prizes and the jury's judgment of environmental improvement features. The results were evaluated at different times on a typically warm day in June, at 10:00, 12:00, 14:00, and 16:00 to assess temporal variations of the environmental conditions in relation to the occupancy patterns of the open space. Four specific spots of the square were selected for the comparison of the simulation results and assessment of the impact of design parameters on the square's microclimate development. The amount of vegetation and trees proposed in the design schemes was found to play a crucial role in improving the cooling potential of the square in summer while the hard pavement was found to increase ambient temperatures and deteriorate comfort conditions. Air velocity increase by unobstructed ventilation also proved to improve summer cooling potential.

Jury Decision-Making Process—Eleftherias Square, Heraklion Heritage Site Competition

The Directorate of Technical Projects & Studies of the Municipality of Heraklion officially engaged in an ongoing effort to regenerate the city's historic center in June 2020, beginning the process toward a panhellenic architectural competition for the redesign of the greater area of Eleftherias Square. This involved not only the general drafting/preparation of the competition brief but also the managerial work for securing the relevant funding for the implementation of the competition by the Green Fund/Ministry of the Environment and Energy. Concerning the wider eastern area within the declared-demarcated archaeological site of the city of Heraklion and delineating the Venetian walls, the competition with a total area of 53,000m2 focuses on bioclimatic interventions toward the reformation of Eleftherias Square and its connection with Daskalogianni, the Archaeological Museum and N. Kazantzakis squares, as well as the redesign of Beaufort Street, part of Dimokratias Street and Ikarou Street, which delimit the wider area of the eastern entrance to the historic center of the city. Launched in June 2021, the competition duration fell within the Covid-19 pandemic period; originally envisaged to last between 09/06/2021 and 09/

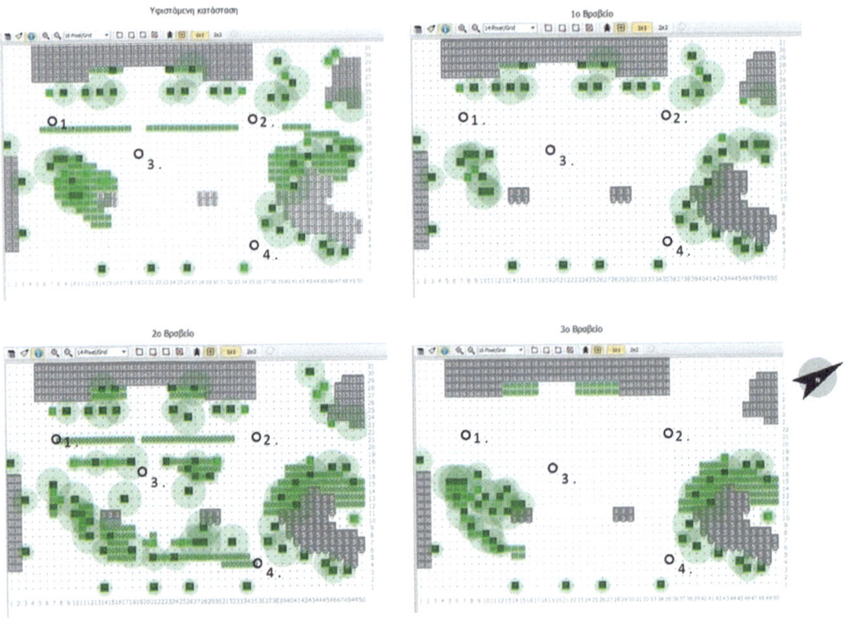

Fig. 6.3 Buildings and green infrastructure in the ENVI-met digital models of Pyrros square. Baseline conditions (top left), and first, second, and third prizes (Solidaki 2019)

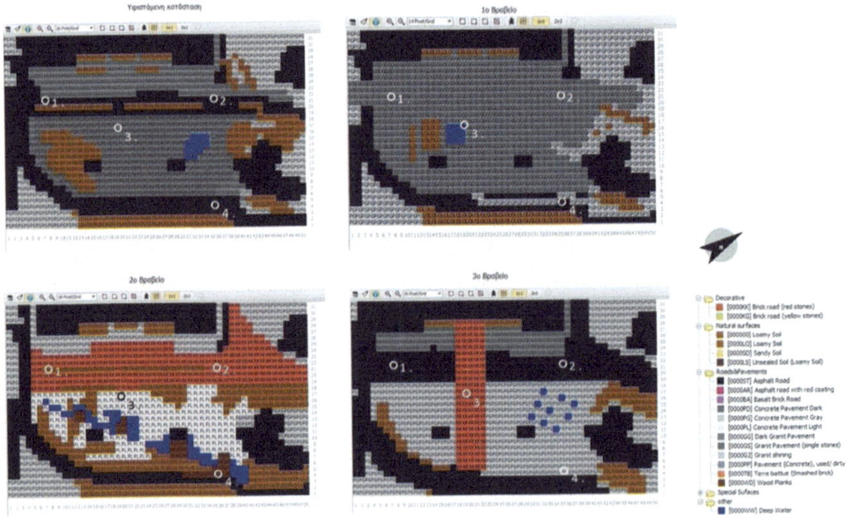

Fig. 6.4 Materials and blue infrastructure in the ENVI-met digital models of Pyrros square. Baseline conditions (top left), and first, second, and third prizes (Solidaki 2019)

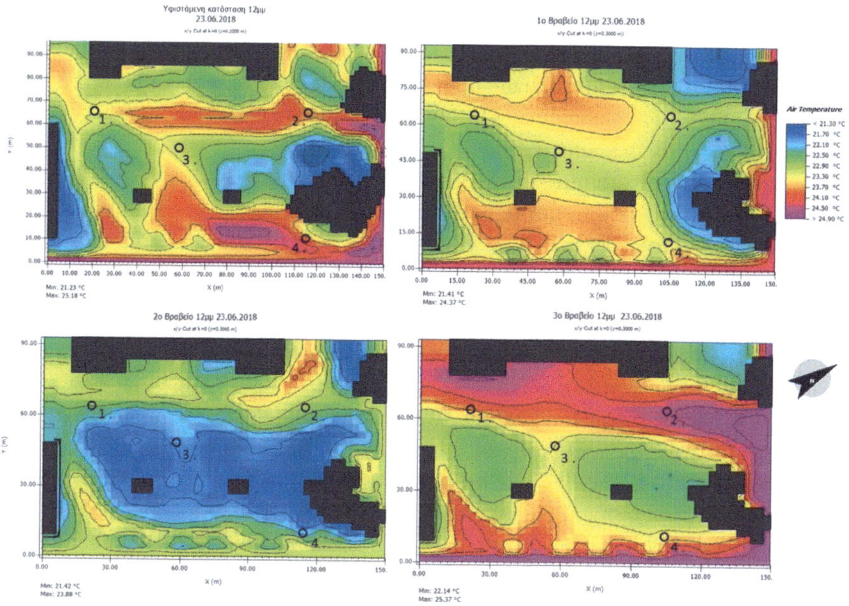

Fig. 6.5 Color maps of air temperature as calculated by the ENVI-met software, corresponding to the existing conditions of Pyrros square (top left), 1st award (top right), 2nd award (bottom left) 3rd award (bottom right), on 23rd June at 12:00 (noon). (Solidaki 2019)

11/2021, its duration was ultimately prolonged by three months, due to disputed issues between the public bodies overseeing the competition.

K. Sakantamis' involvement with the competition starts officially in January 2022, when he is confirmed at the honorary position of the president of the Jury by the Municipality of Heraklion. As prescribed by the legal framework of the competition (Ministry of Environment Energy and Climate Change 2011; Law 4412/2016), the five-member judging panel would be completed by an architect—member of staff of the Municipality, and three further architects drawn from the list of judges compiled by the Ministry of the Environment and Energy. Notably, an expert advisor without voting rights also sat at the Jury; the position was offered by the municipality to a Ministry of Culture representative who was to advise the jury on issues of heritage protection/conservation.

The competition call included a list of criteria which stressed the two dominant issues pertaining to the historic value of the site and its potential for improving the local microclimate, while also referring to issues of traffic/mobility, applicability of proposals, materiality and the scale of interventions, and budgeting. The weighting/enumeration and priority of the above criteria were open for interpretation by the Jury. A site visit and the eager engagement of the expert advisor were enough for the jury to accept that the historicity of Eleftheria Square has been lost through the transformations and redesigns it has undergone; the contemporary city no longer refers to the historical square as it was first created. The reflection and compromise between present and past constituted a strong factor in the committee's judgment. In stark contrast to the historical significance of the area, which was substantiated for competition participants through the relevant plans/documentation and references to historical sources (Tzompanaki 2011; Starida 2012, 2013), the bioclimatic concerns for the area were set in vague terms.

The eleven entries submitted varied in their interpretation/utilization of the main issues raised in the Brief; throughout the judging process, several issues were discussed by the jury:

- The committee did not identify any proposal that clearly addressed all dimensions of the issue of sustainable urban design set within the context/multiple scales of the competition; only one proposal employed simulation results to document microclime improvement;
- The lack of clear goals/indicators in the Brief, for the environmental performance of proposals in relation to a base-case scenario, allowed participants to also respond to issues of sustainability in generic terms—in most cases even lacking reference to local climatic data;
- The committee was being assisted in issues pertaining to the history of the site and the relevant built heritage, through the presence of the expert; however, their role, responsibility, and deliverables were largely undefined; on the other hand, no such assistance was possible for issues of sustainability/microclimate improvement;
- Subsequent phases, after the successful conclusion of the competition and the award of prizes, include consultations with the relevant stakeholders and culture ministry officials for the safeguarding of built heritage; no such practice for the consolidation/development of the environmental performance of the application study is set;
- The discussion on the potential for a more systematic approach of sustainability and heritage in pragmatic terms and taking into account the participation costs allowed jurors to reach a consensus on the necessity of discussing two-stage competitions, whereby shortlisted proposals receive funding to further develop ideas and respond to clearly set targets.

The proposal awarded the first prize was notably also the one that had the clearest substantiation of the microclimate improvement, utilizing simulation tools and local climate data. Unfortunately, to this date—almost one year after the conclusion of the competition, its results have not been presented and/or exhibited to the general public, thus severely restricting the commentary of the public and stakeholders on the proposed designs. Furthermore, the winning design has not been officially contracted to further develop the proposal or to engage in the consultation stages with ephorates and other public authorities.

Discussion and Conclusions: Setting the Framework for Bioclimatic Architectural Competitions at Historic Sites

The three case studies are indicative of the recent framework for "bioclimatic" urban design competitions in Greece. These follow the international norm, whereby environmental benchmarking is avoided or in any case not a frequent practice. More than proving that juries can be wrong about certain parameters of their appreciation of specific urban design proposals in the context of competitions, the methodology used in the study of the first two cases is shown to be able to determine the key parameters influencing the microclimatic conditions and also be able to provide a potential environmental value matrix for the analysis of the different proposals' merits. The study of the case for the Eleftherias Square competition, Heraklion, allows the authors to open the discussion on a reformation of current practices toward employing environmental benchmarking and simulation/analysis in the context of architectural competitions for historic open spaces:

- The role of experts/advisors should neither be confined to Ephorates/Ministry of Culture representatives nor only to the Jury but can be expanded to include representatives of public bodies and independent chartered engineers to assist early in the Briefing process,
- The applied best practice of sufficient data provision in the context of the Briefing process, regarding diagrams, schemes, historic data, photographs, surveys, can be expanded to include the monitoring of environmental conditions to assist in baseline conditions evaluation and software validation and/or to be used as boundary conditions input for initializing simulations,
- Funding bodies such as the Green Fund can be instrumental in creating such pilot applications of the above, setting target thresholds to suit microclimatic conditions, spearheading a more nuanced approach to environmental design of open historic spaces,
- Regarding issues of sustainable design, the appraisal of proposals can and should rely on both measurable and qualitative criteria and assessment methods for microclimate development,
- Two-stage competitions can enable more conceptual approaches to develop to maturity, whereby funding is provided to architects to fine-tune the environmental performance of their proposals; such an approach can accelerate our adaptation to climate change and further serve to protect and preserve built cultural heritage.

Acknowledgements Acknowledgements are due to the two Master Students of HOU Anastasia Tarasi and Maria Solidaki for their contribution. Also, the authors would like to thank the jurors and the organizing committee of the competition of Eleftherias Squares in Heraklion for their openness and availability in exchanging views and ideas.

References

Atsalinou, A. 2020. *Urban blocks' microclimate. Simulation of an urban block at Galatsi, Athens, and microclimate improvement potential investigation through the redevelopment of its open spaces.* MSc Diploma Thesis, MSc Program 'Environmental Design of Cities and Buildings', Hellenic Open University.

Baeza, J.L., J.L. Sievert, A. Landwehr, J. Luft, P. Preuner, J. Bruns-Berentelg, A. Noyman, and J.R. Noennig. 2021. CityScope platform for real-time analysis and decision-support in urban design competitions. *International Journal of E-Planning Research (IJEPR)* 10 (4): 121–137. https://doi.org/10.4018/IJEPR.20211001.oa8.

CRES. 2011. Program for Bioclimatic Upgrade of Public Open Spaces: Implementation Guide (in Greek). Operational Program "Environment and Sustainable Development 2007–2013". Centre for Renewable Energy Sources, Ministry of Environment, Energy and Climate Change. http://www.cres.gr/epperaa/bioclimat_anavathm.htm, http://www.cres.gr/epperaa/pdf/bioclim/odigos_efarmogis_programmatos_vioklimatikwn_anavathmisewn.pdf. Accessed on 01.03.2023.

European Commission, Directorate-General for Environment. 2018. I*ndicators for sustainable cities*, Publications Office. https://doi.org/10.2779/121865

European Commission Directorate-General for Environment. (2019). *Natural capital accounting: overview and progress in the European Union: 6th report,* Publications Office. https://doi.org/10.2779/733830

EIB (n.d.) *Natural Capital Financing Facility. Boosting investment for biodiversity and nature-based adaptation to climate'.* Available at: https://www.eib.org/en/products/mandates-partnerships/ncff/index.htm. Accessed on 01.03.2023.

Gentsidis, A. 2020. *Pocket parks. The contribution of scattered green spaces in the improvement of urban microclimate.* MSc Diploma Thesis, MSc Program 'Environmental Design of Cities and Buildings', Hellenic Open University.

Kalaitzidou, C. 2022. *Green- blue corridors and urban resilience. Urban streets regeneration and possible effects on the microclimate. Case study in the city of Edessa.* MSc Diploma Thesis, MSc Program 'Environmental Design of Cities and Buildings', Hellenic Open University.

Law 4412/2016 (Official Gazette A' 147/08–08–2016). Public Contracts for Works, Supplies and Services (adaptation to Directives 2014/24/EU and 2014/25/EU).

Ministry of Environment and Energy. 2021. Ministerial Decision No ΥΠΕΝ/ΔΜΕΑΑΠ/48505/387/2021. Official Gazette 2239/B/31–5–2021. *New framework for conducting architectural competitions and in general prizes awarding design competitions.*

Ministry of Environment Energy and Climate Change. 2011. Ministerial Decision No ΥΠΕΚΑ/26804/16.06.2011 Official Gazette 1427/B/16.06.2011. *New framework for conducting architectural competitions and in general prizes awarding design competitions.*

Moleiro Dale, M. A. 2021. I*mpact of Digital Planning Tools on the enhancement of the Urban Microclimate Case study: Grasbrook New District, Hamburg.* MSc Research Thesis, MSc Program 'Resource Efficiency in Architecture and Planning' (REAP), Hafencity University.

Nikolopoulou, M., N. Chrisomallidou, K. Steemers, R. Compagnon, J. Kang, N. Kofoed, G. Scudo, L. Katzschner. 2004. Designing open spaces in the urban environment: A bioclimatic approach. RUROS—Rediscovering the Urban Realm and Open Spaces. 2001–2003, Key Action 4, "City of Tomorrow and Cultural Heritage", from the program "*Energy, Environment and Sustainable Development*", within the Fifth Framework Program of the EU. ISBN: 960–86907–2–2.

Papaioannou, T. 2023. "Designer selection", another find of the elite (in Greek: «Επιλογή μελετητή», ακόμη ένα εύρημα των αρίστων. In *EfSyn* 04.02.2023, https://www.efsyn.gr/nisides/377184_epilogi-meletiti-akomi-ena-eyrima-ton-ariston. Accessed in 15.03.2023.

Sakelaridou, R. 2022. Something is up with architectural competitions (in Greek: Κάτι τρέχει με τους αρχιτεκτονικούς διαγωνισμούς). In *Kathimerini* 16.03.2022, https://www.kathimerini.gr/society/561758140/kati-trechei-me-toys-architektonikoys-diagonismoys/. Accessed in 15.03.2023.

Solidaki, M. 2019. *Are Green Awards really "green"?: A comparative study of the environmental response of the prizes of the Panhellenic architectural competition for the redevelopment of Pyrros Square in Ioannina, Greece.* MSc Diploma Thesis, MSc Program 'Environmental Design of Cities and Buildings', Hellenic Open University.
Starida, L. 2012. "*There was a city...The great Castle*", Itanos.
Starida, L. 2013. "*There was a city...Neighborhoods, streets, squares*", Itanos.
Tarasi, A. 2019. *The Contribution of Greenery to the Social Role and Microclimate of a Square. Comparative Evaluation of Proposals for the Redevelopment of Emporiou Square in Serres.* MSc Diploma Thesis, MSc Program 'Environmental Design of Cities and Buildings', Hellenic Open University.
Tsigaki, M. 2019. *The development of microclimate and pedestrian comfort conditions in urban squares through urban design; investigation of the relative influence of design parameters and improvement proposals for a central square at Chania.* MSc Diploma Thesis, MSc Program 'Environmental Design of Cities and Buildings', Hellenic Open University.
Tzompanaki, C. 2011. Khandakas, the city and the walls, Society of Cretan Historical Studies.

Chapter 7
The Identification of Cultural Attributes and Values in Heritage Management as a Key to Resolving the Gap Between Urban Planning and Intervention

Marta García-Casasola and Julia Rey-Pérez

Abstract The main objective of this research is to redefine the importance of the heritage perspective in the scenario of sustainable urban development, updating the methodologies. In this case, by focusing on the implementation of intermediate processes between urban planning and heritage intervention projects, processes associated with the identification of attributes and values. Likewise, progress will be made in the implementation of evaluation mechanisms that make it possible to control the changes, as well as the possible effects and impacts that may occur as a consequence of the intervention processes associated with heritage management.

Keywords Heritage processes · Criteria · Value report · Assessment

Introduction. Cultural Policies for Heritage Preservation

The cultural policies for management and conservation that characterise the current debate in the field of heritage have been formalised, from the second half of the twentieth century to the present day, through a mainly Europeanist vision. During this time, the working methodologies have been constantly updated, refining the phases of the process and implementing measures to ensure compliance.

The approach to the theoretical conceptualisation and the approach to a way of proceeding with objects from the past has been outlined through international texts and charters, taking the Venice Charter (ICOMOS 1964) as a reference point. This serves as a starting point illustrating the evolution of the concepts, criteria and methods that characterise heritage management and conservation. It is a genealogical reading (Fig. 7.1) that considers the following:

M. García-Casasola · J. Rey-Pérez (✉)
Department of Architectural History, Theory and Composition, School of Architecture, Universidad de Sevilla, Sevilla, Spain
e-mail: jrey1@us.es

M. Loren-Méndez et al. (eds.), *Sustainable Heritage Management - Design: Notions, Methods and Techniques*, https://doi.org/10.1007/978-3-031-81970-4_7

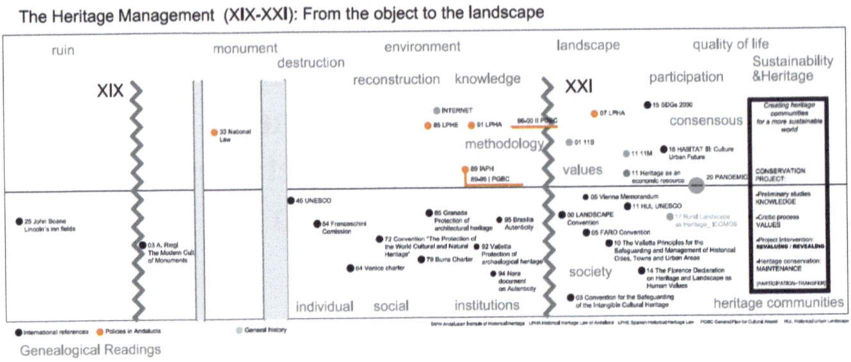

Fig. 7.1 A genealogical reading to approach the heritage management from the object to the landscape, from the end of the nineteenth century to the present. *Source* The authors, 2023

- The evolution of the nomadic concept of heritage (Choay 2007): from object to (cultural) landscape.
- The importance of memory and the commemoration of absence since World War II.
- The recovery of the context, first giving importance to environments and buffer zones and then giving way to the figure of landscape as a working tool.
- The social shift in the heritage approach, fundamentally from the safeguarding of intangible cultural heritage (UNESCO 2003) and its development towards the implementation of participatory processes.
- The growing importance of identifying cultural attributes and values as a key to guaranteeing heritage conservation.

In this context, the main objective of this research is to redefine the importance of the heritage perspective in the scenario of sustainable urban development, again updating the methodologies (González-Varas 2021). Through the revision of *Cultural policies for heritage preservation* from the European perspective, this work approaches *The theory of value and authenticity* as the main issue to manage heritage, always working with *Citizenship and participatory processes.* Both topics are in the purpose of the Faro Convention (2005). The *Methodology for Sustainable Heritage Management* is focused on quality control following the different steps defined from an integral view and with an interdisciplinary team. In this case, the focus is on the implementation of intermediate processes between urban planning and heritage intervention projects, processes associated with the identification of attributes and values. Likewise, progress will be made in the implementation of evaluation mechanisms that make it possible to control the changes, as well as the possible effects and impacts that may occur as a consequence of the intervention processes associated with heritage management. *Working with heritage communities to share knowledge and perceptions, to preserve and add cultural values and to construct significance* is the way to look for a consensus, to make decisions based on knowledge. *Conclusions* proposes the *Evaluation of heritage impacts: implementing processes to guarantee the preservation of cultural assets* as the final task to guarantee good practices in heritage management.

Numerous authors have reflected on the concept of value, constituting a theory of value that has continued to evolve throughout the twentieth and twenty-first centuries. Starting with Riegl who in 1903 published *The Modern Cult of Monuments* (Riegl 1999), the aim is to understand the complexity of the heritage quality of an object or place in which to recognise not so much the monumental values and their historical evolution as the remembrance and contemporary values—values that are understood from the present, inserted in a continuous learning process and which are in permanent transformation. An evolutionary process never ends and requires monitoring mechanisms capable of accounting for successive changes (Jokilehto 2016).

The Theory of Value and Authenticity

The theory of value since Aloïs Riegl, *a modern cult of monuments* understood as an evolutionary process, has been constructed in relation to the concept of authenticity, a concept that the Krakow Charter (2000) defines in its annex as a term that: "means the sum of substantial, historically determined characteristics: from the original to the present state, as a result of the various transformations that have occurred over time".

A definition must incorporate the future in order to have a complete time perspective, because heritage is projected into the future: a way of understanding heritage, from its value, requires a management that faces balanced and justified decisions on change in order to understand the modifications of the identified values, as well as the new values that may be generated from the projected proposals.

The concept of authenticity has been present in all the international reference charters and texts, from the Venice Charter, where it is barely mentioned, to the Nara Charter on authenticity (1994), where the task of understanding the authenticity of cultural heritage, rather than conserving it, is made explicit. The importance of proceeding with rigour, according to a working methodology, is emphasised. However, it is not until the Charter of Brasilia, in 1995, that the concept is addressed in its complexity, placing it ahead of the concept of identity with all its implicit subjectivity, a changing and dynamic concept. It is a call to recognize the importance of the perceptive that would later be reinforced with the Declaration for the Safeguarding of the Intangible Heritage (2004). One scenario in the challenge of heritage conservation will be related to the generation of a collective message, based on the understanding of the pre-existing. These issues will be taken up in the last internationally agreed text of a general nature, the Faro Convention (2005), whose title "On the value of cultural heritage for society" stresses the fundamental issues: making explicit what is of value for communities. The reformulation of the heritage concept at present, from a sustainable point of view, leads to talking about the concept of shared management, applying participation processes in all the phases.

Citizenship and Participatory Processes

The focus on citizens as responsible for the care of heritage values began timidly with the Krakow Charter: Principles for the Conservation and Restoration of Built Heritage (2000). Although it has not been adopted by any international organisation, it has had a considerable impact on the heritage contexts. It incorporates new elements such as the dynamic idea of memory in the construction of identity through the recognition and attribution of values to cultural heritage, the joint responsibility of administrators, heritage technicians and society in heritage management or the difficulty of establishing a single definition of universal and stable cultural heritage due to the variability of values in a process of continuous evolution (Naeyer et al. 2000, art. 11).

The beginning of this new approach takes place, as mentioned above, with the important contribution of the Council of Europe, materialised in the Convention on the Value of Cultural Heritage for Society—Faro Convention, 2005, which insists on cultural heritage as a component of special importance for the quality of life and sustainable development (Council of Europe 2005). The conservation of cultural heritage is no longer seen as an end in itself, since "the conservation of cultural heritage and its sustainable use have human development and quality of life as their goal" (art. 1c). And here appears the concept of heritage communities, which has acquired an important role today in urban heritage management and reflects the Council of Europe's commitment to the democratic management of Europe's growing cultural diversity. It also incorporates the right to heritage on the part of citizens (art. 4).

In 2011, ICOMOS adopted the Valletta Principles for the Safeguarding and Management of Historic Cities, Towns and Urban Areas. As stated in its preamble, it is based on the recognition of the changes that have occurred worldwide in the process of urbanisation and globalisation. It refers to urban identity, characterised not only by its physical and perceptual elements but also taking into account an intangible element such as the spirit of the place: the set of material and immaterial, physical and spiritual elements that give a site its specific identity, meaning, emotion and mystery. It also mentions a methodological renewal in relation to safeguarding and promotes the Management Plan. But it speaks of the key issue of urban governance (ICOMOS 2011a, b, c).

The Florence Declaration on Heritage and Landscape as Human Values in 2014 is based on the ethical commitment of ICOMOS to the protection and vindication of human values that protect the "spirit of place", i.e., the identity of a community in order to improve its quality of life. The formulation of this Declaration invites governments and civil society to participate together, considering landscape, in the improvement and quality of life (ICOMOS 2014).

Methodology for Sustainable Heritage Management

The methodology for the sustainable management of heritage has as its ultimate aim the social use of cultural assets, always on the basis of transdisciplinary knowledge, approached from an integral perspective and making use of the necessary planning/programming based on the recognition of what is valuable.

In the case of Andalusia, Law 14/2007 on Historical Heritage proposes the Conservation Project as the document that defines and orders the conservation, restoration and rehabilitation actions of Andalusian heritage assets. A document, the conservation project, is understood as the tool for intervention in heritage that responds to the methodological development carried out in complex cultural contexts, as is the case in Andalusia, focusing its proposal on the process formulated on the basis of the minimum contents that it must have (article 22):

- Study of the property and its cultural values;
- Diagnosis of its condition;
- Description of the methodology;
- Proposal for action (theoretical, technical and economic);
- Impact on the protected values;
- Maintenance programme.

The management of change or the implementation of the conservation project emphasises two fundamental issues of absolute relevance: the importance of recognising what is of value and the possibility of transforming these values, improving them and even implementing new ones. It should be considered alongside the importance of the project as a tool for projecting into the future what is yet to come. It is a way of working in which decision making is based on knowledge and where the impact on recognised cultural values is controlled. The methodology focuses on the identification of cultural values as a means to ensure the quality of the process: "The process of managing change to a significant place in its setting in ways that will best sustain its heritage values, while recognising opportunities to reveal or reinforce those values for present and future generations" (Conservation Principles 2008).

The issue of heritage management understood as a process (Fig. 7.2) seems to be sufficiently accepted, but it has had to overcome in numerous situations the gap between the guidelines set out in urban planning and the drafting and processes of intervention in the heritage (Gómez-Villa 2019).

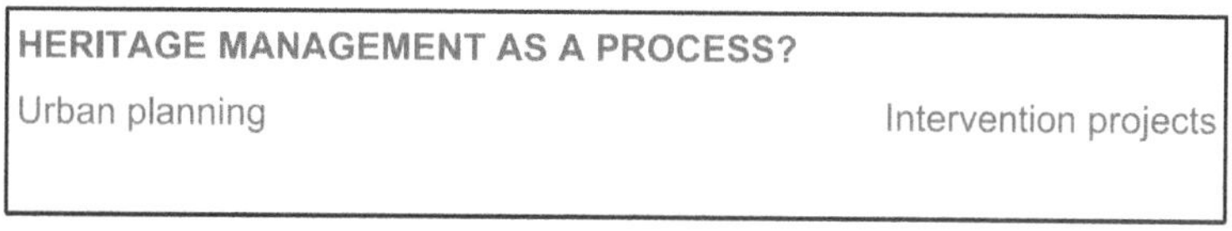

Fig. 7.2 Heritage management as a process: from urban planning to the intervention projects. *Source* The authors, 2023

This need for urban planning and heritage intervention to be managed in a connected way has been highlighted on several occasions by international organisations such as UNESCO or by various researchers (Bandarin and Oers 2015; Pereira Roders and Van Oers 2014; Pereira Roders 2016; Labadi and Logan 2016). In fact, the main objective of the Recommendation on the Historic Urban Landscape is to support inclusive and sustainable planning from the outset of the process, in which decision-making is carried out jointly and transversally between citizens, administration and technicians throughout the process (UNESCO 2011). In the heat of this change of approach, there have been different pilot cases that have tried to implement this new perspective where planning goes hand in hand with conservation (Pereira Roders and Bandarin 2019), with the case of Ballarat in Australia standing out (Murphy et al. 2016; Ballarat 2017, 2015).

In situations where urban planning instruments have a relevant role in the development of cities with a marked heritage character, other instruments have been implemented to cover the leap and provide intervention criteria based on knowledge and the identification of cultural values (Fernández-Baca et al. 2017). These applications have been produced on the basis of the European Landscape Convention (2000) and the *Recommendations on the Historic Urban Landscape* (2011), generating interdisciplinary work dynamics based on the development of rigorous characterisation processes as a prior step to detailed planning (Fernández-Cacho et al. 2022).

At the next level, closer to the object of intervention, the cultural values reports seek to identify and define the keys to the heritage reading of a cultural asset inserted in a landscape. In this way, it is possible to measure the degree of transformation that a given intervention causes on the identified values, trying to minimise the risks for better conservation of the heritage located in its context. These are instruments that facilitate the definition of criteria for intervention on cultural property, advancing guidelines and project strategies aligned with the identified values (Del Espino and Gómez 2022).

These working instruments (Guidance and Values Report) have also been implemented to cover a usually long-time span in which modifications occur in the culturally characterised contexts in which work is being done. In order to account for this evolutionary/transformative process, it is necessary to implement processes of evaluation of the impacts that are occurring on the identified values (ICOMOS 2011a, b, c). This evaluation, understood as quality control of the sustainable management of heritage, makes it possible to show the scope of the proposals that are planned prior to their implementation. It is a working process that must be accompanied by participatory and dissemination initiatives (Fig. 7.3).

SUSTAINABLE HERITAGE MANAGEMENT AS A QUALITY PROCESS
Urban planning - Landscape Guide - Value reports - Intervention projects
IMPACT EVALUATION______PARTICIPATION______DISSEMINATION

Fig. 7.3 Sustainable heritage management as a quality process: working step by step, developing different tools to guarantee the preservation of the values. Implementing the impact evaluation, the participation and the dissemination as a cross-cutting process. *Source* The authors

Identification of Cultural Attributes and Values

Identifying the multiple values and attributes that characterise a given cultural heritage can be complex, as it is a broad and diffuse subject of study that involves multiple actors for its definition and numerous disciplines of a humanistic, scientific, social or legal nature. Given its cross-cutting and wide-ranging nature, it is in fact a "discipline of disciplines" which implies a consensus-building process in which personal or collective assessments must be objectified and decisions justified. The identification of cultural values and attributes is a necessary process to develop intervention criteria to guide the different strategies of action, since as the Burra Charter states: "Conservation of a place should identify and take into consideration all aspects of cultural and natural significance without unwarranted emphasis on any one value at the expense of others" (ICOMOS Australia 1999, art. 5.1).

If values are not objectively given, neither can they be considered fixed and immutable, as was theorised in early modern conservation, where it was argued that heritage values are intrinsic to the property, then limited to historical and artistic values (de la Torre and Mason 2002). The contemporary perspective assumes that values are socially constructed in a given context and time and, therefore, change along with society. As the Nara Document on authenticity states:

> All judgements about values attributed to cultural properties as well as the credibility of related information sources may differ from culture to culture, and even within the same culture. It is thus not possible to base judgements of values and authenticity within fixed criteria. On the contrary, the respect due to all cultures requires that heritage properties must be considered and judged within the cultural contexts to which they belong (ICOMOS 1994, art. 11).

This very personal perception of heritage means that cultural values are often multiple (never just one) and often in conflict (de la Torre and Mason 2002). Assuming that the value of heritage is created by the generations of its time and changes over time implies not only valuing the outcome, but also the process.

This idea, this value, is formed from a series of objects or a cultural manifestation (tangible and intangible), which is where it resides. UNESCO calls these objective elements attributes (Court et al. 2022, p. 33). Both tangible and intangible attributes are considered heritage because of the cultural values assigned by us, as a community, as experts, as residents or as tourists. The moment we stop ascribing values to these attributes, they are no longer considered heritage (Tarrafa Silva and Pereira Roders

2021, p. 24). As Turner states: "Values are authenticated into the World Heritage through their attributes, while they are managed through indicators that measure the state of conservation and monitor change and threats" (Turner 2017, p. 26).

Heritage values can be an ambiguous term that can lead to misunderstandings in interdisciplinary heritage projects. The ambiguity of the term can be explained by the different cultural and ideological models, methods and approaches used by different disciplines, as well as other factors (such as institutional ethics of organisations, legislation, politics, etc.) (UNESCO ICOMOS and ICCROM 2021). From the scientific field, there are different researchers who have contributed different typologies of values, from those who differentiate heritage values from social values (Avrami et al. 2019, p. 11), or those who focus on economic values (Macdonald and Ostergren 2011), or even those who consider the ecological or indeed political question (Pereira 2007). In short, we can find proposals for defining values: historical, age, artistic, social, ecological, economic, landscape, symbolic… Working with a typology of values helps (Mason 2002), but based on the issues mentioned above, these typologies depend on the approach and the context. Therefore, these typologies are also in a continuous process of evolution/transformation. The same is true for attributes. There are different approaches to their identification, e.g., in 2005, the Operational Guidelines for the Implementation of World Heritage already referred to attributes (UNESCO 2005a, b). Nevertheless, ten years later, Loes Veldpaus proposes a fairly comprehensive identification of tangible and intangible attributes, which helps a lot in that disaggregation of landscapes which is sometimes necessary in heritage identification (Veldpaus 2015). The meeting in Fukuoka in 2021 also proposes an Annex C of possible attributes (UNESCO ICOMOS and ICCROM 2021), so all these possible categories are subject to different approaches and proposals, which tells us that it is also in the process of development.

The aim of this breakdown is to convey that the attributes, whether tangible or intangible, are the real objects of protection, both of which can coexist in the same heritage item and represent different values. Distinguishing between what is valued and why it is valued clarifies the approach to the protection of cultural heritage. The definition of cultural significance is broad and abstract, which hinders the scientific process and probably the definition of attributes, objectifying the process (Tarrafa Silva and Pereira Roders 2021). This situation highlights the existence of other values beyond the historical or aesthetic ones, also thanks to the variety of stakeholders involved in this valuation.

What is clear is that the Impact Assessment being undertaken by ICCROM and IUCN in cooperation with the World Heritage Centre and ICOMOS emphasises, as a first step, clearly identifying attributes and establishing a methodology for managing change and new development affecting heritage in urban contexts (UNESCO ICOMOS and ICCROM 2021, p. 2). The definition of cultural significance (identification of attributes and values) should guide all steps/aspects of decision-making affecting heritage conservation practice: identification, conservation policies, intervention strategies, investment, resource allocation… as well as the engagement of all actors involved in its care and enjoyment (Avrami et al. 2019, p. viii). Therefore, how can we correctly attribute attributes and values to heritage in order to understand and manage it and, consequently, establish the criteria that guide the different intervention strategies? Through readings, we can identify values and attributes in isolation, but how can we connect them? How can we work on this relationship?

Working with Heritage Communities to Share Knowledge and Perceptions, to Preserve and Add Cultural Values, to Construct Significance

Participatory planning consists of integrating heritage resources into urban planning, being aware that there are many variables that condition it, such as cultural dynamics or political issues. In the current situation, conservation management is complex as resources are diversified: immovable, movable, tangible, intangible, natural, cultural…; there are multiple disciplines: architecture, urban planning, history, geography, economics, anthropology, archaeology…; different actors: government authorities, professionals, experts, citizens, tourists, financial partners and real estate developers; a variety of values: aesthetic, historical, scientific, social, economic, environmental and political; and different approaches: preservation, restoration, renovation, recycling…

The introduction of social and cultural value from citizenship allows for the official recognition and protection of previously unrecognised types of heritage, as participatory processes go beyond the knowledge of heritage experts. In participatory planning, it is common for communities to identify a set of attributes linked to the collective memory or spirit of place that may challenge the position taken by heritage regulations and technicians. The attributes and values obtained in a participatory process go beyond the limits of materiality and heritage language (Avrami et al. 2019, p. 6). Normally, this approach shows a side of heritage unknown to technicians and administrations. The more important social values evolve into decision-making in heritage conservation, the more important participatory processes become. The heritage-community relationship today is very active.

It is a challenge to find mechanisms to encourage citizen engagement in participation (teaching them to identify attributes and values, to spatialise them…). Another important issue is to train citizens to participate in these workshops, as the cultural outlook of each stakeholder conditions the identification of the different values and attributes. This subjectivity responds to the evolutionary and temporal dynamics, which is why the role of the multidisciplinary team is necessary to mediate in the process and establish decisions based on a scientific approach. This process of identifying attributes and values must take into account the mutable nature of heritage as a process of social construction (Tarrafa Silva and Pereira Roders 2021, p. 24).

From a socio-community and inclusive perspective, a methodological proposal is put forward which, through a participatory approach to planning, incorporates citizens in the different phases of the project. This approach would in itself be a

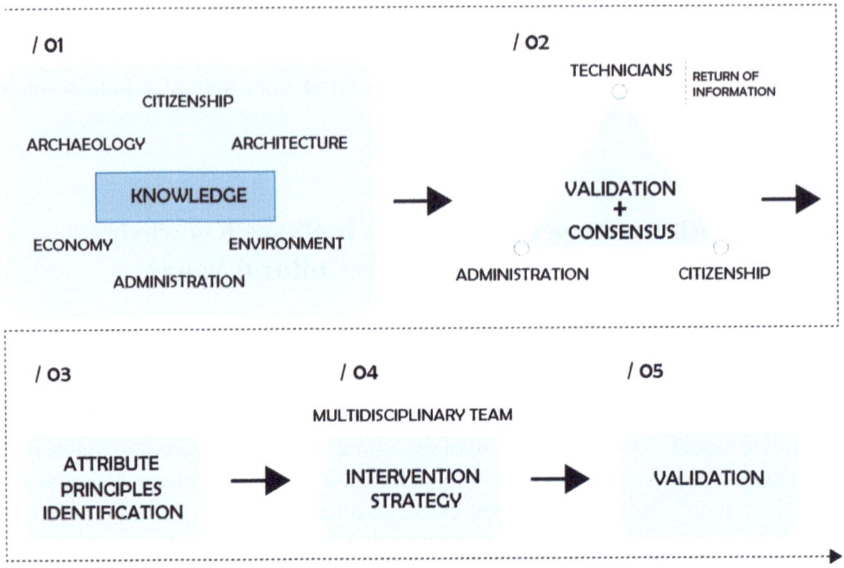

Fig. 7.4 Methodological process to incorporate citizens in the identification of heritage attributes and values from the beginning of the process. *Source* The authors, 2023

sustainable development strategy by guaranteeing the inclusion and integration of all those stakeholders who should be involved in the enhancement of cultural heritage. It is important to consider the heritage community from the diagnosis of the first analytical phase, which identifies its heritage attributes and values, to the design and prioritisation of sustainable intervention proposals aimed at planning the evolution of the property. This approach ensures not only the inclusion of citizens in the whole process of the research project, but also the consideration of the data obtained from citizen participation at the same level as the data provided by the studies of the other disciplines (Fig. 7.4).

This approach defines a new heritage paradigm in which it can usually be clearly seen how citizens value more intensely that which is linked to gastronomy, traditions, the physical environment or public spaces. This type of approach usually leaves built heritage in the background, even though it is the only one with protection regulations (Rey-Pérez et al. 2017). The participatory study usually makes it clear that the valuation of urban heritage by citizens is far removed from what is considered by the administration, hence the need to establish a participatory process at all times.

Conclusions. Evaluation of Heritage Impacts: Implementing Processes to Guarantee the Preservation of Cultural Assets

The identification of values and attributes for heritage conservation is presented as the key task for implementing rigorous heritage management processes and with which it is possible to make assessments. However, it is a complex process and there is a lack of examples to illustrate the heterogeneity of situations in which to apply this methodology. "Values-based approaches are not well researched and formalized" (Avrami et al. 2019, p. 8), although there have been numerous interesting research projects that have addressed the question of heritage in relation to its value (Avrami et al. 2000).

The main difficulty has to do with measuring something that changes over time and pertains to the idea of diversity. ICOMOS started this journey a few years ago (2011) to respond to governance processes at World Heritage properties. The *Guidance on Heritage Impact Assessments for World Cultural Heritage propertie*s represents a milestone in the process of putting in place a system that offers guarantees throughout the processes of change to which these properties and the environments in which they are situated are subject. The assessment of the impact on the values identified in cultural properties has been a concern of UNESCO since the *Convention Concerning the Protection of the World Cultural and Natural Heritage* (1972), which defines the Outstanding Universal Values (OUV) that characterise World Heritage. The experience of ensuring the preservation of these properties has led to the establishment of guidelines for impact assessment. Undoubtedly, a path has begun that starts from a point of no return and which will require institutions and heritage intervention project drafters to demonstrate the benefits of their projects and the non-affection of the cultural values initially identified in the areas being worked on. A working methodology focused on World Heritage, but extrapolated to other cultural assets to be conserved, can be consulted in detail in the Manual recently published (2022) under the title *Guidance and toolkit for impact assessments in a World Heritage Context* (Court et al. 2022).

This approach seeks to define a new way of thinking about heritage management based on its heritage values and citizen participation. However, it is necessary to work in a unified way with the local administration and the rest of the public institutions involved, reinforced by the research work of the academy and with the possibility that all interested parties can participate, especially users. To this end, it is necessary to broaden the generation of spaces for discussion where this new vision of the good is defined together with the citizens.

The path that is opening up focuses on integrating urban conservation into urban planning based on the theory of value, reinforcing the role of communities (Fouseki et al. 2023). Heritage conservation is therefore understood as a social process, thus recovering the true meaning of heritage comprehended as a social construction, this being appreciated as a collective message.

References

ICOMOS Australia. 1999. The Australia ICOMOS Charter for the Conservation of Places of Cultural Significance. http://openarchive.icomos.org/id/eprint/2145/

Avrami, Erica, Susan Macdonald, Randall Mason, and David Myers, eds. 2019. *Values in Heritage Management*. Los Angeles: Emerging Approaches and Research Directions. The Getty Conservation Institute.

Avrami, Erica, Randall Mason, and Marta De la Torre. 2000. Values and Heritage *Conservation: Research Report*. Los Angeles, CA: Getty Conservation Institute http://hdl.handle.net/10020/gci_pubs/values_heritage_research_report

Ballarat, City of. 2015. "Today Tomorrow Together. The Ballarat Strategy Our Vision for 2040." Ballarat. http://www.ballarat.vic.gov.au/media/3277688/ballarat_strategy_-_final_-_july_2015_-_reduced.pdf

Ballarat, City of. 2017. "Our People, Culture and Place. A Plan to Sustain Ballarat's Heritage." Ballarat. https://www.hulballarat.org.au/cb_pages/ballarat_heritage_plan.php

Bandarin, F., and R. Van Oers, eds. 2015. *Reconnecting the City: The Historic Urban Landscape Approach and the Future of Urban Heritage*. Chichester: John Wiley.

Choay, F. 2007. Alegoría del Patrimonio. 1ª ed. 1992. Barcelona: Editorial GG

Conservation Principles Policies and Guidance for the Sustainable Management of the Historic Environment (2008) https://historicengland.org.uk/advice/constructive-conservation/conservation-principles

Council of Europe. 2005. "Convention on the Value of Cultural Heritage for Society (Faro Convention, 2005)." Faro. https://rm.coe.int/1680083746

Court, Sarah, Eugene Jo Richard Mackay, Mizuki Murai, and Riki Therivel. 2022. *Guidance and Toolkit for Impact Assessments in a World Heritage Context*. Edited by Laura Frank and Eugene Jo. París: UNESCO, ICCROM, ICOMOS, IUCN. https://whc.unesco.org/en/guidance-toolkit-impact-assessments/

Del Espino, B., and J.L. Gómez Villa, Coords. 2022. *Guía básica para el uso, la gestión y la intervención en el patrimonio cultural de Andalucía*. e-phc 7. Seville: Regional Department for Culture. Available from: https://www.juntadeandalucia.es/organismos/iaph/servicios/publicaciones/detalle/391729.html. [Accessed: 2023/01/23]

Fernández Cacho, and S. Coord. 2022. *How to Create a Cultural Landscape Guide cultural*. phc 38. Sevilla: Regional Department for Culture. Available from: https://www.juntadeandalucia.es/organismos/iaph/servicios/publicaciones/detalle/384676.html [Accessed: 2023/02/23]

Fernández-Baca Casares, Román, Silvia Fernández Cacho, Pedro Salmerón Escobar. (Lead). 2017. A Guide to Seville's historic urban landscape. An informative document. Seville: Regional Department for Culture. Available from: https://repositorio.iaph.es/bitstream/11532/327459/1/Guide_to_the_Historic_Urban_Landscape_of.pdf [Accessed: 2022/06/29]

Fouseki, Kalliopi, May Cassar, Guillaume Dreyfuss, Kelvin Kah Ang Eng. 2023. Routledge Handbook of Sustainable Heritage. London: Routledg. https://www.routledge.com/Routledge-Handbook-of-Sustainable-Heritage/Fouseki-Cassar-Dreyfuss-Eng/p/book/9780367482749

Gómez-Villa, J.L. (Coed.), M. García-Casasola, (Coed.), del B. Espino Hidalgo, (Coord.). 2019. _re-HABITAR. Project for the technological updating of contemporary heritage: social housing of the Modern Movement. Seville: Regional Department for Culture. Available from: https://www.juntadeandalucia.es/servicios/publicaciones/detalle/79312.html [Accessed: 2022/03/16]

González-Varas, I. 2021. Coord. PH: Boletín del Instituto Andaluz del Patrimonio Histórico, ISSN 1136–1867, Año nº 29, Nº Extra 104. Buenas prácticas en conservación y revitalización del patrimonio cultural desde los Objetivos de Desarrollo Sostenible (ODS). https://doi.org/10.33349/2021.104

ICOMOS. 1964. "International Charter for the Conservation and Restoration of Monuments and Sites." http://www.icomos.org/charters/venice_e.pdf

ICOMOS. 1994. "The Nara Document on Authenticity." www.icomos.org/charters/nara-e.pdf

ICOMOS. 2011. "The Valletta Principles for the Safeguarding and Management of Historic Cities, Towns and Urban Areas." París. http://www.icomos.org/Paris2011/GA2011_CIVVIH_text_EN_FR_final_20120110.pdf

ICOMOS. 2011. "Guidance on Heritage Impact Assessments for Cultural World Heritage Assets" https://bit.ly/3Xs5g0N

ICOMOS. 2014. "The Florence Declaration on Heritage and Landscape as Human Values."

Jokilehto, J. 2016. "Valores patrimoniales y valoración", Conversaciones. En Revista de Conservación 2, Ciudad de México, Instituto Nacional de Antropología e Historia, Secretaría de Cultura, Coordinación Nacional de Conservación del Patrimonio Nacional, pp. 20–34.

Labadi, S., and W. Logan. 2016. Urban Heritage, Development and Sustainability. Edited by S. Labadi and W. Logan. Oxon and New York: Routledge.

Macdonald, Susan, and Gail Ostergren. 2011. "Developing an Historic Thematic Framework to Assess the Significance of Twentieth-Century Cultural Heritage: An Initiative of the ICOMOS International Scientific Committee on Twentieth-Century Heritage. An Expert Meeting Hosted by the Getty Conservation I." Los Ángeles, CA. https://www.getty.edu/conservation/publications_resources/pdf_publications/pdf/develop_historic.pdf

Mason, Randall. 2002. "Assessing Values in Conservation Planning: Methodological Issues and Choices." In Assessing the Values of Cultural Heritage: Research Report, edited by Marta de la Torre, 5–30. The Getty Conservation Institute, Los Angeles. https://www.getty.edu/conservation/publications_resources/pdf_publications/pdf/assessing.pdf

Murphy, Angela, Dahlhaus, Peter and Helen Thompson. 2016. "Historic Urban Landscapes and Visualising Ballarat: Citizen Participation for Sustainable Urban Planning and Design." In CEUR Workshop Proceedings, 1570, edited by A. Both, M. Duckham, and A. Kealy, 41–46. Melbourne: CEUR-WS. https://pdfs.semanticscholar.org/1498/7f170d2a2277755e1e79012ec9c9e085240b.pdf

De Naeyer, André, Salvador Pérez Arroyo, and Javier Rivera Blanco. 2000. "Krakow Charter 2000: Principles for Conservation and Restoration of Built Heritage." Krakow, Bureau Kracow. http://ipce.mcu.es/pdfs/2000_Carta_Cracovia.pdf

Pereira Roders, Ana. 2007. Re-Architecture: Lifespan Rehabilitation of Built Heritage—Basis. https://doi.org/10.6100/IR751759

Pereira Roders, A., and F. Bandarin. 2019. Reshaping Urban Conservation. The Historic Urban Landscape Approach in Action. Edited by Ana Pereira Roders and Francesco Bandarin. Singapure: Springer Nature Singapure. https://doi.org/10.1007/978-981-10-8887-2

Pereira Roders, Ana, and Ron Van Oers. 2014. Wedding cultural heritage and sustainable development: Three years after. *Journal of Cultural Heritage Management and Sustainable Development* 4 (1): 2–15. https://doi.org/10.1108/JCHMSD-04-2014-0015.

Pereira Roders, Ana. 2016. Improved Urban Governance. In Culture: Urban Future. Global Report on Culture on Culture for Sustainable Urban Development, 220–27. París: UNESCO. http://unesdoc.unesco.org/images/0024/002459/245999e.pdf

Rey-Pérez, Julia, Sebastián Astudillo, M. Siguencia Ávila, Juliana Forero, and Silvia Auquilla. 2017. The Application of the Recommendation on Historic Urban Landscape (HUL) in Cuenca—Ecuador. A New App. Edited by Julia Rey Pérez. Cuenca: Universidad de Cuenca. http://historicurbanlandscape.com/index.php?classid=5356

Riegl, A. 1999. El culto moderno a los monumentos. First edition 1903. Madrid: Visor. Dis., S.A. (La balsa de la Medusa 7).

Silva, Ana Tarrafa., and Ana, Pereira Roders. 2021. Taxonomias Do Significado Cultural Do Património: Valores e Atributos. In Gestão Do Património Cultural. Experiências e Desafios, edited by Catarina Almeida Marado, Teresa Valente, and Joao Pedro Bernardes, 10–31. Universidade do Algarve, Centro de Estudios em Arqueología, Artes e Ciencias do Património. https://doi.org/10.34623/vt4s-7r64

De la Torre, Marta, and Randal, Mason. 2002. Introduction. In Assessing the Values of Cultural Heritage. Research Report, edited by Marta de la Torre, 3–4. Los Ángeles, CA: The Getty

Conservation Institute. https://www.getty.edu/conservation/publications_resources/pdf_publications/pdf/assessing.pdf
Turner, Michael. 2017. Culture as an Enabler for Sustainable Development: Challenges for the World Heritage Convention in Adopting the UN Sustainable Development Goals. In Going Beyond. Perceptions of Sustainability in Heritage Studies No. 2, edited by Marie-Theres Albert, Francesco Bandarin, and Ana Pereira Roders, Heritage S, 19–32. Springer International Publishing AG. https://doi.org/10.1007/978-3-319-57165-2
UNESCO, ICOMOS, and ICCROM. 2021. Heritage in Urban Contexts: Impacts of Development Projects on World Heritage Properties in Cities. Final Outcomes. Fukuoka.
UNESCO. 2003. Convention for the Safeguarding of the Intangible Cultural Heritage. http://portal.unesco.org/en/ev.php-URL_ID=17716&URL_DO=DO_TOPIC&URL_SECTION=201.html
UNESCO. 2005. Operational Guidelines for the Implementation of the World Heritage Convention. https://whc.unesco.org/archive/opguide05-en.pdf
UNESCO. 2011. Recommendation on the Historic Urban Landscape, Including a Glossary of Definitions. http://portal.unesco.org/en/ev.php-URL_ID=48857&URL_DO=DO_TOPIC&URL_SECTION=201.html
Veldpaus, Loes. 2015. Historic Urban Landscapes. Framing the Integration of Urban and Heritage Planning in Multilevel Governance. Eindhoven: Eindhoven University of Technology, Eindhoven, the Netherlands. https://pure.tue.nl/ws/portalfiles/portal/3914913/798291.pdf

Part III
Values. Identification, Documentation and Characterization

Some study cases help to demonstrate the application of the methodology described in Part II, focusing on the identification of values and the assessment of impact: how an intervention or proposal affects the identified values.

The contribution, with an example of modern architectural heritage in Cyprus, also makes it possible to update the process, generating a diagram in which it is possible to identify the main values: cultural, environmental, social and economic (E. Siandou). At the same time, it is possible to see the main issues to manage heritage from a sustainable development point of view, indicating the way to improve the quality of life in this context, working with the communities.

From an educational point of view, it is very important to train capacities to be able to meet the requirements of the methodology related to the identification of values. In this sense, one of the first tasks to be carried out at the beginning of the heritage management process is the organization of documentation. Different types of documentation from different sources will be archived in a large FAIR (Findable, Accessible, Interoperable, Reusable) file. For this purpose, the use of traditional, digital and environmental tools is welcome (C. Herakleous, G. Tryfonos, M. Philokyprou, O. Kontovourkis, A. Michael).

Documenting architectural heritage, especially historic buildings and sites, is the first task to do when you have to start the identification of values. In fact, the documentation archive should be understood as an open tool, which never ends to grow. The management of this type of documentation implies working with graphic documentation, also images, which are very important, not only in the documentation process but also in the identification of values.

On the other hand, the experience shared between researchers and educators provides an important theoretical foundation of urban morphology in relation to fieldwork and the identification of in situ values. In this way, the first main steps to be followed in sustainable heritage management—observation, documentation and characterization—can be strengthened through the identification of the object of study, the methodological approaches and the tools used (M. P. Milojević, A. Đorđević, M. Pešić, J. Živković, V. Djokić).

In addition to documentation management, the development of innovative processes for identifying values and monitoring them throughout the intervention project is a major task for practitioners involved in sustainable heritage conservation. To do this, it is proposed to contextualize the process in the revision of the various international charters related to heritage, which are constantly evolving and need to be managed from a contemporary perspective: reconstruction, reuse and resilience are defined as the three aptitudes to approach the sustainable heritage project (E. Sorbo, S. Tonello).

The result tries to be an advance of how artificial intelligence will be the way to manage the different situations, monitoring the actions and the consequences in cultural values identified. New lines of research where the dynamic graphic representation shows the complexity of the task, giving an image of each charter, understood from the perspective of sustainable heritage management.

Chapter 8
Mobilising Heritage for Sustainable Development. Opportunities in Value-Based Approaches for the Protection of Modern Architectural Heritage in Cyprus

Emilia Siandou

Keywords Heritage values · Modernism · Contested heritage · Community involvement · Heritage planning

Introduction

This paper presents the main findings of a PhD research examining the case study of post-world War II schools built in the city of Larnaka, Cyprus, between 1945 and 1963, through a value-based approach. Modern architectural heritage in Cyprus, which developed during politically turbulent periods, became entangled with processes of colonialism, nationalism, and conflict. As such, it constitutes a complex, contested heritage. Mobilising such heritage for sustainable development is challenging. However, this paper argues that new approaches in the heritage field related to heritage values, participation, and sustainability create opportunities for addressing complex histories in conflict-affected contexts.

E. Siandou (✉)
Department of Town Planning and Housing, Republic of Cyprus, Nicosia, Cyprus

M. Loren-Méndez et al. (eds.), *Sustainable Heritage Management - Design: Notions, Methods and Techniques*, https://doi.org/10.1007/978-3-031-81970-4_8

Heritage and Sustainable Development

The concepts of "heritage" and "sustainable development" have been increasingly linked as the heritage field has been undergoing fundamental transformations within the twenty-first century, shifting from the "traditional conservation practice" towards the "new paradigm" as it has been characterised by Araoz (2011)[1]:

> Based on observations and discussions in various international fora, it is argued that alongside traditional heritage places, a new paradigm for heritage sites has emerged whose values no longer rest entirely on material culture, but on intangible concepts for which traditional conservation practice often is neither effective nor applicable.

In 1975, the Council of Europe introduced the term "integrated conservation" in the *European Charter of the Architectural Heritage*. Initially, this referred to integrating conservation into planning and legislation (Council of Europe 1975, 1985). Later, interpretations of "integrated conservation" included the integration of various professional disciplines in the conservation processes (Council of Europe 2005), the integration of all stakeholders in the process including the public (Council of Europe 1975, 2005), the integration of all the steps of the heritage process from identification to safeguarding and maintenance (Council of Europe 1975, 1985, 2005), and an overall more holistic approach to heritage which takes into account its complexity (Council of Europe 1975, 2005).[2] The consideration of more stakeholders in the heritage processes was followed by a numerical explosion of the heritage objects (Araoz 2011). The 2000 Getty Conservation Institute report introduced **values and valuing processes as a key concept to the integration of the field of conservation** (Avrami et al. 2000a, b), precipitating a shift towards value-based approaches (Table 8.1).

In 2011, the UNESCO Recommendation on the Historic Urban Landscape proposed an integrated approach to urban heritage conservation in order to achieve greater goals of overall sustainable development (UNESCO 2011). The above developments were reflected in the 2014 European Commission communication paper *Towards an integrated approach to cultural heritage for Europe*.[3] **The report also reflected the recognition of the role and impact of cultural heritage as a shared resource for sustainable development and refers to heritage values, namely social, societal, economic, and cultural values.** Thus, the official link was made between the two very influential concepts which shifted heritage theory and practice from the "traditional model" to the "new paradigm", namely integrated conservation and sustainable development.

The correlations between heritage and sustainable development evolved to be stronger as heritage became widely accepted as the source of important benefits for

---

[1] Gustavo F. Araoz was the elected President of the International Council of Monuments and Sites between 2008 and 2018.

[2] For further analysis of how the term "integrated conservation" was previously used in various international conservation charters also see (Bell 1997).

[3] Overall aim of the communication paper is supporting progress towards a more integrated approach to heritage at national and EU level.

Table 8.1 Overview of the increasing use of the term "values" in international conservation documents

International charter for the conservation and restoration of monuments and sites (Venice)	European charter of architectural heritage (Amsterdam)	Burra charter	Burra charter	ICOMOS New Zealand charter for the conservation of places of cultural heritage value	The Paris declaration on heritage as a driver of development	
1964	1975	1979	1999	2010	2011	
				Tangible values	Intangible values	
Historic	Spiritual	Aesthetic	Aesthetic	Aesthetic	Commemorative	Cultural
Aesthetic	Cultural	Historic	Historic	Archaeological	Historical	Historical
Archaeological	Social	Scientific	Scientific	Architectural	Social	Environmental
	Economic	Social	Social	Functional	Spiritual	Aesthetic
			Spiritual	Landscape	Symbolic	Memory
				Monumental	Traditional	Social
				Scientific		Human
				Technological		Economic

Mason, Getty conservation institute	Guidelines documentation fiche 2003	Writing statements of significance	Conservation principles	Guidelines on cultural heritage
2002	2003	2006	2008	2012
Historical	Technical	Aesthetic	Evidential	Historic and aesthetic
Cultural/symbolic	Social	Historic	Historical	Scientific or research
Social	Aesthetic	Cultural	Aesthetic	Social and spiritual
Spiritual/religious		Scientific	Communal	
Aesthetic		Social		
		Spiritual		

society, culture, and economy. In 2005, the Faro Convention emphasised "the value and potential of cultural heritage wisely used as a resource for sustainable development and quality of life in a constantly evolving society" in one of the first acknowledgements of the value of cultural heritage to society in international documents (Council of Europe 2005). Since then, significant research efforts have supported the much-needed articulation of the role, impact, and potential of cultural heritage for sustainable development by providing quantitative and qualitative evidence for this.[4]

Since 2002, Historic England on behalf of the Historic Environment Forum conducts the annual audit of England's heritage, named Heritage Counts (Historic England n.d.). A series of reports published annually comment on different issues every year. In 2014, the report explored *The Value and Impact of Heritage* on many factors including growth, the economy, wellbeing, and sense of place attempting to answer the question "can heritage really contribute to sustainable development?" (Historic England 2014). The report uses an economic approach of measuring the value of heritage, while at the same time, it examines three distinct types of heritage impacts: individual impacts, community impacts, and economic impacts. The report concluded that:

> Heritage has the potential to achieve a range of positive impacts for individuals, communities and the wider economy, although we need to know more about the mechanisms by which heritage might improve wellbeing, say, or bring people close together.

On the parallel, yearly studies analysing specifically the importance of heritage to individuals and communities (Historic England n.d.), as well as the economic impacts of heritage in England (Historic England n.d.), are published within this scope. The data gathered and analysed, over this period of twenty years, provide a significant body of evidence, which promotes not only the local understanding of the social and economic benefits of heritage, but also the international understanding of these issues.

In 2013, the two-year EU-funded research project *Cultural Heritage Counts for Europe: Towards a European Index for Cultural Heritage* was launched (CHCfE Consortium n.d.). The research was conducted by the Cultural Heritage Counts for Europe (CHCfE) Consortium comprised of six partner institutions, namely Europa Nostra, the European Network on Cultural Management and Cultural Policy Education (ENCATC), the European Association of Historic Towns and Regions (Heritage Europe), the International Cultural Centre in Krakow (ICC), the Raymond Lemaire International Centre for Conservation (RLICC) at KU Leuven, and The Heritage Alliance, England.

---

[4] This research acknowledges the fleeting nature of the concept of "sustainability", as this is reflected in discourses of political ecologies (Worster 1993). The use of the term within the framework of this paper focuses on how the term has infiltrated cultural heritage policy documents based mostly on the definition of sustainable development coined by the "Brudlant Report" (United Nations 1987). Sustainable development is defined by the report as follows: "Sustainable development is development that meets the needs of the present without compromising the ability of future generations to meet their own needs" (United Nations 1987).

The project aimed to collect and analyse existing and accessible evidence-based research and case studies regarding the economic, social, cultural, and environmental impacts of cultural heritage in Europe. In 2015, the *Cultural Heritage Counts for Europe* executive summary and strategic recommendations were published, supporting the wide-ranging benefits of cultural heritage in Europe as a key component in attracting investment, producing revenue through cultural tourism, creating jobs, and driving heritage-led regeneration (CHCFE Consortium 2015). The research highlighted heritage's benefits for identity, creativity, energy efficiency, quality of life, lifelong learning, social cohesion, and integration (Fig. 8.1).

The importance of this research was paramount since based on its results, the EU Council of Ministers, on the 21st of May 2014, recognised "cultural heritage as a strategic resource for a sustainable Europe" (Council of the European Union 2014). EU Council's conclusions recognised that ""cultural heritage plays a specific role

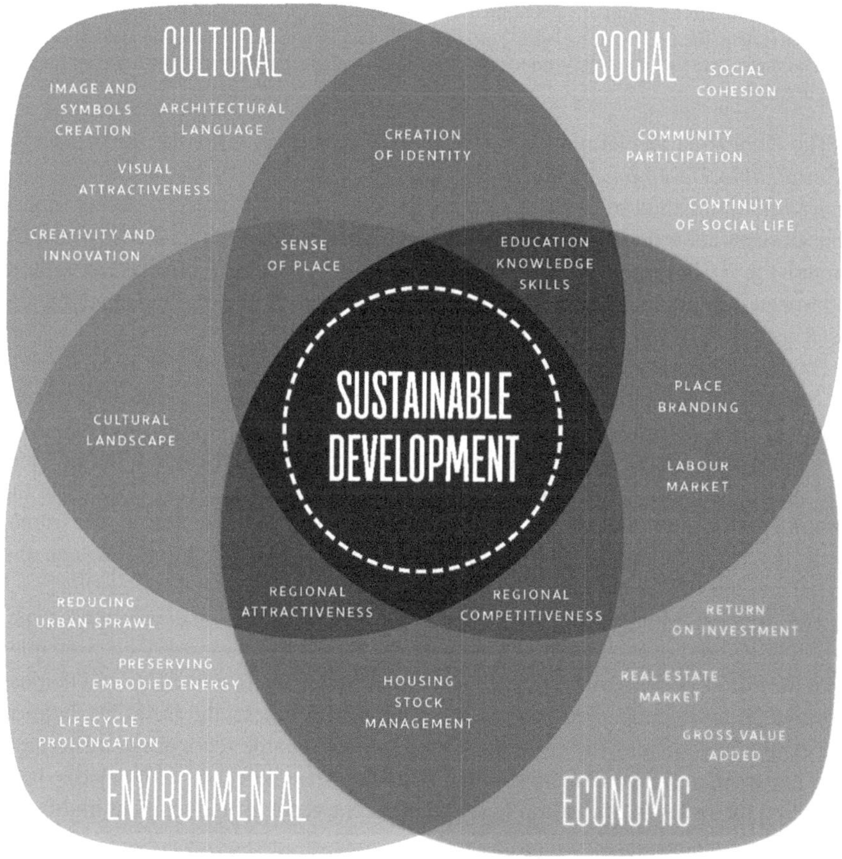

Fig. 8.1 The project cultural Heritage Counts for Europe (CHCFE) (2013–2015) was coordinated by Europa Nostra through a Consortium composed of all project partners. © CHCFE Consortium

in achieving the Europe 2020 strategy goals for a "smart, sustainable, and inclusive growth" because it has a social and economic impact and contributes to environmental sustainability"". The Council called for the EU member states and the Commission to recognise heritage's impact, the need for dialogue with the stakeholders, integrated approaches, awareness-raising, and supportive policies (Council of the European Union 2014).

Furthermore, this research informed the 2018 European Year of Cultural Heritage and the 2019 publication *European Framework for Action on Cultural Heritage* (European Commission 2019).

At the same time, there were developments in the further recognition and wide acceptance of the value of cultural heritage for society as a key resource for sustainable development internationally, further than the English and European context. In 2013 UNESCO adopted the Hangzhou Declaration stressing the need for:

> New approaches [towards mounting international challenges] should fully acknowledge the role of culture as a system of values and a resource and framework to build truly sustainable development, the need to draw from the experiences of past generations, and the recognition of culture as part of the global and local commons as well as a wellspring for creativity and renewal.

The Hangzhou Declaration was succeeded by the *Policy for the Integration of a Sustainable Development Perspective into the Processes of the World Heritage Convention* published by UNESCO in 2015. The policy adopted the three dimensions of sustainable development from the conceptual framework accepted by the UN through the 2030 Agenda for Sustainable Development, namely, inclusive economic and social development and environmental sustainability, complemented by the fostering of peace and security (United Nations 2015).[5]

For the integration of cultural heritage in the context of the UN 2030 Agenda for Sustainable Development, ICOMOS has also published, in 2016, the Concept Note *Cultural Heritage, the UN Sustainable Development Goals, and the New Urban Agenda.* The Concept Note argues for the positive integration of culture and cultural heritage into urban development plans and policies as a way to enhance the sustainability of urban areas through heritage (ICOMOS 2016). Within this framework, ICOMOS published, in 2017, also an Action Plan for Cultural Heritage and Localizing the UN Sustainable Development Goals (SDGs) (ICOMOS 2017a).

The 2017 ICOMOS *Delhi Declaration* recognised heritage and democracy as vital to people-centred sustainable development. The Delhi Declaration reflects principles from the Faro Convention in being founded on respect for human rights and democracy and promoting a people-centred approach. Nevertheless, the Delhi Declaration reflects the widening of the concept of heritage, reflects the recognition of heritage as a fundamental right and responsibility of all, and also reflects the knowledge built regarding the benefits of cultural heritage in recent years. The document highlights

[5] The UN 2030 Agenda for Sustainable Development has introduced the fifteen Sustainable Development Goals (SDGs), which have been currently adopted by many organisations as a way of evaluating sustainable development.

the link between involvement of the community in the planning process in the framework of an integrated approach to conservation and the pursuit of sustainability as did also other international documents during the last decade (UNESCO 2011; Council of the European Union 2014; Council of Europe 2017). Nevertheless, ICOMOS through this document goes one step further to recognise heritage conservation as "a condition of sustainable development" while it emphasised that "The protection and sustenance of heritage resources should be the basis of development policies and planning programmes, integrating heritage-conservation strategies within the larger goals of sustainable development" (ICOMOS 2017b).

Modern Architectural Heritage in Cyprus

As heritage has been increasingly accepted as the source of important benefits to society, and participation and access to heritage is now recognised as a human right, the need to protect complex layers of our heritage and to mobilise them in favour of sustainable development is today more urgent than ever.

Modern architecture in Cyprus developed during the most turbulent periods of the twentieth century, amid decolonisation, modernisation, Westernisation, and nation-building, has been used as a form of representation, carrier of ideologies, and ideas of national identity, often conflicting. Cyprus' modern architecture is tied to major twentieth-century historical periods: colonial rule until 1960, post-independence pre-1974, and post-1974 division. The development and expression of modern architecture locally was influenced by the colonial rule, the anticolonial fight, the Cypriot Independence, local and international ideas of progress and utopias, the tensions between the two larger local communities, Greek Cypriots and Turkish Cypriots, and the Turkish military operation which resulted in the ongoing division of the island. At the same time, it has also been physically involved in the conflict (Demetriou 2012; 2015; Given 1997; Sioulas and Pyla 2015). In such a manner, modern architecture in Cyprus constitutes a complex layer of the island's architectural heritage with many contested aspects.

Recent scholarship regarding modern architecture in Cyprus has made great steps in situating it on the map of the international development of the modern movement, in unfolding its history and in promoting understanding of its local development, about the processes which created and sustained it (Fereos and Phokaides 2006; Given 1997; Pyla 2009; Phokaides and Pyla 2012, 2016; Pyla and Phokaides 2009, 2011; Tzirtzilakis 2010; Georghiou 2013, 2018).[6] Concurrently, the intersections between modern architecture and the island's colonisation, decolonisation, nation-building, modernisation, conflict and division, its meaning and symbolism

---

[6] In evaluating local modernism as heritage, the referenced PhD study considered modern architecture in Cyprus within the framework of alternative or multiple accounts for modern architecture which developed in the periphery of the West, but also considers the manners in which this historic view of the "modern" and the "Western" as advanced influenced the local development of modern architecture and defined the built environment in Cyprus in the twentieth century.

as heritage are being further examined by sociological, anthropological, and ethnological research (Papadakis et al. 2006; Constantinou et al. 2012; Demetriou 2012; 2015). This scholarship has indicated the wealth and the complexity of this kind of heritage.

Recent designations by the Department of Preservation of the Town Housing and Planning Department of the Republic of Cyprus show growing recognition, but modern buildings still comprise only around 1% of listed heritage. According to information provided by the Preservation and Cultural Heritage Sector so far, Preservation Orders were issued for approximately 70 modern architecture buildings. The low level of statutory protection of modern architecture in Cyprus, according to information from the Preservation and Cultural Heritage Sector, of the Department of Town Planning and Housing of the Republic of Cyprus, relates to the following issues: (i) the lack of extended bibliography on the subject of the modern architectural heritage of Cyprus, (ii) lack of awareness regarding the value of modern architectural heritage among people, professionals, and authorities, (iii) the fact that in Cyprus, modern architecture buildings are in most cases individual buildings dispersed all over the island.[7] An additional challenge to addressing it as heritage and to mobilising it towards sustainable development is its contested aspects.

PhD Research on Heritage Values of Post-world War II Schools in Larnaka

This paper presents the main findings of a PhD research examining the case study of post-world War II schools built in the city of Larnaka, Cyprus, between 1945 and 1963, through a value-based approach (Siandou 2019).

Schools (1945–1963) were selected as a case study as a prevalent modern building typology used by all communities, integral to Cyprus' modern architecture. The important role of school buildings in the development of the modern movement in Cyprus has been a point of agreement between researchers on the history of modern architecture in Cyprus (Sierepeklis 1997; Fereos and Phokaides 2006; Pyla and Phokaides 2009, 2011; Tzirtzilakis 2010; Georghiou 2013, 2018; Docomomo Cyprus 2014). Furthermore, Fereos and Phokaides argue, "Through the new school buildings, modern architecture established its pioneering role as a catalyst of modern education" (2006). The history and development of educational buildings in Cyprus in the twentieth century reflect the socio-political discourses of that period. The social and political tendencies of the main historical periods of the twentieth century in Cyprus, as well as the shifts from one period to the other, all found representation in the school buildings themselves. The historical and geopolitical context of Cyprus played a major role in the colonial policies which were implemented, especially in the development of educational matters.

---

[7] Interview with Ms Yiola Kourou, Senior Officer of the Preservation and Cultural Heritage Sector, and Ms Kyriaki Kalava, Officer of the Preservation and Cultural Heritage Sector, 5 March 2020.

Education was historically segregated between the Greek Cypriot and Turkish Cypriot communities. During British colonial rule in the island, the colonial educational policy did not remain static, it was changing in response to the evolving local conditions. Three major factors influenced education in Cyprus during the colonial period: (i) the colonial educational policy implemented in Cyprus, (ii) the Greek Cypriot nationalistic movement which had been developing since the nineteenth century, and (iii) the rising nationalistic movement of the Turkish Cypriot community. School architecture in Cyprus in the twentieth century constituted a form of representation and conveyed rival Cypriot identities.

During the twentieth century, conflicting nationalisms of the two main communities shaped Cyprus's contested self-image and identity. Within this context, the most important problems in Cyprus during the twentieth century were the problem of identity and otherness. The political and educational discourse was focused on this problem which also influenced the production of educational buildings (Persianis 2006). Over and above, schools became physically involved in the conflicts of the twentieth century in Cyprus. Modern school buildings were also physically involved in the local conflicts by being targeted and attacked because of their association with the colonial government amidst the rising public anticolonial feelings, or by being used for the accommodation of displaced populations both in 1963 and in 1974. In these multiple ways, modern school buildings constitute a building type explicitly involved in the local conflicts of the twentieth century in Cyprus.

Thus, the study of modern school buildings in Cyprus has the prospect to highlight the complex and contested aspects of modern architecture as heritage and to highlight how these aspects are reflected in its value as heritage. A value-based approach for modern school buildings in Cyprus has the prospect to highlight how the modern architectural style held different values for different groups of people during the same period. At the same time, it was considered interesting to examine if and how the same architectural vocabulary held different social meanings during the colonial and postcolonial periods.

Efforts were made to select a coherent group of buildings within the wider category of school buildings of the twentieth century in Cyprus as a case study. In such a manner, it would be possible to have comparability and to extract generalisations based on the case study findings. Hence, a group of buildings was identified through the examination of various categorisations of educational buildings of this period, namely typological (elementary schools, secondary schools, technical schools, higher education institutions, private schools, public schools, communal school, etc.), geographic (urban, peri-urban, rural areas), and chronological (colonial period, independence period, and their historic subdivisions).

Taking into account all the above, it was decided to focus on the **communal school buildings built in Larnaka between 1945 and 1963**. The urban areas of Cyprus at the time concentrated more types of schools in relation to the rural areas while at the same time urban areas were more linked with the processes of creating images, narratives, and representations of statehood. The city of Larnaka provided a sample of schools diverse enough, yet manageable enough to be analysed as a case study within the framework of this PhD study. The research examined operating primary

Table 8.2 School buildings included in the case study

No	Schools	Education level
1	Agios Georgios elementary school (AGES)	Primary education
2	Agios IOANNIS elementary school (AIES)	Primary education
3	Drosia middle school (DMS)	Secondary education
4	Agios Georgios high school (AGHS)	Secondary education
5	Prodromos elementary school (PES)	Primary education

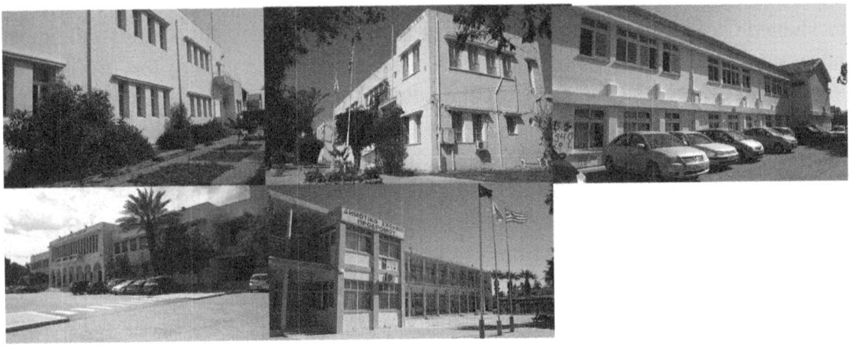

Fig. 8.2 The five schools included in the case study. AGES (top left), AIES (top middle), DMS (top right), AGHS (bottom left), PES (bottom middle)

and secondary schools from 1945 to 1963, a historically coherent period bounded by the 1963 bicommunal strife. After 1963, the Turkish Cypriot community withdrew to enclaves in Larnaka (and all the other cities of Cyprus). This meant that the schools, which were situated out of the enclaves, were abandoned while the community did not have the possibility to build any more schools between 1963 and 1974 due to the financial hardships caused by the isolation. During the same time, the Greek Cypriot community entered a new phase in education with the creation of the Ministry of Education in 1964 and the appointment of a head architect for the Technical Services of the Ministry, which was responsible for the design of all Greek Cypriot school buildings in the period between 1964 and 1974. The case study includes five schools, as listed in Table 8.2 and pictured in Fig. 8.2.

Methodology

It is often argued that the introduction of the concept of cultural significance has slowly but fundamentally transformed the field of heritage from the last quarter of the twentieth century until today (Avrami et al. 2000a, b; Araoz 2011; Fredheim and Khalaf 2016). However, one could also claim the opposite; it is the fundamental changes in the heritage field which created the need to define cultural significance, and hence heritage values.

This research analysed heritage literature and policies, identifying opportunities in value-based approaches for mobilising contested heritage in Cyprus for sustainable development, as these are mentioned below.

Firstly, value-based approaches encourage the involvement of all stakeholders in all stages of the heritage process (Figs. 8.3 and 8.4). Access to knowledge and active participation by each individual or group who holds an interest in heritage has been stressed as very important in making heritage relevant to society within the framework of the heritage discourse. In line with the above, value typologies are currently required to be inclusive and democratic (Avrami et al. 2000a, b; De la Torre 2002; Stephenson 2008; Worthing and Bond 2008; Fredheim and Khalaf 2016). In such a manner, values of all stakeholders involved with a heritage site, even conflicting, can be recognised and through a value-based approach, difficult questions such as "whose heritage?" and "for whom?" might be answered. The involvement of all the related stakeholders and the consideration of all their values in the heritage processes is even more critical in a conflict-affected environment where heritage may carry different meanings and interpretations for each of the conflicted local communities and where the ultimate service of heritage would be the promotion of peace and democracy.

Furthermore, in the framework of value-based approaches, the consideration of intangible values has been evaluated as an important component (Australia ICOMOS 1979; 1999; 2013; Gibson et al. 2009; Tomaszewski 2005; Worthing and Bond 2008; UNESCO 1972). This allows for the consideration of values, meanings, and associations related to notions of identity, nation or nationality, conflict, etc.

Value-based approaches allow considering aims like peace and reconciliation, which have been recognised as key for sustainability in Cyprus, from the start (Fig. 8.4). In the context of Cyprus, peace and reconciliation are considered as key to sustainable development and vice versa. Hence, through value-based approaches, it would be possible to consider the aims of peace and reconciliation from the outset of the design of a heritage process. Within this framework, the identification of common values between stakeholders can be used as an opportunity for the promotion of peace in line with the Council of Europe Recommendation for Cultural heritage in crisis and post-crisis situations (Council of Europe 2015). The Parliamentary Assembly through the above-mentioned recommendation stresses "the importance of 'common' cultural heritage and its enshrined values as key factors in the process of sustainable reconciliation and conflict resolution" (Council of Europe 2015). Within the same framework, conflicting values can be acknowledged, addressed, and monitored in a risk mitigation process in a crisis and a post-crisis scenario. The need for the identification of threats and opportunities in value-based approaches has been highlighted by experts (De la Torre 2002). This can be considered also in the framework of an integrated conservation approach where heritage management and planning regulations should have the capacity to identify, assess, and balance potentially conflicting views and interests of the multiple stakeholders, currently involved in the conservation process.

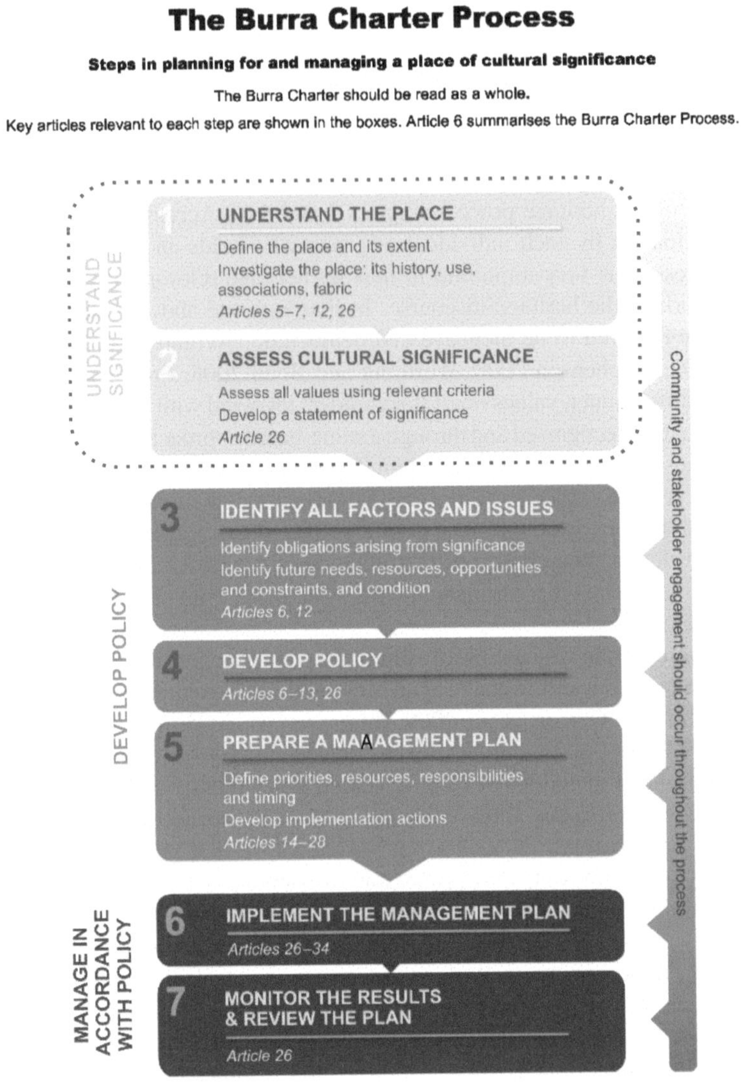

Fig. 8.3 Understanding and assessing significance as the first recommended steps in The Burra Charter Process (Australia ICOMOS 2013)

In addition, heritage values have been recognised to be time- and context-specific (De la Torre 2002; Gibson et al. 2009; Fredheim and Khalaf 2016). The consideration of the factors of time and context in contested sites is important since shifts in the socio-political context or events in the history of the sites can have an impact on the associations and symbolic meanings related to architectural heritage. The need to include these factors in value-based methods has taken an important place in the heritage values discourse (De la Torre 2002; Gibson et al. 2009).

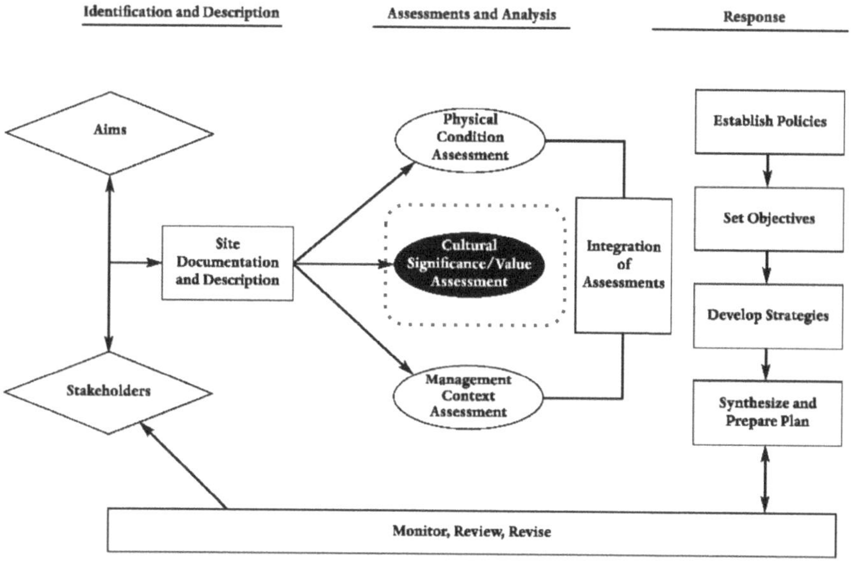

Fig. 8.4 Value assessment in the heart of the planning process methodology proposed by Mason (De la Torre 2002) ©2002 J. Paul Getty Trust

Overall, a value-based approach for modern architectural heritage in Cyprus has the potential to provide a holistic and inclusive method of identifying the variety of values modern architecture carries in Cyprus and this is why it has been selected as the methodology of this PhD research.

This case study focused on the identification and description of the heritage values linked with the modern schools constructed between 1945 and 1963 in Larnaka. The assessment of the significance of heritage sites was proposed within the first steps of a value-based planning process methodology both by Mason (De la Torre 2002) and by ICOMOS Australia in The Burra Charter Process (2013) (Figs. 8.3 and 8.4).

The value-based methodology used for the purposes of this thesis has been developed in line with the international conservation recommendations and benefiting from international experience and research. Through a critical analysis of the value-based approaches, a "toolbox approach" was used and a variety of qualitative methods were selected in order to address the research questions of this case study. Mason (2002) describes the distinction between sociocultural, economic, and ecological values, as the three major categories of values which are based on different conceptual frameworks and hence require different tools for their articulation and assessment (Mason in De la Torre 2002: p. 11). This distinction is adopted within the framework of this research and the research focused on the category of sociocultural values.

Taking into account the current discourse on value-based methods, it was decided for this study to divide the case study research into two parts. The first part is an expert analysis conducted by the author, and the second is the analysis of the users' interviews. For the expert analysis, only primary and secondary sources were used, and a statement of significance was produced for each of the buildings.

PhD Research Findings: The Range and Nature of Heritage Values Associated with Modern Schools in Larnaka

The case studies showed that **modern schools in Larnaka carry multiple heritage values as landscapes, architectural ensembles, singular buildings, and building networks. These values have different levels of importance, namely personal, local, regional, national, and even international importance and they are linked to specific physical elements of the buildings.** The identified heritage values, tangible or intangible, fit into several heritage value categorisations—e.g., architectural/aesthetic, historic, evidential, social, communal, technical, use, emotional, and others.

Furthermore, the systematic analysis of the research findings provided information regarding the nature of the heritage values linked with the school buildings.

It was demonstrated that these values are not static but dynamic, have changed through time, and thus are **time-dependent.** This was very evident in the analysis of the results of the interviews. Even in the cases of schools which generated continuous appreciation in depth of time, the interviews have highlighted variations in the heritage values linked with the school buildings in the past and at present. These changes are related to the socio-political context of the main historic periods of the twentieth century locally (colonial, postcolonial, pre-1963 independence period) and internationally (post-world War II period), and are **linked with specific events**, such as the division of Cyprus that followed the 1974 war.

At the same time, the study has demonstrated the effect of time on a building's significance. Schools, which enjoyed a continuous relationship with their users through many generations, accrued historic values, evidential values, and emotional values. Nevertheless, in some occasions, the architectural value of the buildings is currently dismissed due to a lack of awareness regarding the value of modern architecture as heritage or due to extensive alterations to the original school buildings which compromised their architectural values. Interviews showed stakeholders also value reuse prospects.

Furthermore, the findings highlighted that **the local context influenced the schools' heritage values**. For example, the study highlighted that the heritage values linked with the school buildings during the period of the 1950s had much to do with the value of education for Cypriots at the time. Additionally, the research findings highlighted how heritage values are linked with the representational attributes of modern school architecture in the context of colonial and postcolonial Cyprus.

Over and above, the case study has demonstrated how the context impacted how past and current users value the schools. This is most evident in the cases of the former Turkish Cypriot schools. It was evident in the interviews that their Turkish Cypriot users highly value the schools even after more than 40 years that they have been displaced, and they still associate them with the Turkish Cypriot identity. Furthermore, the Greek Cypriot community also associates these schools with the Turkish Cypriot community even today. The research findings further highlighted that the Turkish Cypriot schools, which switched hands after 1963 or after 1974, as the AIES and the DMS, were less appreciated by their Greek Cypriot users. This highlights the **association of the schools with the identities of the two communities and the impact of the conflict between the two communities to their valuing**. It is important to note that although it took years for the Greek Cypriot users of the two schools to value them, today their users have a very positive approach towards their preservation and expressed willingness to participate in decision-making processes regarding the building.

What the case study has further highlighted is the fact that some of these schools were physically involved in the conflict and most specifically in the war. One of the past users of the AGES described fighting scenes she witnessed during the coup d'état and the war in 1974 at the school which was overtaken by people involved in the coup. The interviewee described a scene of fighting between the two schools AGES and AIES, which was overtaken by Turkish Cypriot fighters. This is one of the many stories of conflict in the city of Larnaka which are not widely known and have not been documented so far.

Furthermore, as it has been documented in the interviews of users of the DMS and it was also supported by historical documentation, the DMS was used for the internment of the Turkish Cypriot men during the 1974 war for several weeks until they were finally exchanged with Greek Cypriot war captives (Fig. 8.5).

Furthermore, after 1974, all the schools in the case study were involved in mitigating the impact of the conflict for the accommodation of the needs of the refugees. As it was also clear in the case study, the heritage values of the buildings were defined by the dramatic change in the context after the 1974 events. The schools are valued for their contribution to the needs of the displaced population and are linked with the experiences of the refugees. This was identified by the interviews as one of the important reasons for which school buildings of that period are valued. In these multiple ways, the findings of the case study have demonstrated that modern school buildings in Larnaka are linked with local conflicts both physically and by association. Furthermore, it has been demonstrated that this fact affects the way these school buildings are valued currently by different groups of people.

Conclusion: Value-Based Methodologies as Tools for Mobilising Heritage Towards Sustainable Development

The knowledge built on the range and nature of heritage values linked with the case studies, a small, but nevertheless consistent group of modern school buildings in

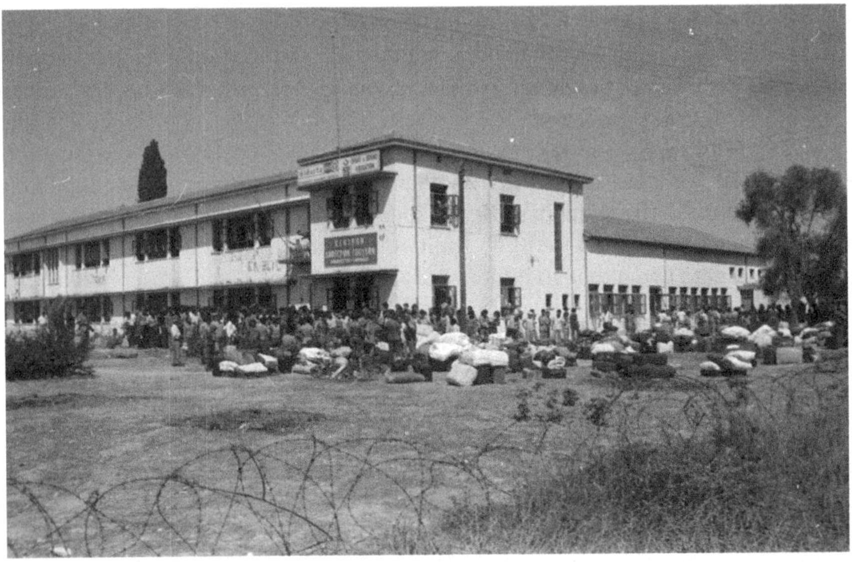

Fig. 8.5 Turkish Cypriots being withheld at the Drosia Middle School in Larnaka, 23 September 1974 @ Press and Information Office of the Republic of Cyprus

Larnaka, can be used as a starting point, **a provisional typology**, as recommended by Mason (2002) suitable for the evaluation of modern architecture in Cyprus.

At this stage, it is important to reiterate the need for robust thematic studies on modern architecture in Cyprus. **Thematic studies and value-based assessments** can enable comparative analysis and systematic prioritisation in heritage planning, considering significance at personal to international levels. At the same time, **linking values with specific tangible or intangible elements and their attributes** could allow for strategic conservation interventions for heritage buildings, ensembles, and sites.

The findings of this research have demonstrated how factors of time and context have influenced the values, associations, and symbolic meanings related to modern architectural heritage in Cyprus. It is necessary for any value-based approach to include a **monitoring mechanism for heritage values**. In a conflict-affected environment, monitoring possible changes in the set of meanings and associations related to heritage can be critical to achieving or sustaining peace.

The findings indicated that involving more disciplines would benefit the evaluation.

The main finding of this research has been that **community involvement in heritage evaluation** through value-based methods is **key for sustainable development**. Despite recognised as essential, effectively involving all stakeholders remains a core challenge. Access to knowledge and active participation by each individual or group who holds an interest in heritage has been stressed as very important in making heritage relevant to society and, in recent years, has come to be recognised as a human right. In such a manner, the values of all stakeholders involved with a heritage site, even conflicting, can be recognised through a value-based approach. The involvement of all the related stakeholders and the consideration of all their values in the heritage processes is even more critical in a conflict-affected environment where heritage may carry different meanings and interpretations for each of the conflicted local communities and where the ultimate service of heritage would be the promotion of peace and democracy. Through the introduction of user interviews in the value assessment process, this research highlighted some of the benefits of including stakeholders in these early stages of the heritage planning process. This research comprises interviews with current and past users of the school buildings, people who use and value the building but currently have no leverage in the local heritage planning process for public buildings. It was one of the methodological aims to highlight the benefits of bringing light to the values and views of these groups excluded by the "authorised heritage discourse" around modern architecture in Cyprus, and the benefits of the democratic inclusion of these groups of stakeholders in the evaluation processes.

User interviews enriched historical knowledge about the schools. Oral testimonies and personal archives provided key primary sources, especially for former Turkish Cypriot schools where other sources were limited. Interviewees who witnessed historic events supplied context about the buildings unavailable elsewhere.

Furthermore, the user interviews validated the expert's assessment and ensured important values not identified by the expert were captured, while concurrently demonstrated whether expert-identified values reflect those of society. The user interviews overall enriched the knowledge about the buildings' values. Moreover, the user interviews were useful in highlighting different perspectives regarding the value of the schools as architectural heritage. As Avrami, Mason, and De la Torre stress in the GCI, 2000 report:

> Heritage is valued in myriad and sometimes conflicting ways. These different means of valuing influence negotiations among various stakeholders and thus shape conservation decision making.

Interviews revealed common and conflicting values. The AIES and DMS held communal, but also emotional values, for both Turkish Cypriot and Greek Cypriot, as both communities used them.

Within this framework, the identification of common values between stakeholders can be used as an opportunity for the promotion of peace, while concurrently conflicting values can be acknowledged, addressed, and monitored in a risk mitigation process. Over and above, through the user interviews, contested aspects of the schools, which were not previously known, were highlighted. Incorporating excluded views in the value assessment process revealed hidden contested aspects, aiding prioritisation processes for strategic heritage planning. The value-based assessment of the schools provided an overall image of the school's significance, including how the schools are valued, or not valued, by their users and the users' attitudes related to the buildings' protection. The last two points constitute important information for decision-making and prioritisation in the framework of strategic heritage planning for modern architectural heritage in Cyprus.

The users' interviews further provided information about the perception of modern architecture in Cyprus as heritage. For example, the fact that the interviews highlighted a high level of aesthetic appreciation for modern school buildings, especially linked with the initial school buildings and building elements (and not the later additions), constitutes an interesting finding since the lack of aesthetic appreciation for modern buildings is considered one of the obstacles in raising awareness about their heritage value.

This study's expert and user analyses revealed that modern school buildings hold diverse heritage values on multiple scales, from personal to international significance. The identified values are dynamic, context-dependent, and connected to specific physical elements. Critically, incorporating stakeholder perspectives through interviews provided additional insights into contested histories and priorities for preservation. Overall, the value-based approach enabled holistic understanding of the buildings' entanglement with complex socio-political processes in Cyprus. This methodology can inform strategies for participatory, equitable mobilisation of difficult heritage towards sustainability and peace-building. Despite widespread acceptance of heritage's benefits (ICOMOS 2017b), unprotected modern architectural heritage assets represent missed opportunities for sustainable development in Cyprus. Safeguarding all manifestations of the local heritage remains an urgent, collective duty.

Acknowledgements This paper is in part based on material from my PhD thesis on "A value-based approach to modern architectural heritage in Cyprus: Schools in Larnaka, 1945–1963" at University of Westminster, UK, carried out under the supervision of Dr John Bold, Dr Davide Deriu, and Dr Panayiota Pyla. I would like to express special thanks to the A.G. Leventis Foundation for financially supporting my research for three years.

References

Araoz, Gustavo F. 2011. Preserving heritage places under a new paradigm. *Journal of Cultural Heritage Management and Sustainable Development* 1 (1): 55–60.

Australia ICOMOS. 1979. *Charter for the Conservation of Places of Cultural Significance*. Burra: ICOMOS.

Australia ICOMOS. 1999. *The Burra Charter: The Australia ICOMOS Charter for Places of Cultural Significance*. http://australia.icomos.org/wpcontent/uploads/BURRA_CHARTER.pdf. Accessed 15 February 2023.

Australia ICOMOS. 2013. *Practice Note: Understanding and Assessing Cultural Significance*. https://australia.icomos.org/wp-content/uploads/Practice-Note_Understanding-and-assessing-cultural-significance.pdf. Accessed 15 October 2023.

Australia ICOMOS. 2013. *The Burra Charter: The Australia ICOMOS Charter for Places of Cultural Significance*. https://australia.icomos.org/wp-content/uploads/The-Burra-Charter-2013-Adopted-31.10.2013.pdf. [Accessed 15 October 2023.

Avrami, E., R. Mason, and M. De la Torre. 2000. Values and Heritage Conservation. Los Angeles, California: The Getty Conservation Institute. Aydintik, S., Pulhan, H. and Uraz, T. (2016). Adaptive Reuse of Modern School Building in PostColonial Cyprus. In: Docomomo 14 th international conference proceedings.

Avrami, Erica, Mason, Randall and De la Torre, Marta. 2000. *Values and Heritage Conservation.* Los Angeles: The Getty Conservation Institute.

Bell, D. 1997. *Technical Advice Note 8: The Historic Scotland Guide to International Conservation Charters.* Edinburgh: Historic Scotland.

CHCFE Consortium. 2015. *Cultural Heritage Counts for Europe.* http://blogs.encatc.org/culturalheritagecountsforeurope/outcomes/. Accessed 15 February 2023.

CHCfE Consortium. no date. *Cultural Heritage Counts for Europe.* http://blogs.encatc.org/culturalheritagecountsforeurope/. Accessed 15 February 2023.

Constantinou, Costas M., Olga Demetriou, and Mete Hatay. 2012. Conflicts and uses of cultural heritage in Cyprus. *Journal of Balkan and near Eastern Studies* 14 (2): 177–198.

Council of Europe. 1975. *European Charter of the Architectural Heritage [Amsterdam].* https://www.icomos.org/en/resources/charters-and-texts/179-articles-en-francais/ressources/charters-and-standards/170-european-charter-of-the-architectural-heritage. Accessed 15 February 2023.

Council of Europe. 1985. *Convention for the Protection of the Architectural Heritage of Europe [Granada].* https://www.coe.int/en/web/culture-and-heritage/granada-convention. Accessed 15 February 2023.

Council of Europe. 2005. *Council of Europe Framework Convention on the Value of Cultural Heritage for Society [Faro].* https://rm.coe.int/1680083746. Accessed 15 February 2023.

Council of Europe. 2015. *Parliamentary Assembly (Recommendation 2071): Cultural heritage in crisis and post-crisis situations.* http://semanticpace.net/tools/pdf.aspx?doc%3DaHR0cDovL2Fzc2VtYmx5LmNvZS5pbnQvbncveG1sL1hSZWY%20vWDJILURXLWV4dHIuYXNwP2ZpbGVpZD0yMTc4NSZsYW5nPUVO&xsl%3DaHR0cDovL3Nlb%20WFudGljcGFjZS5uZXQvWHNsdC9QZGYvWFJlZi1XRC1BVC1YTUwyUERGLnhzbA%3D%3D&xsltparams%3DZmlsZWlkPTIxNzg1. Accessed 15 February 2023.

Council of Europe. 2017. Ministers' Deputies (Recommendations CM/Rec (2017)1): Recommendation of the Committee of Ministers to member States on the European Cultural Heritage Strategy for the 21st century. https://rm.coe.int/16806f6a03%0A, https://www.coe.int/en/web/culture-and-heritage/-%20/new-strategy-21-website-how-to-make-heritage-management-inclusive. Accessed 15 February 2023.

Council of the European Union. 2014. Council conclusions of 21 May 2014 on cultural heritage as a strategic resource for a sustainable Europe (2014/C 183/08). *Official Journal of the European Union (C 183/36).* https://eur-lex.europa.eu/legal-content/EN/TXT/?uri=CELEX:52014XG0614%2808%29. Accessed 15 February 2023.

Docomomo Cyprus. 2014. Cyprus 100 [most] important buildings, sites and neighbourhoods. https://issuu.com/docomomo.cyprus/docs/_importantbuildingsdocomomocy_18_04. Accessed 15 February 2023.

De la Torre, Marta. 2002. *Assessing the Values of Cultural Heritage.* Los Angeles: The Getty Conservation Institute. https://www.getty.edu/conservation/publications_resources/pdf_publications/values_cultural_heritage.html. Accessed 15 February 2023.

Demetriou, Olga. 2012. The militarization of opulence: Engendering a conflict heritage site. *International Feminist Journal of Politics* 14 (1): 56–77.

Demetriou, Olga. 2015. Grand Ruins: Ledra Palace hotel and the rendering of 'conflict' as heritage in Cyprus. In *War and Cultural Heritage: Biographies of Place,* ed. Sørensen, Marie Louise S., and Viejo-Rose, Dacia: 183–207. Cambridge: Cambridge University Press.

Historic England. 2014. The value and impact of heritage. In *Heritage Counts.* https://historicengland.org.uk/content/heritage-counts/pub/2014/value-impact-chapterpdf/. Accessed 15 February 2023.

European Commission. 2019. *European Framework for Action on Cultural Heritage.* Luxemburg: Publications Office of the European Union. https://op.europa.eu/en/publication-detail/-/publication/5a9c3144-80f1-11e9-9f05-01aa75ed71a1/language-en/format-PDF/source-101251729. Accessed 15 February 2023.

European Commission. 2014. *Communication from the Commission to the European Parliament, the Council, the European Economic and Social Committee and the Committee of*

Regions: Towards an Integrated Approach to Cultural Heritage for Europe (COM (2014) 477). Brussels. https://ec.europa.eu/assets/eac/culture/library/publications/2014-heritage-communication_en.pdf. Accessed 15 February 2023.

Fereos, Stefanos, and Petros Phokaides. 2006. Architecture in Cyprus between the 1930s and 1970s. The search for modern heritage. *DOCOMOMO Journal* 35: 15–19.

Fredheim, Harald L. and Khalaf, Manal. 2016. The significance of values: heritage value typologies reexamined. *International Journal of Heritage Studies*, 22 (6): 466–481. Routledge. https://doi.org/10.1080/13527258.2016.1171247. Accessed 15 February 2023.

Georghiou, Costas. 2013. *British Colonial Architecture in Cyprus*. Nicosia: En Tipis Publications.

Georghiou, Costas. 2018. *The Architecture of the Cypriots during British rule: 1878–1960*. Nicosia: En Tipis Publications.

Gibson, Lisanne, and John R. Pendlebury, eds. 2009. *Valuing Historic Environments*. Surrey: Ashgate Publishing Limited.

Given, Michael. 1997. Star of the Parthenon, Cypriot melange: Education and representation in colonial Cyprus. *Journal of Mediterranean Studies* 7 (1): 59–82.

Historic England. no date. *Heritage Counts*. https://historicengland.org.uk/research/heritage-counts/. Accessed 15 February 2023.

ICOMOS. 1998. *Declaration of ICOMOS Marking the 50th Anniversary of the Universal Declaration of Human Rights [Stockholm]*. https://www.icomos.org/charters/Stockholm-e.pdf. Accessed 15 February 2023.

ICOMOS. 2016. *Cultural Heritage, the UN Sustainable Development Goals, and the New Urban Agenda*. http://www.usicomos.org/wp-content/uploads/2016/05/Final-Concept-Note.pdf Accessed 15 February 2023.

ICOMOS. 2017a. *ICOMOS Action Plan: Cultural Heritage and Localizing the UN Sustainable Development Goals (SDGs)*. http://www.icomos.org/en/what-we-do/involvement-in-international-conventions/un-sustainable-development-goals-2. Accessed 15 February 2023.

ICOMOS. 2017b. *Delhi Declaration on Heritage and Democracy [Delhi]*. https://www.icomos.org/images/DOCUMENTS/Charters/GA2017_Delhi-Declaration_20180117_EN.pdf. Accessed 15 February 2023.

ICOMOS. 2018. *Buenos Aires Declaration marking the 70th anniversary of the Universal Declaration of Human Rights [Buenos Aires]*. https://www.icomos.org/en/about-icomos/governance/general-information-about-the-general-assembly/list-of-general-assemblies/20th-general-assembly-2018/52583-declaration-de-buenos-aires-marquant-le-70e-anniversaire-de-la-declaration-universelle-des-droits-de-l-homme-2. Accessed 15 February 2023.

Mason, Randall. 2002. Assessing values in conservation planning: Methodological issues and choices. In De la Torre, M. ed. *Assessing the Values of Cultural Heritage*. Los Angeles: The Getty Conservation Institute.

Papadakis, Yiannis, Nicos Peristianis, and Gisela Welz, eds. 2006. *Divided Cyprus: Modernity, History, and an Island in Conflict*. Bloomington: Indiana University Press.

Phokaides, Petros, and Panayiota Pyla. 2012. Peripheral hubs and alternative modernisations: Designing for peace and tourism in postcolonial Cyprus. In *Proceedings of the 2nd international conference of the European Architectural History Network*, eds. Heynen, Hilde, and Gosseye, Janina.

Phokaides, Petros, and Panayiota Pyla. 2016. Leisure-scapes and conflict-scapes: The Famagusta modern coastline. In *Adaptive reuse: the modern movement towards the future*, eds. Tostoes, Ana, and Ferreira, Zara. Lisbon: Docomomo International.

Pyla, Panayiota, and Petros Phokaides. 2009. Vitrual tour: Twentieth century architecture in Cyprus. *EAHN Newsletter* 2: 36–49.

Pyla, Panayiota, and Petros Phokaides. 2011. Ambivalent politics and modernist debates in postcolonial Cyprus. *Journal of Architecture* 16 (6): 885–913.

Pyla, Panayiota, 2009. Modernism, Modernization and Mid-20th Century Middle East. *European Heritage Days 2009: Learning from the Heritage of the Modern*: pp. 32–39. Nicosia: Ministry of Interior, Department of Town Planning & Housing and Cyprus Architects Association.

Siandou, Emilia. 2019. *A Value-Based Approach to Modern Architectural Heritage in Cyprus: Schools in Larnaka, 1945–1963*. PhD thesis. London: University of Westminster.

Sierepeklis, Zinonas. 1997. Diadromes tis metapolemikis arxitektonikis sti kypro. *Phileleutheros*.

Sioulas, Michalis, and Pyla, Panayiota. 2015. Technical schools for inter-communal crisis management in mid-20th century Cyprus. In *EAHN 2015 International scientific thematic conference: entangled histories, multiple geographies:* 271–277. Belgrade: European Architectural History Network, University of Belgrade.

Stephenson, Janet. 2008. The Cultural values model: An integrated approach to values in landscapes. *Landscape and Urban Planning* 84 (2): 127–139.

Tomaszewski, Andrej. 2005. Tangible and intangible values of cultural property in Western tradition and science. In *Proceedings of the 15th general assembly and scientific symposium of ICOMOS, Xi" an, 17–21 October, 2005*. http://openarchive.icomos.org/id/eprint/465/. Accessed 15 February 2023.

Tzirtzilakis, Giorgos. 2010. Topio, architektoniki kai geopolitiki: Mathenontas apo tin Kypro. *Kaput*. 08: 1–7.

UNESCO. 1972. *Convention Concerning the Protection of the World Cultural and Natural Heritage [Paris]*. https://whc.unesco.org/en/conventiontext/. Accessed 15 February 2023.

UNESCO. 2011. *Recommendation on the Historic Urban Landscape [Paris]*. https://whc.unesco.org/uploads/activities/documents/activity-638-98.pdf. Accessed 15 February 2023.

UNESCO. 2013. *The Hangzhou Declaration: Placing Culture at the Heart of Sustainable Development Policies*. Hangzhou, People's Republic of China. https://unesdoc.unesco.org/ark:/48223/pf0000221238. Accessed 15 February 2023.

UNESCO. 2015. *Policy for the Integration of a Sustainable Development Perspective into the Processes of the World Heritage Convention*. https://whc.unesco.org/en/sustainabledevelopment/. Accessed 15 February 2023.

United Nations. 1987. *Report of the World Commission on Environment and Development: Our Common Future (Brundtland Report)*. https://www.are.admin.ch/are/en/home/sustainable-development/internationalcooperation/2030agenda/un-_-milestones-in-sustainable-development/1987--brundtlandreport.html. Accessed 15 February 2023.

United Nations. 2015. *Transforming Our World: The 2030 Agenda for sustainable development*.https://sustainabledevelopment.un.org/post2015/transformingourworld/ Accessed 15 February 2023.

Worster, Donald. 1993. The shaky ground of sustainability. In *Global Ecology: A New Arena of Political Conflict*, ed. Sachs, Wolfang, 133–145. London and New Jersey: Zed Books & Halifax: Fernwood Publishing.

Worthing, Derek, and Stephen Bond. 2008. *Managing Built Heritage: The Role of Cultural Significance*. Oxford: Blackwell Publishing Ltd.

Chapter 9
A Holistic Teaching Approach for Recording and Documenting Cultural Heritage Buildings, Through the Implementation of Traditional, Digital and Environmental Tools

Chryso Heracleous, George Tryfonos, Maria Philokyprou, Odysseas Kontovourkis, and Aimilios Michael

Abstract Teaching the various methods of recording and documenting heritage buildings is an important way of gathering valued information about architectural heritage. Diachronically, there are various courses, at both undergraduate and graduate levels, on traditional methods of documenting heritage buildings, which in turn, have recently been enriched by advanced methods and technologies. This chapter introduced a new graduate course, through which theoretical and applied knowledge is gained by recording and documenting cultural heritage buildings using various methodologies (traditional, digital, and environmental). The main requirement of the course is the preparation of a project, with a historic church as a case study. The incorporation of environmental recording, together with traditional and digital architectural documentation of cultural heritage buildings, is an innovative teaching element that is introduced in this course, which furthermore, is necessary for the holistic understanding of built heritage.

Keywords Historic structures · Educational strategy · Documentation · Monitoring equipment · Laser scanning

Introduction

Recording and documenting architectural heritage is very important, as through this process valuable information about the various heritage buildings is gathered. According to Khalil et al. (2021), the documentation of heritage buildings is the preliminary action to deal with any problem related to built heritage. The procedure

---

C. Heracleous (✉) · G. Tryfonos · M. Philokyprou · O. Kontovourkis · A. Michael
Department of Architecture, University of Cyprus, Nicosia, Cyprus
e-mail: chrysohe@gmail.com

M. Loren-Méndez et al. (eds.), *Sustainable Heritage Management - Design: Notions, Methods and Techniques*, https://doi.org/10.1007/978-3-031-81970-4_9

of documentation requires a very diverse range of data (quantitative and qualitative). Khalil et al. (2021) suggested four main categories that cover the whole documentation data areas: geometry surveys and subsurface materials documentation, structural integrity investigations, data concerning the building performance and historic records concerning the building's morphology over time. All this data can help to create in-depth knowledge about the heritage building's status and performance, while creating a solid base for any required restoration and retrofitting processes (Khalil et al. 2021).

The preservation and documentation of architectural heritage has a long tradition in European history. The first thoughts regarding recording methods were formulated in the middle of the seventeenth century. At this time, the Royal School of Architecture in Paris considered Palladio's drawings unreliable and decided to send groups of French architecture students to record the historic buildings of Rome, even giving instructions on practical matters of equipment. For the first time, accurate impressions of monuments were made. At the same time, Antoine Babuty Desgodetz (1653–1728), a young, trained French architect, travelled to Rome and prepared the first really accurate drawings of Rome's monuments. But his views were not accepted among the architects of the time, who ultimately preferred to follow Le Roy's instructions, giving attention to general proportions and typology rather than the accuracy of measurements. Since then, these views regarding the protection of cultural heritage have varied greatly.

According to Klein (2014), art historians started the systematic documentation of cultural heritage around 1900. The most influential in Central Europe were Georg Dehio in Strasbourg, and Alois Riegl and Max Dvořák in Vienna. They left printed census annuals and art-topographies, which documented the built heritage and gave the public access to heritage information.

The importance of documenting the existing situation of a structure before any intervention, as well as the necessity for public accessibility to recording reports and drawings, are both underlined in international charters on conservation such as Venice Charter, Granada Convention, ICOMOS reports, Faro Charter etc. (Smars 2008).

Thus, the involvement of universities in the process of documenting heritage buildings is necessary and has many benefits. Students gain expertise during their academic education, learning to work on-site and experience the knowledge of the past through accurate observation, documentation and detailed analysis. The reports, detailed sketches and drawings prepared are the base for evaluation and protection, as well as the repair and adaptation of the structure for contemporary use.

Various tools and methods are used diachronically in the process of documentation. Documenting heritage structures using traditional tools has many advantages. For example, it offers a close relationship with the architectural heritage under investigation. In addition, it incorporates the use of very low-cost equipment, which is easy to acquire and can be used without any particular training. Despite this, the introduction of digital tools is also very important. This is because their use accelerates the speed of documentation, helping to facilitate access to many inaccessible parts of the buildings, while also making 3D representation of the structures both possible and easy (Yilmaz et al. 2008). According to Arayici (2007), cultural heritage needs more advanced techniques to support conservation and preservation. Recently, in the teaching of corresponding courses at university level, in addition to the traditional methods of recording and documenting historic buildings, advanced methods and technologies are also being introduced, including laser-scanning and photogrammetric surveying.

Today, another category of tools has been added—environmental tools—in order to proceed to a holistic documentation of heritage structures. Through the environmental evaluation of historic buildings, it is easier to understand the impact of architectural design and building construction on their environmental performance. Thus, the incorporation of environmental tools offers a holistic approach to the recording of heritage structures.

The aim of this paper is to show the importance of a holistic documentation of architectural heritage, involving different methods and tools for its architectural and environmental documentation. This is achieved through the presentation of an innovative course on the documentation of historic structures, offered to graduate students within the framework of a graduate programme on conservation at the University of Cyprus. At the same time, similar courses from other universities are presented for comparative reasons. In addition, the scientific results of the students' projects carried out as part of the aforementioned course are briefly discussed and compared with information derived from literature review.

Literature Review

Traditional methods of documenting heritage buildings are incorporated in many undergraduate and graduate courses at various institutions, thus showing the significance of this field of investigation. Specifically, many graduate conservation programmes in European universities include courses related to the analysis and documentation methodologies for heritage buildings. An example is a course called 'Analysis, Documentation and Recording Monuments and Historic Complexes' at the Technical University of Athens, that includes in-depth knowledge on the methodology of recording, documenting and evaluating historical buildings, as well as the analysis of historical structural techniques. In this course, the sources of information for the historical documentation of heritage buildings are presented. This includes contemporary recording tools (geodetic, photogrammetry), the methodology for documenting their historical, morphological and construction status, as well as their pathology.

Recently, the incorporation of digital technologies and advanced methods of laser-scanning and photogrammetric surveying have increased the teaching of corresponding existing or newly established courses at undergraduate and postgraduate university levels. Such technologies include laser-scanning, and 2D/3D photogrammetric surveying with aerial image capturing for interior and exterior digitisation. This is due to the ever-increasing impact that technology has on the cultural sector and the need for easy, fast and accurate reproduction, display and dissemination of both art and material heritage in general, as well as of architectural heritage in particular (Hess et al. 2017).

In literature, the critical assessment of digital technologies in art and cultural heritage is discussed in the work of Hess et al. (2017), through an object-based teaching and learning module that was applied within the framework of the course 'Technologies in Arts and Cultural Heritage', at University College London (UCL). Similarly, Hess et al. (2019) introduced the concept of critical assessment and object-oriented solutions in digital technologies used for heritage conservation, by establishing a Master's degree at the University of Bamberg, which involved and focused on specific areas of digital technology (Hess et al. 2019). In a similar direction, Bern University of the Arts started a Master of 'Applied Sciences BFH in the Preservation of Digital Art & Cultural Heritage' in 2013. This graduate programme aims to combine different fields, such as conservation and restoration, information science and media technology (Lurk 2014).

Another programme that focuses on cultural heritage research with an emphasis on the application of digital technology, is the MSc degree in 'Digital Cultural Heritage' at The Cyprus Institute. This programme aims to take an interdisciplinary approach, where issues such as cultural heritage research, conservation, management, and public engagement, are examined through digital methods and applications (The Cyprus Institute 2022). The programme focuses on photogrammetry and the sensor-based 3D documentation of tangible cultural heritage assets, leading to the 3D processing analysis and surface characterisation.

Although examples of the application of digital technology in recording and documenting cultural heritage buildings at undergraduate or postgraduate level are few, gradually they are being introduced as modules within a wider teaching framework of architectural curricula that refer to cultural heritage. Specifically, at the Polytechnic University of Milan as part of the Architectural Preservation Studio course, a module called 'Advanced Survey and Representation Techniques' has been introduced. Through the 'Learning by Doing' principle, this module aims to offer students the possibility of implementing critical skills and operational tools in the case studies under investigation (Lavori 2008; Orario 2015). In another course at the Polytechnic University of Milan (Politecnico di Milano, 2020), digital technology is applied within a cultural heritage context, to enhance its lighting design through 3D modelling and lighting simulation testing (Siniscalco et al. 2021).

In order to gain a holistic overview and documentation of a heritage building, recording its environmental data is also necessary, in addition to its architectural and structural survey drawings. The environmental investigation approach to cultural heritage buildings is mainly based on building physics, introducing the fundamentals of the thermal and visual aspects of building performance. Despite the aforementioned necessities, it is noted that until now, only a limited number of courses regarding the environmental perspective of heritage buildings have been introduced into the higher education of architecture and conservation. The combination of environmental and architectural documentation is even more scarce in education programmes on conservation. The Master's programme on Sustainable Heritage at

UCL includes the module 'Strategies for Sustainable Heritage Buildings and Collections', which aims to explore strategies for supporting the environmental sustainability of heritage buildings and collections. These include the control of environmental conditions, energy performance, and the use of renewable energy. The objective is to identify interconnections between building condition, building performance and collection condition, as well as strategies to enhance conservation. Furthermore, the 'Sustainable Building Conservation Skills' course offered by the Department of Archaeology at the University of York, focuses on the sustainable adaptation and retrofit of cultural heritage buildings, highlighting the environmental perspective. Similarly, Boston Architectural College's module, 'Sustainable Design and Preservation', tries to identify the features of cultural heritage buildings that conserve energy and create comfortable environments, as well as to explore ways to design energy improvements that meet historic preservation requirements. As observed, the aforementioned courses mainly cover the theoretical background of environmental perspective, and do not usually involve environmental monitoring and tools.

Case Study: 'Special Topics on Recording and Documenting Historic Buildings and Sites'- University of Cyprus

General Description of the Course

The University of Cyprus recently developed the course: '*Special Topics on Recording and Documenting Historic Buildings and Sites*' (ARH 550) as part of the graduate programme *Conservation and Restoration of Historic Buildings and Sites*. This new course was introduced due to the growth of new digital techniques and tools for the architectural and environmental documentation of cultural heritage buildings. It involves close collaboration between teachers from different backgrounds (in the field of traditional and digital tools, as well as in the field of environmental recording). The integration of traditional and digital architectural documentation, together with environmental monitoring, constitutes an innovative element of this course, not found in other similar courses.

This course provides basic and specialised knowledge of recording and documenting heritage buildings and includes the use of both conventional and contemporary digital methods. It introduces technologies that refer to issues of recording spatial organisation, the morphology and construction of buildings, as well as their 3D representation and documentation in the digital environment. In addition, it refers to the quantitative recording and analysis of a building's thermal and visual comfort conditions and energy efficiency, as well as the recording of external and internal environmental data.

The course aims to introduce research tools and methodological approaches for the on-site recording and documentation of buildings and sites, as well as individual structural and morphological building elements. It also includes methodologies for categorising, evaluating and processing recorded data.

Main Objectives of the Course

One of the main objectives of the course is recognising the importance of documenting architectural heritage, in order to have a holistic image of a structure before any conservation action is carried out. In addition, this course underlines the importance of the process in carrying out a detailed study and analysis of a historic structure (architectural, morphological, structural, environmental).

Through lectures and especially through case study investigations, students are familiarised with various methods and tools. They learn to combine different methods, in order to prepare all drawings necessary for the representation of a historic structure, using both 2D (plans, sections, elevations) and 3D models. Using traditional methods, students work very closely with the heritage building, thus managing to gain valuable information through observation and detailed in situ study and measurements.

In parallel, the application of digital methods through field work and hands-on experience, leads to data capturing and processing, thus improving students' knowledge on the application of various digital techniques, as well as on traditional documentation in survey projects. Furthermore, the course focuses on an in-depth analysis of the building geometry and its relation to the materials, building phases and the broader environment.

With regards to environmental monitoring, enhancing the students' abilities in the recording and analysis of comfort conditions through internal and external environmental data, can help them understand the contribution of design and structure in the overall performance of historic buildings.

Simultaneously, through the course, the students learn how to work in groups, cooperating with members from different backgrounds and skill sets, as well as contributing individually in different ways.

Methodology of the Course

This course includes theoretical teaching, as well as fieldwork. The teaching is based on lectures, tutorials and classroom presentations by both the professors and students, as well as discussions incorporating student participation. At the same time, field visits to support a better understanding of the course content are organised.

The major requirement of the course is the preparation of a project which uses a historic church as a case study (Fig. 9.1). Ecclesiastical buildings are considered suitable to achieve the goals of the course. This is because they follow some common typological characteristics, serve the same functional needs and are easily accessible

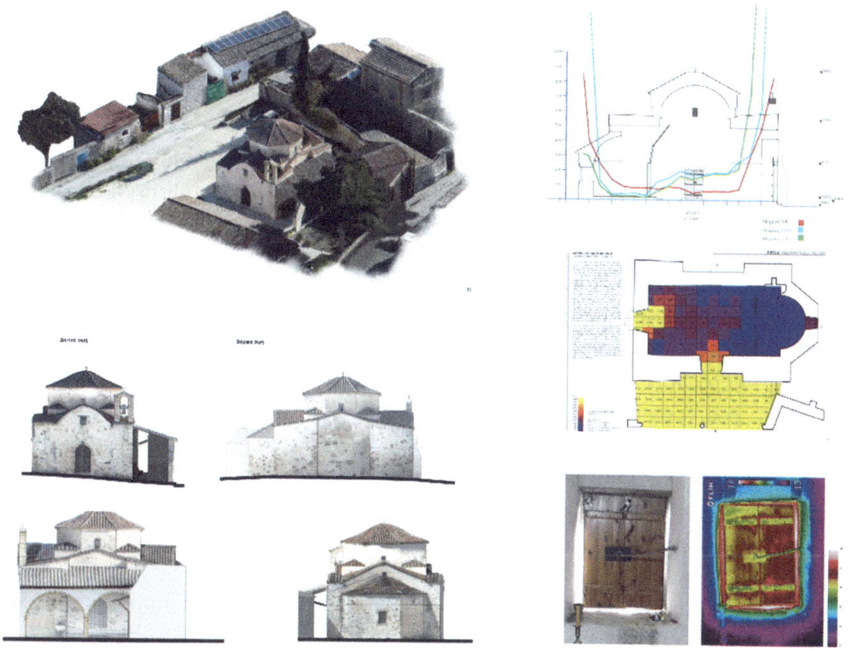

Fig. 9.1 Church of Archangel Michael, Pera Orinis. Views of a students' project showing the textured 3D mesh model exported from the photogrammetric process (top Left), four elevations of the church (left), one plan and one section of the building presenting the lighting levels inside the building (right), and thermography of a window (right/below) (Athanasiou et al. 2021)

for public use. Simultaneously, these buildings show variations in terms of location, size and typology. In general, churches follow different typologies with common characteristics: varying number of aisles (bays), as well as different types of roofing.

Within the framework of the course, various types of churches have been studied so far, such as single- and double-aisle, vaulted or cross-vaulted, as well as churches with a central dome. The selection of churches with different typologies, sizes and locations, but with common characteristics (eastern orientation of the chancel, main access from the west) helps to draw interesting conclusions. These conclusions not only relate to the teaching methodology, but also to the typology of the study buildings, as well as to their environmental behaviour. This in turn allows for comparison between them, which is considered an innovative element of the course.

The methodology followed in the preparation of the project includes three main stages:

(a) Preparation of floor plans and sections using mainly traditional surveying methods.
(b) Preparation of elevations, sections, 3D models and details using traditional and digital tools (terrestrial laser scanning and photogrammetry via Unmanned Arial

Vehicle (UAV) and cameras). The collected data is processed in 3D modelling programs such as Rhino3D (2023) and SketchUp (Trimble 2023).

(c) Recording of environmental data, including temperature, humidity, lighting levels, CO_2 concentrations, and thermography.

Initially, a site visit to the church under investigation for the study is carried out using traditional documentation tools and observations. This is followed by digital data collection using laser scanning (FARO 2023a, b) and a UAV (DJI Enterprise 2023) for image capturing. The laser scanner produces individual point clouds, which are registered together in the FARO SCENE Software (FARO 2023a). A georeferenced unified point cloud is then created, which is georeferenced at Cyprus Geodetic Reference System 1993, CGRS93/Cyprus Local Transverse Mercator (LTM) (EPSG:6312). For the georeferenced process, ground control points (GCP) are placed around the site of the church.

In parallel, an aerial photogrammetric method is applied using the aforementioned UAV, with a camera mounted at the bottom of it. Then, all images are imported to the Agisoft Metashape software (Agissoft 2023) for further processing. This includes high-quality alignment, dense point cloud creation, mesh geometry generation, orthophotos and 3D texture mapping.

The data from the laser scanning and photogrammetry are imported in 3D modelling environments for further processing and 3D model development. The 3D model, together with the accompanied 2D projected drawings developed, are used to verify the accuracy of the 2D line drawings. In addition, the use of textured orthophotos and views allows detailed 2D sections and elevations to be created. Finally, the 3D mesh of the church building can be exported and 3D printed as a scale model.

The environmental monitoring mainly involves the evaluation of thermal comfort conditions and lighting levels in the church under investigation, which in turn affects the energy performance of the building. More specifically, EL-USB-2-LCD temperature and humidity data loggers are given to the students, to be placed in selected positions inside and outside the church. The results are evaluated considering the building typology and morphology, orientation, shading devices, height, and exposure to the external environment. The temperature and relative humidity are recorded for the time span of one week from the end of March to the end of April. The internal spaces of the buildings under study have no technical support for heating or cooling during the aforementioned period.

At the same time, the distribution of light inside the buildings is examined using monitoring equipment. Natural lighting measurements are conducted on-site at 09:00, 12:00 noon and 15:00, near the equinox date. The sky conditions are recorded. The equipment TECPEL 536 Light Meter RS232 Data Logger is employed. Lighting data is taken from specific points of the church, in a grid of 0.50–1 m in both directions, which coincides with the centre of the grid at the openings of the church. The lighting levels are taken at a height of about 0.75 m from the church's finished floor level. During the recording, the buildings are free of users, artificial lighting is not used, and the shading systems remain opened.

Outcomes

The combination of theoretical teaching, assignments, and projects is a pedagogical approach that leads to students gaining an overall knowledge of the subject, as well as essential practical skills. Through this course, students are familiarised with the use of different conventional and digital tools, and methods of documentation. At the same time, they gain knowledge regarding the architecture (morphology, structure, typology) and environmental behaviour of ecclesiastical buildings.

Architectural Outcomes

After the site visits, field work and preparation of detailed survey architectural drawings, the students are able to understand and extract conclusions about the architecture of historic churches in Cyprus. Some common architectural characteristics of the churches observed are the following: they are stone-built structures, having walls of 50–100 cm in width and stone-built vaults for roofing. The number and size of openings is small as a rule. All the churches have a small east opening in the central apse, a main entrance to the west and auxiliary entrances in the south and west. The survey showed the existence of pairs of churches with almost exactly the same plan and proportions leading to the assumption that they were erected by the same builders, using the same architectural plans.

Environmental Outcomes

The evaluation of the churches' environmental characteristics allows for the implementation of research-based learning (RPL), where students can develop competencies and analysis, reflection, and argumentation skills, along with their training.

Some of the most common environmental characteristics of the churches under investigation include the use of building materials with high thermal mass, as well as the use of solid timber planks in the openings, which leads to a small temperature fluctuation inside the building. The cross arrangement of the openings of most of the churches under investigation helps towards the ventilation of the buildings during the cooling period. The existence of small openings high up, above doors and windows and on the domes, helps towards the extraction of hot air and the cooling of the space.

The field measurements of the churches' daylighting levels were largely conducted under a mostly clear or partly cloudy sky. The lighting levels were relatively low in all typologies of churches due to the small openings that characterised the local ecclesiastical architecture. Specifically, the average illuminance of single or double-aisle churches is around 100 lx during the morning (09:00), around 150–200 lx at noon (12:00) and around 100 lx in the afternoon (15:00). The lighting levels of

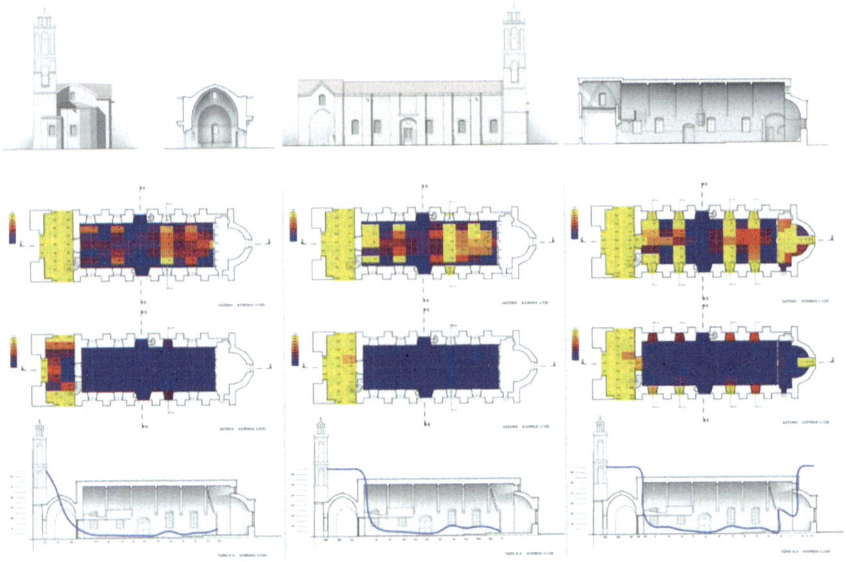

Fig. 9.2 Church of Ayios Ioannis, Nicosia. Views of a students' project showing some of the survey methods followed, four elevations of the church (above) (Gavalis et al. 2018). Church of Ayios Antonios, Nicosia. Views of a students' project showing plans and sections of the building presenting the lighting levels inside the building (below) (Hadjipetrou et al. 2018)

double-aisle churches are slightly lower due to a deeper plan compared to the single-aisle churches. Churches with domes provide higher lighting levels due to additional openings around the dome. It is interesting to mention that some churches have a semi-open space in front of one façade (mostly west or south). These spaces show sufficient lighting levels, but on the other hand may reduce the lighting levels which reach the windows of the specific façade (Fig. 9.2).

The investigation of temperature and relative humidity aims to assess the contribution of semi-open spaces, the impact of orientation, openings, and the height of the buildings, as well as the typology of church to the thermal performance of the structure. With regards to semi-open spaces, the results indicated that these spaces have higher diurnal fluctuations compared to internal spaces, as they are directly affected by the external environment. However, the temperature fluctuations in these semi-open spaces are lower than those of the external environment, as they work as sun-shading elements. Higher internal temperature fluctuations were observed near the openings of the buildings. This shows the poor airtightness of openings, allowing the external environment to influence the internal. The poor airtightness of openings was also confirmed by thermography which illustrates inhomogeneous temperature distribution in the same material, and increased temperature in the cracks and gaps (Fig. 9.1). The temperature difference along the height of the building i.e., from the lower layers (at about 1.10 m) to the higher layers above 3.50 m, is 0.5–1 °C. As a result of heat transfer on the higher levels, the domed churches presented slightly lower temperatures at the level of ground floor, compared to churches without domes.

The temperature distribution inside the churches at the end of March / early April was relatively low in all cases, ranging between 12 and 16 °C, and therefore not meeting the comfort zone limits. On the other hand, during mid-April to the end of April, the churches reached 20–25 °C, as the external temperature rose and fell within the comfort zone (Fig. 9.3). It is interesting to mention that the cross arrangement of openings enhances ventilation and thus comfort conditions during the intermediate and summer periods. Another important outcome is the fact that in some cases, a time lag of 4–5 h was observed due to the thick stone masonry walls, which delay the interior maximum temperature, thus improving thermal comfort during all periods. With respect to the relative humidity, generally all churches met the norms of 40–70%.

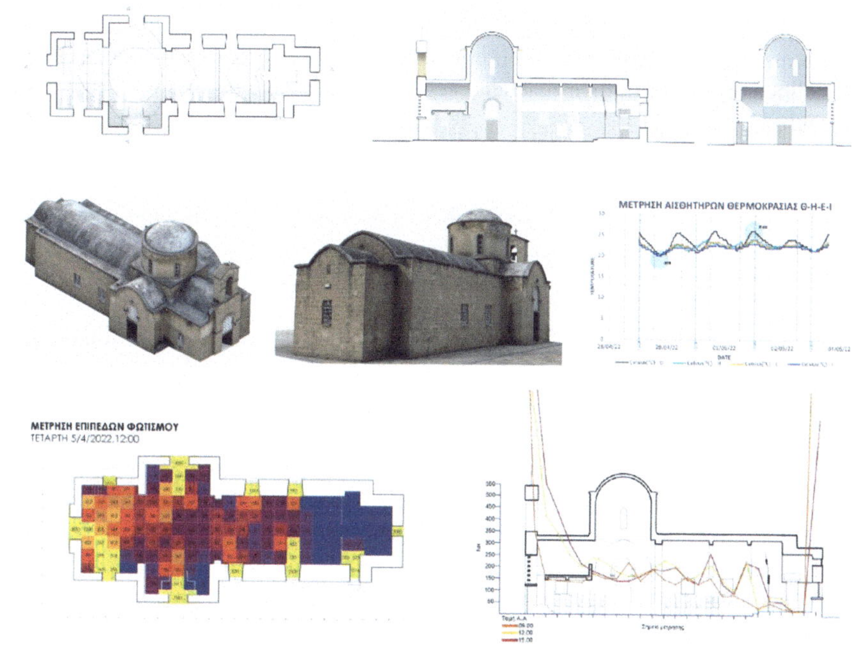

Fig. 9.3 Church of Ayios Pavlos, Nicosia. Views of a students' project showing some of the survey methods followed, plan and sections (above), digital scanning merged with photogrammetry results (middle), temperature distribution of the church (middle), one plan and one section of the building presenting the lighting levels inside the building (below) (Demou et al. 2022)

Critical Overview of the Course

Education and research significance

The theoretical and practical parts of the course are developed simultaneously in order to allow the 'Learning by Doing' method. This is a hands-on approach that allows students' interaction with their environment to adapt and learn, thus promoting experiential learning (Study.com 2022). This approach also helps the teachers to learn from students, while giving all participants in the learning experience an equal voice.

The implementation of the research-based course introduces students to professional research activities early in their career. Research-based learning (RBL) aims to promote and develop student competencies related to research practice, and to benefit students through activities linked to research (Noguez and Neri 2019). This technique implies the application of learning and teaching strategies that link research with teaching. In addition, through project-based learning, the students investigate various aspects of the case study building, actively interacting with their co-workers in small teams, and with their teacher, exchanging ideas and discussing progress in each subject.

Also, the students can use different methodological approaches for the completion of their project. This allows comparison of the research results, aiming at identifying possible divergences and convergences between them and ensuring the validity of the outputs.

Assessment of the results by students and by professors

The evaluation of the results by students and professors is important for improving the course over the years. According to the students′ evaluation reports, this course is an essential element of their studies in conservation, as they gain the necessary theoretical and practical knowledge on the field of documentation, which can help them on their professional career. At the same time, they become familiar with research methods and tools when following an academic path. According to their comments, the combination of architectural and environmental documentation of heritage buildings constitutes a challenge for them, as this is the first time they have combined different methodological tools in the same project.

Recording and documenting architectural heritage through different methodological approaches is very important, as it enables students to learn useful tools necessary for other courses, as well as for professional work and experience.

The introduction of environmental monitoring of heritage buildings is important to evaluate the effectiveness of their interior environmental conditions. Specifically, through the course, students and professors can interpret the impact of architectural characteristics of cultural heritage on the environment and disseminate valuable research results, adding to existing knowledge through participation in international conferences and writing articles in international scientific journals.

Discussion and Conclusions

Several courses on the traditional and digital methods of documenting heritage buildings are incorporated in undergraduate and graduate programmes on architecture and conservation. However, to the best of our knowledge, existing courses do not cover the environment in correlation with the architectural documentation. The coexistence of environmental monitoring with architectural documentation, is an innovative element of the course and at the same time essential for the holistic understanding of a heritage building. With the inclusion of environmental data in the education of architectural conservation, a multidisciplinary and integrative approach is ensured (Santamouris and Mumovic 2009). Also, the environmental study of cultural heritage buildings, in combination with their architectural and historic analysis and documentation gives valuable and necessary information about their thermal and visual aspects (Kakoulli et al. 2022).

The course allows students to develop deep knowledge, as well as skills such as critical thinking, collaboration, creativity, and communication. Moreover, it allows deeper interaction with the learning content, while enhancing autonomy, time management and valuable career insights. Additionally, as a consequence of the course's research-oriented approach, the students gain a better understanding of research methods. Thus, they develop research, writing and organisational skills. Furthermore, they can demonstrate a mastery of the information through their own reflection and reporting of their findings. By synthesising knowledge in final research reports, they also manage to understand and communicate the work of others. Finally, students are familiarised with various contemporary documentation and monitoring tools, which can be useful for their professional career in the field of architecture, engineering, archaeology, and conservation.

Acknowledgements This paper has been developed based on the teaching experience gained from the graduate course 'Special Topics in Recording and Documenting Historic Buildings and Sites' (ARH 550). We would like to sincerely thank all students who attended the course and prepared outstanding projects, some parts of which are presented in this paper.

References

Agisoft. 2023. *Discover Intelligent Photogrammetry with Metashape*. https://www.agisoft.com/. Accessed 1 December 2022.

Arayici, Y. 2007. An approach for real world data modelling with the 3D terrestrial laser scanner for built environment. *Automation in Construction* 16 (6): 816–829.

Athanasiou, D., S. Dimitriou, I. Kyriakou, and I. Panteli. 2021. *Church of Archangel Michael, Pera Orinis. Documentation and Recording. Graduate Study in the framework of the course ARH 550*. Special Topics on Documenting and Recording Historic Buildings and Sites. https://www.ucy.ac.cy/conservation/student-assignments/. Accessed 3 October 2023.

Demou, A., A. Kapetanios, P. Liassi, M. Patsias, and A. Psoma. 2022. Church of Ayios Pavlos, Nicosia. *Documentation and Recording. Graduate Study in the framework of the course ARH*

550. Special Topics on Documenting and Recording Historic Buildings and Sites. https://www.ucy.ac.cy/conservation/student-assignments/. Accessed 3 October 2023.

DJI Enterprise. 2023. *PHANTOM 4 RTK*. https://enterprise.dji.com/phantom-4-rtk. Accessed 1 December 2022.

FARO. 2023. *FARO Focus Laser Scanners*. https://www.faro.com/en/Products/Hardware/Focus-Laser-Scanners. Accessed 1 December 2022.

FARO. 2023a. *FARO® SCENE Software*. https://www.faro.com/en/Products/Software/SCENE-Software. Accessed 1 December 2022.

Gavalis K., M. Georgiou, S. Hadjisotiriou, and M. Yiapa. 2018. Church of Ayios Ioannis, Nicosia. *Documentation and Recording. Graduate Study in the framework of the course ARH 550*. Special Topics on Documenting and Recording Historic Buildings and Sites. https://www.ucy.ac.cy/conservation/student-assignments/. Accessed 3 October 2023.

Hadjipetrou, T., N. Kotsoni, M. Kounnapi, and M. Solomou, 2018. *Church of Ayios Antonios, Nicosia. Documentation and Recording. Graduate Study in the framework of the course ARH 550*. Special Topics on Documenting and Recording Historic Buildings and Sites. https://www.ucy.ac.cy/conservation/student-assignments/. Accessed 3 October 2023.

Hess, M., D. Garside, T. Nelson, S. Robson, and T. Weyrich. 2017. *Object-based teaching and learning for a critical assessment of digital technologies in arts and cultural heritage*, in International Archives of the Photogrammetry, Remote Sensing and Spatial Information Sciences—ISPRS Archives, pp. 349–354. https://doi.org/10.5194/isprs-archives-XLII-2-W5-349-2017

Hess, M., C. Schlieder, A. Troi, O. Huth, M. Jagfeld, J. Hindmarch, and A. Henrich. 2019. *Digital Technologies in Heritage Conservation*. Methods of teaching and learning this M.SC. Degree, unique in Germany, in Advances in Intelligent Systems and Computing. Springer Verlag, pp. 53–63. https://doi.org/10.1007/978-3-030-12240-9_7

Kakoulli, C., A. Kyriacou, and M. Michaelides. 2022. A review of field measurement studies on thermal comfort, indoor air quality and virus risk. *Atmosphere* 13: 191.

Khalil, A., S. Stravoravdis, and D. Backes. 2021. Categorisation of building data in the digital documentation of heritage buildings. *Appl Geomat* 13: 29–54. https://doi.org/10.1007/s12518-020-00322-7.

Klein, Caroline Jaeger. 2014. *Documentation and Inventory of Architectural Heritage*. The European model including the role of universities within. UBT International Conference. 16. https://knowledgecenter.ubt-uni.net/conference/2014/all-events/16. Accessed 3 October 2023.

Lavori, S.U. 2008. Scheda docente, pp. 2008–2008. https://www4.ceda.polimi.it/manifesti/manifesti/controller/ricerche/RicercaPerDocentiPublic.do?evn_didattica=evento&k_doc=118320&aa=2019&lang=EN&jaf_currentWFID=main. Accessed 8 December 2022.

Lurk, T. 2014. *Preservation of Digital Art & Cultural Heritage*. A graduate program at Bern University of the Arts, CeROArt, (HS). https://doi.org/10.4000/ceroart.4301

Politecnico di Milano. 2020. *Architectural Preservation Studio*. https://www4.ceda.polimi.it/manifesti/manifesti/controller/ManifestoPublic.do?EVN_DETTAGLIO_RIGA_MANIFESTO=evento&aa=2019&k_cf=222&k_corso_la=1195&k_indir=BE1&codDescr=051566&lang=EN&semestre=1&anno_corso=1&idItemOfferta=146159&idRiga=249066. Accessed 3 August 2022.

Noguez, J., and L. Neri. 2019. Research-based learning: A case study for engineering students. *International Journal on Interactive Design and Manufacturing* 13: 1283–1295. https://doi.org/10.1007/s12008-019-00570-x.

Orario, D. 2015. *Architectural Preservation Studio*, pp. 2–3. https://www4.ceda.polimi.it/manifesti/manifesti/controller/ManifestoPublic.do?EVN_DETTAGLIO_RIGA_MANIFESTO=evento&aa=2019&k_cf=222&k_corso_la=1195&k_indir=BE1&codDescr=051566&lang=EN&semestre=1&anno_corso=1&idItemOfferta=146159&idRiga=249066. Accessed 3 August 2022.

Santamouris, M., D. Mumovic. 2009. *A Handbook of Sustainable Building Design and Engineering: An Integrated Approach to Energy, Health and Operational Performance*. Earthscan, James & James, London.

Siniscalco, A., L. Bellia, and L. Marchesi. 2021. Teaching lighting design for cultural heritage in the digital and pandemic era: Experiencing new and old techniques. *Heritage* 4 (4): 3731–3748. https://doi.org/10.3390/heritage4040205.

Smars, P. 2008. *Documenting Architectural Heritage, Looking for a Balance Between Automation and Control*. Journal of Cultural Conservation Science, Taiwan, pp. 3–10. http://smars.yuntech.edu.tw/papers/jccs0.pdf. Accesed 3 October 2023.

Study.com. 2022. *John Dewey on Education: Impact & Theory*. https://study.com/academy/lesson/john-dewey-on-education-impact-theory.html. Accessed 3 November 2022.

The Cyprus Institute. 2022. *Digital Cultural Heritage*. https://www.cyi.ac.cy/index.php/education/masters-programs/digital-cultural-heritage/masters-dch-program-overview.html. Accessed 3 August 2022.

Trible. 2023. SketchUp. https://www.sketchup.com/. Accessed 3 October 2023.

Yilmaz, H.M., M. Yakar, and F. Yildiz. 2008. Documentation of historical caravansaries by digital close range photogrammetry. *Automation in Construction* 17 (4): 489–498.

Chapter 10
Field-Based Learning in Urban Morphology: Observation, Documentation and Characterization

Milica P. Milojević, Aleksandra Đorđević, Mladen Pešić, Jelena Živković, and Vladan Djokić

Abstract This chapter explores relations between heritage, sustainability, and urban morphology by demonstrating the importance of field-based learning in urban morphology for the recognition of in situ values. In line with a contextually sensitive approach, in situ values are recognized as valuable assets that could be found in a specific place both as cultural values that are already present and recognized and patterns that can become values through observation, documentation, and characterization. Accordingly, this study advocates for fieldwork as a framework for developing interest and professional responsibility for the in situ values of the local cultural context.

Keywords Urban form · In situ values · Fieldwork · Urban settlements · Multiscale approach

Introduction

Urban morphology as a specific field of study experienced growth from 1990s onwards (Oliveira 2016), specifically after establishing International Seminar on Urban Form—ISUF in 1994 as an organization for researchers and practitioners of urban form and launching the journal *Urban Morphology* in 1997. Specific focus within this field is given to researching urban form in an interdisciplinary context of architecture, geography, planning, and history, without aspiration to diminish or forget historical research and thinking of physical manifestations, as stated by the founder of the association (Whitehand 2016). Consequently, the multidisciplinary nature of urban morphology requires wider insight to the multifaceted meaning of in situ derived from the original Latin idiom referring to 'on site' or 'in position'. Along with derived meanings that can be interpreted as 'locally', 'on the premises',

M. P. Milojević · A. Đorđević (✉) · M. Pešić · J. Živković · V. Djokić
Faculty of Architecture, University of Belgrade, Belgrade, Serbia
e-mail: aleksandra.dj@arh.bg.ac.rs

M. Loren-Méndez et al. (eds.), *Sustainable Heritage Management - Design: Notions, Methods and Techniques*, https://doi.org/10.1007/978-3-031-81970-4_10

'in place' when emphasizing the position where something is located or where something took place, in general, in situ means that something is situated in its natural or original position, or connected to the specific place and surroundings, and relates to characteristics of place and space in various forms. Nevertheless, the distinctions of meaning change slightly in relation to the discipline or field within which in situ specificities are discussed. Within natural sciences, in situ can describe the way a measurement is taken, that is, in the same place the phenomenon is occurring without isolating it from other systems or altering the original conditions, or it can denote that an experiment is done in the field rather than in the laboratory. In archaeology, within which it is most recognizable and most often used, the term refers to an artefact that has not been moved from its original place of disposition and as such is valuable for the interpretation of the artefact and accordingly of the culture which formed and used it. Work of art made especially for the host site, or that takes into account the site in which it is installed or exhibited could also be regarded as in situ.

If all these vast meanings are taken into account, in the discourse of urban morphology one can relate in situ values as: (1) valuable features that are inseparable from their original environment and crucial for interpretation of the urban patterns and culture which formed them (from archaeology), (2) activities performed on the site—fieldwork activities (from natural sciences), and (3) site-specific and contextual solutions (from the art field). More specifically, in situ values may be used to designate measurable and non-measurable characteristics of the place and its surroundings, and hence to determine potentials and limitations for anticipation of spatial transformation. In situ values are not always easy to define, but it is expected that during intervention in the built environment, they are discovered, highlighted, or enhanced with the minimization of negative consequences.

Depending on the type of value (economic, social, cultural, environmental, or other), and the context (spatial characteristics of the built environment) the process of identification of in situ values and their possible application and meaning could vary. Value may be regarded as 'worth'—as an assessment of whether something (an object, an idea, a state of affairs) matters or not, and how it matters (Chiaradia et al. 2017). Accordingly, 'value' may be aligned with 'meaning' or more precisely with 'articulated meaning' which indicates that if meaning cannot be communicated, then there is no value. Consequently, besides positive inputs and valuable assets that could be found in a specific place both as tangible and intangible cultural values, in situ values are critical to the interpretation of the urban patterns.

To develop adequate descriptive mechanisms for dealing with physical, spatial, and configurational characteristics that are the essence of in situ values, the process of fieldwork is of great importance. Fieldwork could be perceived as instrumental in producing knowledge about the built environment in urban and architectural design, architectural history, and other related disciplines. Historically, fieldwork was closely connected both with academic and practical endeavours within the field of architecture and urbanism and it has put pragmatic challenge and importance for both fields. Role of fieldwork as a part of educational programmes in architecture and urbanism has also been highlighted as a form of knowledge and cultural production (Fekete 2021; Rendell 2010), as a part of studio teaching, study trips, site surveys, and site analysis. Moreover, Peter Larkham highlights the tendency of planning practice and developers to rely more on virtual reality and gaming technology rather than facing the uncertainties of site visits and argues for engagement with real-world urban form that can result in "better understanding of smaller features that cumulatively create a character, the factors that shape the lived experience and the genius loci" (Larkham 2018). For practice, significance of fieldwork is illustrated through internships and reflective design processes that can accumulate existing knowledge and experience, and are complemented by spatial analysis. Fieldwork has been used in other disciplines such as—social sciences, anthropology, geography, history, etc., where it employs associated techniques and skills that include systematic observation, measuring and analysing form, formation, functioning, and use through direct experience. Fieldwork should be an intrinsic part of spatial research of in situ values since it informs and at the same time confronts the practitioners or researchers with the built environment in real life, shows results of implemented plans and projects, enables different experiences, and provides direct contact with local context, people and history.

Linking Urban Morphology to in Situ Values and Fieldwork

The purpose of urban morphology according to the definitions and statements collected by Steven and Olgu (2011) is to (a) find out principles or rules of urban design (Gebauer and Samuels 1981), (b) analyse a city's evolution from its formative years to its subsequent transformations, identifying and dissecting its various components (Moudon 1997), and (c) providing explanations or developing explanatory framework and/or determining the modalities according to which the city should be planned or built in the future (Gauthier and Gilliland 2006). In relation to the role of urban morphology, identification can be perceived as an act of finding out values of local culture through observation, documentation, and characterization of cultural patterns that are loaded and read from the morphological characteristics of the built environment. The values that urban morphology is concerned about are always seen in relation to the groups of patterns, inseparable from their original environment (natural, social, cultural).

Urban morphology is traditionally associated with (1) having physical elements on various scales as a subject of its study, (2) having several different theoretical approaches and research methods that are specific to various schools of thought, and (3) applying various tools for research, transferring knowledge in an academic environment and acting in practice. Accordingly, urban morphology provides a complex framework valuable for discovering and emphasizing values of urban heritage and ensuring their continuity that one can find by researching on and about the site.

The role of urban morphology, as Peter Larkham (2005) highlights it, is to understand the ways of towns growth and development by "understanding the physical complexities of various scales, from individual buildings, plots, street blocks, and the street patterns that make up the structure of towns". Various spatial components and configurational elements are subject of different morphological researches, from landscapes and townscapes (Gu 2019; Whitehand 2021), through urban tissues (Ünlü 2018) and building and street types (Oliveira 2016, p.136, Ünlü 2018, Samuels 2021) to plot dimensions (Slater 1990) and character areas (Samuels 2019; Larkham and Morton 2021). These spatial elements are always perceived into temporal components such as changes and transformation (Conzen 2018; Larkham 2018), conditional/relational elements such as boundaries, limits, divisions (Oliveira 2016, p. 133), and functional-operational characteristics such as agents and agency, uses and planning zones (Oliveira 2016, p. 104, Vitor and Silvia 2012, Oliveira 2021), density (Berghauser Pont, Meta 2018), and land ownership (Gu 2019).

Within urban morphology, several approaches are acknowledged: historico-geographical approach, the process-typological approach, space syntax, and pertinent types of spatial analysis (Whitehand 2016; Oliveira 2016). Rooted in German human geography, the historico-geographical approach was built on the basis of seminal work of MRG Conzen in the UK, further developed by J.W.R. Whitehand and MP Conzen, with specific focus on fringe-belt, the morphological region, and the burgage cycle (Oliveira 2016). The process-typological approach is structured around the work of Saverio Muratori, and his successor Gianfranco Caniggia, developed predominantly in Italian context and focused on the research and endeavour to preserve continuity of types and typological processes. Space syntax and the various forms of 'spatial analysis' are structured around research of Bill Hillier and Julienne Hanson in the UK. Main role of the space syntax in relation to heritage is seen in tracing effects of the layout geometry, understanding behaviour flow and movement pattern on different scales, understanding the interdependence between the two and providing comprehension of the evolution of settlement morphologies and changes that occurred over time.

Each of the previously described approaches brings its own tools. For the purpose of this paper, the most prominent ones related to the field of architecture and urbanism are highlighted. Morphogenesis refers to the analysis of continual and perpetual historical processes in which urban tissue is observed both at the time of their emergence and during the development and transformation in time (Djokić 2009a). The tools within this approach are applied to (a) reveal order within complex urban form, (b) understand and elaborate dynamics of urban form (Oliveira 2021), (c) discover and present transformation of urban form in different historical periods (Đokić 2009b), (d) establish connection with history and interpret it in the present (Sanders and Baker 2016), and (e) add value to the design and place making (Davies in Sanders and Baker 2016). Other important methods refer to methods of morphological regionalization and metrological analysis (Oliveira 2016). Within process-typological school of thought, Caniggia was the figure who contributed the most for operationalization and translating of Muratori's philosophical ideas, perceiving typological process as the most fundamental mode for built environment transformation (Gauthier 2005). Space syntax is both a theory and method based on four components:

Representations of space (through convex map, visibility graph and other representation model such as isovists, agents), spatial form analysis (Segment angular analysis—choice analysis, integration analysis, multiple scale analysis, other measure, and other methods VGA and agent analysis, spatial function analysis (movement, land use, and other methods), and interpretative models (movement model, land use model, and other).

Methodology

The research was conducted in three related and mutually conditioned phases so that the first phase includes building of a theoretical framework to be used as main input for the second and third parts of the research. Theoretical framework is predominantly focused on the content analysis of four books solely related to urban morphology, issued by Springer Publishing from 2016 to date within the Urban Book Series (Oliveira 2016, 2018, 2019, 2021). These books were identified as the most prominent sources that gathered contributions of the leading figures highlighting roles and addressing challenges of urban morphology in relation to education, practice, and research. Different spatial components and configurational elements derived from this research phase were used as constituent components of the subject of the study within Table 10.1. Second phase is systematic by nature and focused on building the relationship between domains of urban morphology (subject of the study, method/approach, tools) and identification of in situ values with its generative activities (observation, documentation, and characterization). Third phase starts from Michael Barke's diagram from his seminal text "Further thoughts on research and practice in urban morphology" (Barke 2015) deals with diverse interests of urban morphologists that act in academic and practical environments, with the intention to understand perspectives of interest between the two. Both phases take fieldwork as its focal point and question its role for identifying in situ values. The focus has been put to the challenges of the fieldwork as research/practice interface, since previous research (Đorđević et al. 2022; Djokić et al. 2021) has testified about the importance of the fieldwork in education for notions and tools related to urban morphology and for securing multiscale approach when dealing with heritage. Hence, the paper starts from the assumption that fieldwork is crucial for identification of in situ values and as such, needs to be equally addressed both in academic and practical environment. In the final step, conclusions were drawn by interrelating the results from the second and third phases.

Table 10.1 Theoretical foundation of urban morphology in relation to fieldwork and identification of in situ values

		Identification of types, processes, agents and agency, phenomena, behavioural patterns		
		A >	< B >	< C
		Observation	Documentation	Characterization
1. Subject of the study		Urban form; urban patterns; typologies; spatial relations (types representatives)	Built environment patterns; genesis; historic urban landscape; urban heritage; types characteristics; distribution of types	Culture of built form; urban form; townscape and landscape; character area; urban tissue; urban pattern, urban block; routes, town square; urban grain (street, plot and building pattern)
	Fieldwork	Better understanding and application of theoretical approaches; learning from experience	Linking ideas and codes of her-their representation; Recognizing types and patterns	Confirmation or expansion of knowledge and verification of theoretical models; Recognition, questioning of dominant paradigms
2. Method/ approach		Dependent on approach: historico-geographical, process-typological and space syntax and pertinent types of spatial analysis	Typo-morphological classification; morphogenesis; space syntax maps; followed by induction, deduction and/or analogy	Combined—quantitative and qualitative approach
	Fieldwork	Analogy; data selection and collection on urban form and spatial relations (under-represented or unavailable in other sources, tracing behavioural patterns;	Data systematization; Deduction and induction when getting acquainted with the phenomena; Conducting visits to archives or other collections of data (private or public); Conducting interviews; Mapping	Experience; sensibility; walking and being at the site—perception and experience of spatial relations and microclimatic conditions

(continued)

Table 10.1 (continued)

		Identification of types, processes, agents and agency, phenomena, behavioural patterns		
		A >	< B >	< C
		Observation	Documentation	Characterization
3. Tool		Mapping; drawing; recording; taking photos/framing; town-plan analysis, cadastral map analysis, site-analysis; morphometric tools	Mapping (data basis for structurization of information); GIS (ArcGis, QGis…) and augmented reality; Morphological map; Photo documenting, maps, drawings, interviews, design guidance	Visualization; 3d modelling, drawing; GIS, 3D GIS mapping; Building typology
	Fieldwork	Walking; sketching; recording; taking photos/framing; mapping; talking	Systematization of recording, interviewing and mapping results, structuring of questionnaires, structuring of guidelines	Preparation of the database, selection of information and visualization techniques. Photo editing and processing, characteristic samples and patterns, or specific frames

Research

Phase 1: Relationship Between Domains of Urban Morphology and Identification of in Situ Values

Research findings were systematized and summarized in the form of a comprehensive table where three main rows correspond to research and practical domains of urban morphology while dealing with the built environment: (1) Subject of the study; (2) Method/theoretical approach; (3) Analytical tools. Three columns follow activities undertaken by urban morphologists in the process of identification of urban patterns, types, and processes: (A) observation, (B) documentation and (C) categorization. It is worth mentioning that each activity of the identification (observation, documentation, and characterization) develops in every domain of urban morphology, there is interdependence and consequentiality among them (< >).

Within each of the rows, the specific role of fieldwork in cognitive processes developed and questioned through education, scientific research, and practice of urban morphology has been listed and elaborated. Learning from experience is a cognitive process for which fieldwork is of specific importance. Accordingly, systematisation in a form of a table enables reading the potential and importance of fieldwork for identification of in situ values while performing activities of observation, documentation, and characterization in all three domains of urban morphology. Moreover, the activities that precede the fieldwork, that is, those that are necessary for the on-site research to produce relevant results and activities that rely on insights from the fieldwork, are indicated (Table 10.1).

From the analytical lenses of fieldwork, the set of terms noted in the first shaded row illustrates the subject of the study within urban morphology, as the highest level of cognitive process which, through connecting the subject of observation (field 1A in Table 10.1) enables further process of documenting (field 1B in Table). Further on, based on the processes and procedures of documenting, new layers of cognition about the subject of study of urban morphology can be defined through characterisation of the subject of study, in the form of a specific discursive formation of urban morphology—notions ((field 1C in Table).

Fieldwork activities enable new insights into what can be learned by identification of patterns, types and processes, specifically through observation, documentation, and characterization. Observation during the fieldwork enables for pre-known, learned, or new ideas to be recognized. By documenting the field, ideas, concepts, phenomena, tacit knowledge, and values are connected with their (visual) spatial representations in the form of types, patterns and spatial–temporal relations. Lastly, performing characterization on the field, enables knowledge confirmation or expansion and puts in question previously known/learned theoretical models. In other words, fieldwork enables urban morphologists to contribute to the confirmation, recognition, questioning of dominant paradigms and encourages and provokes critical thinking of the city and the built environment in order to search for the seeds of new paradigms.

Regarding method/approach, currently, as mentioned, there are three dominant approaches that methodologically contribute the most to the study of urban form: historico-geographical approach, process-typological approach, and space syntax and pertinent types of spatial analysis. Depending on the approach applied in observation activities (field 2A in Table 10.1), the result of the documentation process can be: morphogenesis, typological classification, and space-syntax maps (field 2B in Table 10.1). These results are used as input for a combined qualitative and quantitative method applied for the characterization of urban phenomena (field 2C in Table 10.1) or, more precisely, in the description of the illustrations of the discursive formations of urban morphology (field 1C in Table 10.1). In the domain of urban morphology, it is significant to note the importance of mapping in the cognitive process since certain spatial phenomena can only be explained when their arrangement in space is identified (Nikezić 2007).

Activities that are performed on site are relevant for the identification of urban patterns, types, and spatial processes (shaded fields related to 2A, 2B, and 2C in Table 10.1). During observation, selection of existing and collection of new data are performed. This data usually relates to the characteristics of the site, built structures, or spatial relations that are under-represented or missing from the documentation retrievable from the archive, municipal institutions, or other sources such as cadastral or topographical maps. Specifically, to the space syntax and pertinent types of spatial analysis, behavioural patterns are the ones relevant among other information for the documenting activity. During documentation, previously prepared questionnaires and interviews are conducted and tested, even adjusted (Field 2B in Table 10.1). For characterization, perceived as the most valuable result of identification of in situ values, personal engagement and experience is crucial. By being at the site and walking in real microclimatic conditions, spatial relations are perceived, experienced, and understood. Additionally, one can experience sensitization (of local communities and) of future experts for the discovery of in situ values in real time and space.

When dealing with tools in relation to observation, documentation, and characterization, urban morphology balances between traditional analogues tools such as walking, sketching, recording, photographing, talking, etc., and digital, constantly growing tools (GIS, Augmented Reality (AR), 3D modelling). Based on the results of the observation phase, predominantly dealing with analogues tools (Field 3A in Table 10.1), structuring and systematization of results are performed (Field 3B in Table 10.1). In the final step, characterization is performed through selection of information and visualization techniques (Field 3C in Table 10.1). Before entering the process of data visualization and image processing, it is worth mentioning that fieldwork can contribute to the data selection, sampling, and framing.

Phase 2: Mutual Interest of Urban Morphologists in Fieldwork

Based on the relations between the practitioners and the academic urban morphologists described by Michael Barke (2015, 2018), the position of fieldwork was identified as being capable to connect for both sides in areas of sites of interests, groups of interests, capacity building, building arguments, contextual approach, and responsibility (Table 10.2).

Table 10.2 Mutual interest of practitioner and academic urban morphologists in fieldwork (adopted based on Barke 2015)

Practitioners >	< Mutual interest > mutual interest in fieldwork	< Academic urban morphologists
	Urban form/townscape	
Operate within legislative frameworks	Sites of interest	Operate within personal/research group and peers' agenda
Reconcile perspectives of developers, protagonists and public	Contact with groups of interest	Recognize established knowledge: but challenge it
Operate within established consensual paradigm	Capacity building	Probe the horizons of knowledge: seek new paradigms
Answerable to a local political bureaucracy	Building arguments	Answerable to self/own intellectual integrity
Tangible outputs—implementation of a plan/strategy	Contextual approach and responsibility	'Ideas' main output—academic publications
	Impact Agenda?	

Practitioners tend to document patterns and principles to be applied in the scope of the legislative framework both in implementation of the current plans and in the process of conceptualization and creation of new ones. Following the perspective of academic urban morphologists', working on the field enables identification of site limits and allows reconsideration of initially selected sites and its pertinence in relation to researchers' agenda. In practice, fieldwork is used for tracing changes and processes with the purpose of updating and enriching site-related data, identified as valuable. Based on the fieldwork results and experience, practitioners and academics identify sites of interest. These sites are perceived as parts of the territories exposed to transformation or being pressured to be transformed, regardless of whether these changes are positive or negative. Alignment of sites of interest is carried out while simultaneously collecting additional data relevant for the further research steps. Fieldwork can be organized in a manner that aligns sites of interests of practitioners and academics by close cooperation in the preparatory phase and effective communication. The very communication is usually based either on personal or institutional connections, or through projects that encourage cooperation between multiple institutions from academic, scientific, private, civil, and public sectors. Inclusion of urban morphologists in the work of these institutions enables this communication to be built on mutual values and understanding of fundamental principles, reliability, and applicability of results. It is worth mentioning that there is a need for reaching consensus regarding safety, cost and time efficiency, and accessibility for fieldwork and establishing standards of observation, documentation, and characterization of the area).

When it comes to the different perspectives of developers, protagonists, and the public, direct contact enables one to get insight into different views and aspirations. Only when perspectives of development actors outside the field are combined with information gathered through questionnaires, interviews, and informal talks on the site, practitioners can form a full understanding of how to reconcile perspectives through specific plan, project, or technical documentation. Although this information can be gathered online, personal contact provides a direct, informal approach and hence possibility to articulate interests at risk to be forgotten in online forms. Consequently, practitioners develop responsibility both towards groups of interest with whom they interact (citizens, local communities, landowners) and towards their perspective in future actions. From the perspective of an academic urban morphologist, formal and informal contact, and interaction with various actors during the fieldwork challenges established academic knowledge regarding its applicability in making adequate solutions and confirms the need to engage academic knowledge. On the site, there is an opportunity to directly open up unexpected research questions and develop different directions of action. Even though there is an immense importance for academic urban morphologists to interact with various experts on the field, mutual interest is seen in establishing contact with local communities that can develop their interest for engaging urban morphologists as experts that respect and try to preserve site-specific-in situ values. On the other side, fieldwork can additionally make productive links between academic and practical fields of action.

At the relation between established and new paradigm, workshop can be identified as a leading format for exchange of experience and reacting on a specific issue or site. Even though workshops depend on the methodology adopted (Dragutinovic et al. 2023) and can be successful even in an online format (Milovanović et al. 2020), workshops on the site allow for this exchange to be immediate, intense, and empowering by providing capacity building both individually and for local institutions. Field workshops organized for specific research enables exchange of experience and build researchers capacity to share and understand different attitudes and increase communicative skills. All of the skills are relevant for building tactical skills and flexibility for established or future professionals. Field workshops with local experts enable the exchange of experience between practitioners and academics, expanding the scope of action for both. The condition for this to happen is that through the duration of the workshop mutual trust is gained and prejudices about the irreconcilability of academic and practical action are overcome. This exchange process creates an opportunity to provide the necessary élan for professional action.

Seeking for the answers within existing regulatory framework and personal aspirations, fieldwork enables conditions for building argumentation for decision making. Accordingly, practitioners build contextually based arguments about local conditions and procedures, and become able to recognize the reasons for certain actions and measures within the framework of bureaucratic protocols. At the same time, fieldwork enables academic urban morphologists' to build argumentation for future research steps and actions and to become answerable to self/own intellectual integrity based on direct insight and collected field data. Fieldwork realized in cooperation between practitioner and academic urban morphologist strengthens the argument and encourages critical thinking but does not necessarily lead to consensus.

In regard to the form and type of the output ranging from implementation of the plan to communication of ideas, practitioners and academics are faced with various challenges. Fieldwork in practice enables monitoring of plan and project implementation. Practitioner is faced with challenges of implementation and develops interest for activities and solutions with tangible outputs for local context. Having in my importance of ideas as main outputs, academic urban morphologists develop personal research sensibility and encourage divergent thinking to secure diversity of ideas. Responsibility towards local values and identification of in situ values are the most important characteristics of contextual approach. Consequently, fieldwork can be perceived as a form of work which enables direct personal experience of local culture without which research is possible but scarce from the point of view of personal insight and relationship to the context.

Discussion and Conclusion

The conclusion that can be drawn out from the research is structured on two levels—first, that relates to the results from relating urban morphology, fieldwork, and the process of identification of in situ values, and second, that reconsiders possibilities of the fieldwork as a link within the academic and practical field of action. Urban morphology through fieldwork enables the identification of in situ values on all three levels: subject of study, method, and tools. Additionally, fieldwork allows recognition or questioning of dominant paradigms and seeks new paradigms, approaches, and tools expanding knowledge about in situ values of the built environment.

Regarding the second level, there is a need for developing an agenda for the fieldwork to secure the highest potential of this type of research both in practice and academia, relevant both for urban morphology practitioners and academics. Agenda is structured in five points that relate to five topics perceived by Michael Barke as key distinction in the perspectives of urban morphologists acting in these two fields—academia and practice.

Based on identified mutual interest of urban morphologists engaged either in practice or research and academia, the agenda was structured to include both sides:

1. Identify sites of interest,
2. Establish contact with groups of interests,
3. Organize experts' workshops to enable exchange of experience and induce capacity building,
4. Organize discussions and presentations on the site to build argumentative base,
5. Secure conditions for being and working on the site in order to apply context approach and gain professional responsibility.

Fieldwork components are formulated in a manner to indicate the importance of fieldwork for strengthening the professional capacities and competencies of urban morphologists in the identification of In Situ values. Interest and responsibility for the in situ values of the local cultural context is the common basis through which contextual approach and professional responsibility are developed.

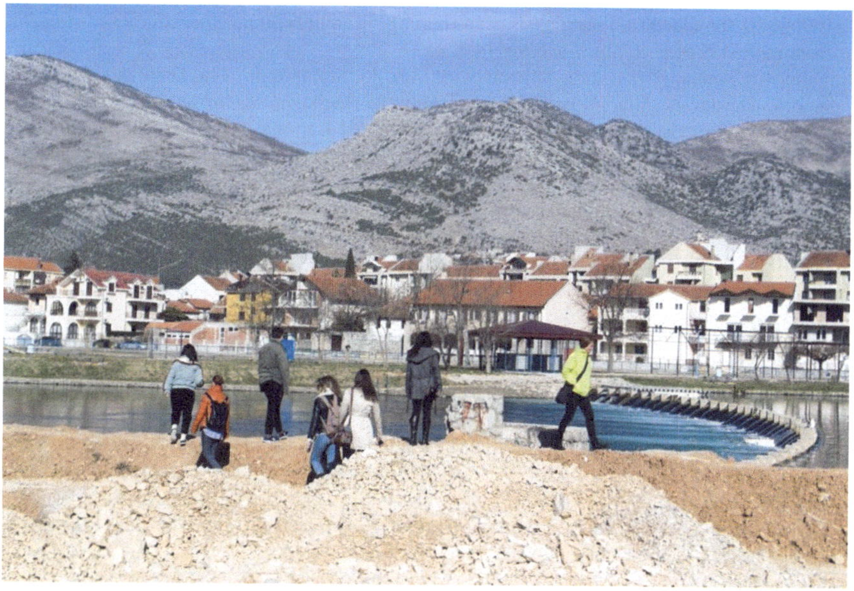

Fig. 10.1 Course: Studio 06U, Filed work team: Vladan Djokić, Milica Milojević, Mladen Pešić, Aleksandra Djordjević, 2019 Trebinje

References

Barke, Michael. 2015. Further thoughts on research and practice in urban morphology: A British perspective. *Urban Morphology* 19 (1): 96–99.

Barke, Michael. 2018. The importance of urban form as an object of study. In *Teaching Urban Morphology*, ed. Vitor Oliveira, 11–30. Cham: Springer.

Berghauser Pont, Meta. 2018. An analytical approach to urban form. In *Teaching Urban Morphology*, ed. Vitor Oliveira, 101–119. Cham: Springer.

Chiaradia, Alain J., Louie Sieh, and Frances Plimmer. 2017. Values in urban design: A design studio teaching approach. *Design Studies*. 1 (49): 66–100. https://doi.org/10.1016/j.destud.2016.10.002.

Conzen, Michael P. 2018. Core concepts in town-plan analysis. In *Teaching Urban Morphology*, ed. Vitor Oliveira, 123–143. Cham: Springer.

Djokić, Vladan, Aana, Nikezić, Mar, Loren-Méndez, Kostantinos, Sakantamis, Maria, Philokyprou, Emanuela, Sorbo, Jose, Péral-López. 2021. *Statements: Teaching through Design for Sustainability of the Bult Environment and Heritage Awareness*. Belgrade: University of Belgrade—Faculty of Architecture. Available at: https://hersus.org/statements-for-teaching-through-design-for-sustainability-of-the-built-environment-and-heritage-awareness/. Accessed on 11 February 2023.

Đokić, Vladan. 2009b. *Urbana Tipologija—Gradski Trg u Srbiji [Urban Tipology—City Square in Serbia]*. Beograd: Arhitektonski fakultet Univerziteta u Beogradu.

Đokić, Vladan. 2009a. Morphology and typology as a unique discourse of research. *SAJ Serbian Architectural Journal* 12: 107–130.

Đorđević, Aleksandra, Aleksandra, Milovanović, P. Milica, Milojević, Ana, Zorić, Mladen, Pešić, Jelena, Ristić Trajković, Ana, Nikezić, and Vladan, Djokić. 2022. Developing methodological

framework for addressing sustainability and heritage in architectural higher education—insights from HERSUS project. *Sustainability* 14.8: 4597.
Dragutinovic, Anica, Aleksandra, Milovanovic, Mihajlo, Stojanovski, Tea, Damjanovska, Aleksandra, Đorđevic, Ana, Nikezić, Ana, Ivanovska Deskova, & Jovan, Ivanovski. 2023. Approaching extracurricular activities for teaching and learning on sustainable rehabilitation of mass housing: Reporting from the arena of architectural higher education. *Sustainability* 15.3: 2476.
Fekete, Albert, and Martin van den Toorn. 2021. Teaching fieldwork in landscape architecture in European context. *Some Backgrounds and Organisation Land* 10: 237. https://doi.org/10.3390/land10030237.
Gauthier, Pierre. 2005. Conceptualizing the social construction of urban and architectural forms through the typological process. *Urban Morphology* 9 (2): 83–93.
Gauthier, Pierre, and Jason Gilliland. 2006. Mapping urban morphology: A classification scheme for interpreting contributions to the study of urban form. *Urban Morphology* 19: 41–50.
Gebauer M, I. Samuels. 1981. *Urban Morphology: an Introduction. Joint Centre for Urban Design. Research Note 8.* Oxford: Oxford Polytechnic.
Gu, Kai. 2019. Urban morphological regions: Development of an idea. In *JWR Whitehand and the Historico-geographical Approach to Urban Morphology*, ed. Vitor Oliveira, 33–46. Cham: Springer.
Larkham, Peter J. 2018. The importance of observation: Urban morphology in the field.". In *Teaching Urban Morphology*, ed. Vitor Oliveira, 265–279. Cham: Springer.
Larkham, Peter J., and Nick Morton. 2021. Conservation and (Sub) Urban Form: Reviewing Policy in Stratford Upon Avon, 2004–2019. In *Morphological Research in Planning, Urban Design and Architecture*, ed. Vitor Oliveira, 43–68. Cham: Springer.
Milovanović, Aleksandra, Miloš Kostić, Ana Zorić, Aleksandra Đorđević, Mladen Pešić, Jovana Bugarski, Neda Sokolović, and Andrej Josifovski. 2020. Transferring COVID-19 challenges into learning potentials: Online workshops in architectural education. *Sustainability* 12 (17): 7024.
Moudon, Anne Vernez. 1997. Urban morphology as an emerging interdisciplinary field. *Urban Morphology* 1: 3–10.
Nikezić, Zoran. 2007. *Građena sredina i arhitektura. [Built Environment and Architecture].* Beograd: Univerzitet u Beogradu, Arhitektonski fakultet.
Oliveira, Vítor. 2016. *Urban Morphology: An Introduction to the Study of the Physical form of Cities.* Cham: Springer.
Oliveira, Vítor., ed. 2018. *Teaching Urban Morphology.* Cham: Springer.
Oliveira, Vítor., ed. 2019. *JWR Whitehand and the Historico-Geographical Approach to Urban Morphology.* Cham: Springer International Publishing.
Oliveira, Vítor., ed. 2021. *Morphological Research in Planning, Urban Design and Architecture*, 2021. Springer: Cham.
Peter, Larkham. 2005. Understanding urban form? *Urban Design* 93: 22–24.
Rendell, J. 2010. Afterword: Working (through) the field. *In Architecture and Field/Work, ed. Suzanne Ewing, Jeremie Michael McGowan, Chris Speed, Victoria Clare Bernie,* 186–191. London: Routledge.
Samuels, Ivor. 2019. Research and Practice. In *JWR Whitehand and the Historico-geographical Approach to Urban Morphology*, ed. Vitor Oliveira, 115–131. Cham: Springer.
Samuels, Ivor. 2021. Towards an eclectic urban morphology. In *Morphological Research in Planning, Urban Design and Architecture*, ed. Vitor Oliveira, 71–96. Cham: Springer.
Sanders, Paul, and Douglas Baker. 2016. Applying urban morphology theory to design practice. *Journal of Urban Design* 21 (2): 213–233.
Slater, Tonny R. 1990. *The Built form of Western Cities.* Leicester: Leicester University Press.
Steven, Marshall, and Çali.şkan Olgu. 2011. A joint framework for urban morphology and design. *Built Environment* 37: 409–426.

Ünlü, Tolga. 2018. Planning Practice and the Shaping of the Urban Pattern. *In Teaching Urban Morphology*, ed. Vitor Oliveira, 31–49. Cham: Springer.
Vitor, Oliveira, and Sousa Silvia. 2012. Urban morphology in planning practice. *Urban Morphology* 16: 80–82.
Whitehand, J.W.R. 2021. Conzenian research in practice. In *Morphological Research in Planning, Urban Design and Architecture*, ed. Vitor Oliveira, 19–42. Cham: Springer.
Whitehand, J.W.R. 2016. Preface. *In Urban morphology: an introduction to the study of the physical form of cities*, ed. Vitor Oliveira. Cham: Springer.

Chapter 11
International Charters. Research and Values for Cultural Heritage

Emanuela Sorbo and Sofia Tonello

Introduction. International Charters: A Dialogue Tool

The Università Iuav di Venezia is conducting experimental research on utilising International Charters as a tool for fostering multidisciplinary dialogue at the global level, aimed at advancing the knowledge and conservation strategies for Cultural Heritage. This investigation traces the evolution of concepts and definitions articulated in key Charters, beginning with the Athens Charter and extending to the most recent international frameworks. The study highlights the dynamic relationship between International Charters and shifting social, cultural, and economic conditions, including the impacts of war, geopolitical reconfigurations, and socio-political transformations. Notable examples such as the ICOMOS Charters of 1982 and 2000, the *Dresden Declaration* (ICOMOS 1982), the *Riga Charter* (ICOMOS 2000), the *San Antonio Declaration* (ICOMOS 1996), the *Québec Declaration* (ICOMOS 2008a), the *Delhi Declaration* (ICOMOS 2017), and the *Nara Document* (ICOMOS 1994) are examined for their evolving responses to issues like territorial boundaries, the conservation and restoration of architectural heritage, and the implications of shifting global power structures. These documents address technical conservation practices and reflect broader concerns, including the principles of interpretation and identification of cultural heritage, as seen in documents such as the *Florence Declaration* (ICOMOS 2014b) and the *ICOMOS Charter for the Interpretation and Presentation of Cultural Heritage* (ICOMOS 2008b). By analysing the evolving frameworks of the Charters, the research positions them as critical instruments for investigating the complex relationships between cultural heritage and communities. The study emphasises their role as privileged observers of the intersection between heritage preservation, global socio-political changes, and the increasing interactions between local and global contexts.

E. Sorbo (✉) · S. Tonello
Department of Architecture and Arts, Università Iuav di Venezia, Venice, Italy
e-mail: esorbo@iuav.it

M. Loren-Méndez et al. (eds.), *Sustainable Heritage Management - Design: Notions, Methods and Techniques*, https://doi.org/10.1007/978-3-031-81970-4_11

Starting from the conditions mentioned in *The Nara Document on Authenticity* (ICOMOS 1994) which recalls and extends the objectives of the *Venice Charter* (ICOMOS 1964), what becomes clear is the need to define common principles for recognition and collective responsibility to protect Cultural Heritage in the light of the evolution of sensitivity and of a critical point of view through which contemporary problems can be observed (Stovel 2000, 397–398). Furthermore, in 2008, *The Charter for the Interpretation and Presentation of Cultural Heritage Sites* (ICOMOS 2008b) identifies places, the natural landscape, settlement areas, architectural structures and archaeological sites as places of historical and cultural importance, drawing strategies and methods to define and evaluate the Cultural Heritage.

The *Charter for the Interpretation and Presentation of Cultural Heritage Sites* (ICOMOS 2008b, p. 3) is based on seven key principles that represent the evolution of Brandi's recognition process for Cultural Heritage:

- Principle 1: Access and understanding;
- Principle 2: Sources of information;
- Principle 3: A focus on the setting and context;
- Principle 4: Preservation of authenticity;
- Principle 5: Sustainability planning;
- Principle 6: Concern for inclusion;
- Principle 7: Importance of research, training and evaluation.

Starting from these premises, the study, carried out by the Iuav research group, in the context of the activities of *Intellectual output 05* of the HERSUS International project, investigates the aspects related to multiculturalism and the definition of ways of interacting and sharing topics related to Cultural Heritage in groups with different geographical and social backgrounds. Furthermore, it suggested a critical-interpretative approach applied to International Charters as a tool of interaction and cultural exchange.

Research: Glossaries and Definitions

The analysis of glossaries and definitions within International Charters suggests that an intercultural and metalinguistic understanding of issues related to Cultural Heritage is possible. With this premise, an interpretative reading of the actions related to the Cultural Heritage project was carried out, excluding the key terms of restoration such as conservation, renovation and protection and using the three terminological references of the HERSUS project: *Reconstruction, Reuse, Resilience* as parameters for checking the use of International Charters as a multicultural dialogue tool.[1]

The first step of the research was to evaluate the Charters selection method. The parameters identified were: the Promoting Bodies, the places connected to the issuing of the Charters, the topics addressed and the historical period of reference. This operation made it possible to select 55 Charters over a period of one century (from 1931 to 2021), taking into consideration those shared by international bodies such as UNESCO, ICOMOS and the European Community.

Charters monitoring was carried out through a research that combined interpretive analysis, data analysis software and artificial intelligence. Each of these three terms was then investigated both in terms of its potential definition and as a reference to the concept expressed therein. Part of the analysis of the International Charters was focused on building a statistical database in which to identify the most frequently used words in relation to the three main topics within the Charters. With this method, in addition to the interpretative analysis, it was possible to identify how other headwords have evolved and potentially have been associated with the three terms of the pyramid (*Reconstruction, Reuse* and *Resilience*). This condition of terminological evolution has made it possible to monitor over time a number of semantic and conceptual elements that are in connection with the three main terms.

Following the identification of the semantic references (codes) by the team, the statistical analysis was then conducted using a data management and analysis application (MAXQDA) which made it possible to identify, among the analysed data, a statistical population (consisting of approximately 25 headwords) that was representative of the three topics. The analysis activity can be summed up in four steps:

- **Identification**: preliminary phase of preparation for Charters reading and application of search codes for headwords for analysis softwares.
- **Headwords Analysis and Data Interpretation**: selection of link-words to be associated with the headwords of the Hersus project—*Reconstruction, Reuse, Resilience.*
- **Graphic Rendering**: analysis of the relationships between the headwords of the project and the link-words through a relational artificial intelligence software (Yewno) and construction of a graphical output.
- **Experimental Verification**: application of International Charters as a tool for sharing concepts in a sample of subjects from cultural, social and geographical contexts at a global level.

[1] The Iuav team is coordinated by Emanuela Sorbo (Scientific coordinator of the project for Iuav) and consists of Sofia Tonello (Junior researcher) and Eleftherios Mantzouneas (Erasmus + trainership). The research activities were carried out in support of the investigations of the HERSUS project for the Intellectual Output 5 —Book of Courses.

Identification. The Composition of a Reading Matrix

The first phase allowed the research team to identify the terms used in HERSUS: *Reconstruction, Reuse, Resilience* and the link-words associated with them in a matrix that associates them with International Charters to understand their transformation and development over time in relation to social, cultural and natural evolutions. The matrix (Fig. 11.1) thus composed consists of:

- Rows: the first row represents the sequence of headwords—*Reconstruction, Reuse* and *Resilience*—and the secondary row the link-words associated with them.
- Columns: the columns chronologically identify the International Charters and the places in which they were drawn up.

The matrix thus composed allowed to perform an interpretative reading on the evolution of the terms *Reconstruction, Reuse* and *Resilience* over time and in the geo-cultural contexts of application and then to create their graphitisation.

Analysis and Data Interpretation

Headword: "Reconstruction"

The link-words are important to understand the evolution of the concept "Reconstruction", they are identified through the matrix. It was in fact possible to recognise that the headword "Reconstruction" takes on multiple connotations and facets of meaning, and is associated with the link-words "Anastylosis", "Rebuild-Rebuilding", "Replication-Replicate", "Re-creation", "Replacement-Replace", "Recovering", "Integrity". These terms generally refer to the process of repairing, improving or restoring something to a previous state or to a better condition (Musso 2023). The specific term used may depend on the context in which the activities are carried out and on their typology. For example, it is possible to note that the headword "Anastylosis", present in the *Athens Charter* of 1931 (International Congress of Architects and Technicians of Historic Monuments 1931), is almost absent in all the subsequent ones, starting from the *Venice Charter* (ICOMOS 1964) in which it is important, with respect to the historical period, to focus attention on the meaning that "Reconstruction" assumes with respect to the term "Anastylosys", the validity of which is only admitted on specific occasions such as in relation to the archaeological field.[2] This concept changes over time according to the causes that underlie the "reconstruction" action and aim to limit and identify the opportunities by using (even conceptually) the

---

[2] *Venice Charter*. ICOMOS 1964.

Art. 15 of the *Venice Charter* states with regard to reconstruction actions that these must be limited and that: "All reconstruction work should however be ruled out 'a priori'. Only anastylosis, that is to say, the reassembling of existing but dismembered parts can be permitted".

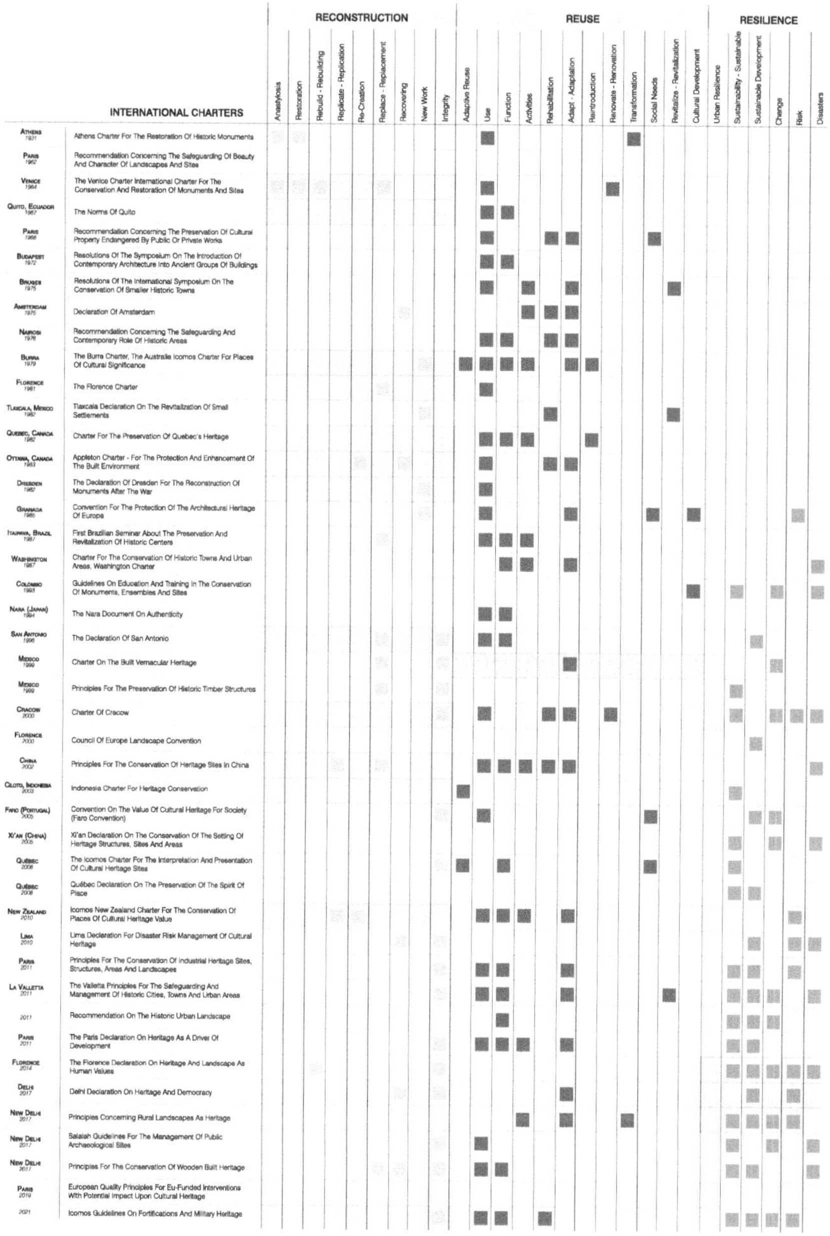

Fig. 11.1 The reading matrix. The matrix is a graphical output by the authors. It represents the main International Charters and helps to monitor the relationship with headwords and link-words (Image by the authors)

terms "Rebuild-Rebuilding", "Replication" or "Replacement" favouring actions of "Recovering" based on the needs and the state of "Integrity" of the artefact (*Declaration of Amsterdam - Congress On The European Architectural Heritage*, COE 1975a, b; Petzet 2009, 30–32). For example, the *Dresden Charter of 1982 on the Reconstruction of Monuments*[3] (ICOMOS 1982) highlights the cultural-identity problem of the communities due to the loss of Cultural Heritage caused by the war and suggests "Reconstruction" as a means to recover the integrity of the main monuments. In 2004, the *Charter for the Conservation of Unprotected Architectural Heritage* (ICOMOS 2004) recognises a role for "Rebuilding" practices in cases in which communities have lost large parts of their heritage due to war or natural disasters and (explicitly) limits their feasibility to cases in which it is possible to respect the overall spatial and volumetric composition of the historical contexts.[4]Three other link-words find a definition and distinction in the same Charter: "Restoration/ Replication/ Rebuilding". "Restoration" is considered an appropriate conservation strategy to restore the integrity or complete the fragmented whole of the architectural heritage or site[5], while "Reconstruction" is assimilated to the term "Rebuilding" which is used to understand the original meanings of the cultural asset in the contemporary context and to improve its link with society[6].

Some Charters, such as the *Appleton Charter—For protection and enhancement of the built environment* (ICOMOS 1983) or the *Icomos New Zealand Charter for the Conservation of places of cultural value* (ICOMOS 1993), try to limit and define the conditions of validity of the actions of "Re-creation" of elements of the Cultural Heritage that have been lost or that cannot be recognised anymore. Furthermore, the latter Charter confirms what is indicated by the *Riga Charter on authenticity and historical reconstruction in relationship to Cultural Heritage* (ICOMOS 2000) that

---

[3] *Declaration of Dresden*, ICOMOS 1982.

Art. 1 states that "The task of social development after the war, the reconstruction of towns and villages, and the resulting task of the protection of monuments constitutes a single entity. The spiritual values of monuments and the desire to acknowledge them both intellectually and politically were the reasons for initiating their reconstruction".

[4] *Charter for the Conservation of Unprotected Architectural Heritage*. ICOMOS 2004.

Regarding the matter of "Rebuilding", the *Charter for the Conservation of Unprotected Architectural Heritage* defines, in Art. 3 that "Conjectural restoration or rebuilding" as an action that "must nevertheless respect the overall spatial and volumetric composition of historic settings. The parameters of the historical setting should be defined through comprehensive urban design studies. These parameters should also guide new urban development in the vicinity of heritage buildings and sites."

[5] *Charter for the Conservation of Unprotected Architectural Heritage*. ICOMOS 2004

In the *Charter for the Conservation of Unprotected Architectural Heritage*, the definitions of "Restoration" is "an appropriate conservation strategy to reinstate the integrity or to complete the fractured 'whole' of the architectural heritage/site"; "replication" that "can be accepted as an appropriate strategy not only to conserve unprotected historic buildings, but especially if such replication encourages historic ways of building".

[6] *Charter for the Conservation of Unprotected Architectural Heritage*. ICOMOS 2004.

The *Charter for the Conservation of Unprotected Architectural Heritage* presents in the glossary definitions of "reconstruction/rebuilding" that "can provide the impetus to develop a parallel market for local buildings materials and new opportunities for the use of alternative systems of building".

prefers actions of maintenance and repair of Cultural Heritage, operated in a recognisable and neutral manner, rather than the "Replication" of a formal vocabulary and means of expression used in the past[7]. This term has critical issues, highlighted since 1994 in the *Nara Document Authenticity* (ICOMOS 1994) of Cultural Heritage in the transmission of historical, artistic and cultural values. The notion of "Authenticity" (see *Nara Document on Authenticity*, ICOMOS 1994; *Nara + 20: on heritage practices, cultural values, and authenticity*, ICOMOS 2014a) of heritage is closely linked to the notion of "Integrity" which is evaluated and reconsidered in terms of materials, form and identity (Jokilehto 2000, 137–139). The notion of "Reconstruction" is also understood in a broad sense as indicated by the *Faro Convention* (COE 2005), referring to the dynamics concerning the recovery of the social, cultural and economic dimension of places, as well as the "Resilience" promoted by "Sustainable Development" projects which aim to satisfy social needs through the "Adaptation" of places, sites and buildings. The *European Framework for Action on Cultural Heritage* (COE 2018) associates the term "Reconstruction" with issues related to "Resilience", such as the "Sustainable reconstruction" of Cultural Heritage through concrete and cost-effective adaptation and mitigation strategies, methods, plans and other corrective tools to protect and preserve Cultural Heritage[8].

Headword: "Reuse"

The analyses carried out on the term "Reuse" have shown that this term begins to be available in the International Charters starting from Art. 5 of the *Charter on the Built Vernacular Heritage* (ICOMOS 1999). Prior to this charter, the connotation assumed in international documents (the *Report of the World Commission on Environment and Development: Our Common Future,* World Commission on Environment and Development 1987, 156) is linked to the "Reuse" of resources and objects, understood as "Recycling" aimed at reducing consumption and waste. Only starting from the first

---

[7] *Riga Charter on Authenticity and Historical Reconstruction in Relationship to Cultural Heritage.* ICOMOS 2000.

On the matter of "maintenance and repair of Cultural Heritage", the *Riga Charter on Authenticity and Historical Reconstruction in Relationship to Cultural Heritage* in the premises declared that these "should be the primary focus of current conservation work, recognising that each historical period has its own particular style which does not replicate previous used formal vocabulary and means of expression".

[8] *European framework for action on Cultural Heritage.* CoE 2019.

In the text of the *European framework for action on Cultural Heritage* in the *CLUSTER OF ACTIONS 9: Protecting Cultural Heritage against natural disasters and climate change*, the following is read: "to improve the resilience of historic settlements and foster their sustainable reconstruction in case of disasters". Furthermore, it is underlined that the project "contributes to the development and dissemination of evidence-based and cost-effective adaptation and mitigation strategies as well as methods, planning and other remedial tools to safeguard Europe's Cultural Heritage".

few years of the twenty-first century has the concept of identifying new uses for buildings that, due to their historical and cultural value, are characterised by problems of compatibility between adaptation and new functions. In International Charters, terms such as "Adaptation", "Use", "Function", "Rehabilitation", "Refurbishment", "Urban Regeneration", "Reintroduction", "Renovate-Renovation", "New Work" are related to the historical, cultural and social value of the heritage built inherited from the past with the current needs of the contemporary community[9]. The selection of link-words has therefore allowed the research to acquire a mapping of the meanings that "Reuse" has assumed over time. These terms generally refer to the act of reusing something, both in its original form and after a modification, and transversal reading of International Charters over time has confirmed what is written in the *Venice Charter* (ICOMOS 1964) that in Art. 5 places the "Use" of monuments, for a socially useful purpose, as a fundamental factor for conservation and transmission. The centrality of the theme of "Use" was also confirmed in later texts such as the *Resolutions of the International Symposium on the Conservation of Smaller Historic Towns* (ICOMOS 1975) that suggest looking for new "appropriate uses" for empty buildings that would otherwise be threatened by decay. *The Convention for the Protection of the Architectural Heritage of Europe* (COE 1985) states the need to guarantee the use of protected buildings in the context of contemporary life and adaptation, where appropriate, of old buildings to new uses. The most recent charters, including the *Principles for the Conservation of Industrial Heritage Sites, Structures, Areas and Landscapes* (*TICCIH-ICOMOS* 2011)[10] and *the Guidelines on Fortifications And Military Heritage* (ICOMOS 2021),[11] are important for cultural recognition that these types of Heritage have received over time and to the possibility that the new "Functions" that allow their "Reuse" are compatible with the sedimentation of cultural, natural and landscape meanings and values of which they have

---

[9] CoE 2010.

In the documents of the Council of Europe of Monitoring the Granada Convention, the role of the interest of the stakeholders in the reuse activities is highlighted in the section *The Netherlands Pilot of the Granada Case Study Modules*: "There was an interesting discussion about the elevation of training for non-technical conservation-related skills, which was seen as relevant to the context in The Netherlands, with a recent policy move from conservation of monuments to adaptive reuse and incorporation in development, placing high tension on the different interests [...] so we have to develop the different concept of what we value in the building as different interest groups".

[10] *Principles for the Conservation of Industrial Heritage Sites, Structures, Areas and Landscapes.* TICCIH-*ICOMOS* 2011.

In *Principles for the Conservation of Industrial Heritage Sites, Structures, Areas and Landscapes,* it is underlined in Art. 8 that with reference to the use, "it must be recognised that their continued use and function might carry some of their heritage significance and provide adequate conditions for their physical and economic sustainability as a living production or extraction facilities".

[11] *Guidelines On Fortifications And Military Heritage.* ICOMOS 2021.

Regarding the "change", in Art. 3.5 of *Guidelines On Fortifications And Military Heritage,* the following is found: "The changing nature of military operations often means that fortifications cannot be reused for the specific purpose for which they were originally built. Fortifications were designed to prevent entry and today this presents obvious challenges for accessibility and the meeting of current use requirements".

become collectors over time. In particular, the *Amsterdam Declaration - Congress On The European Architectural Heritage* (COE 1975a, b) suggests the introduction of new "Activities" in declining areas, in order to control depopulation and thus prevent the deterioration of historic buildings. Similarly, the *Principles Concerning Rural Landscapes As Heritage* (ICOMOS—IFLA 2017) applies the topic of "Function" to the landscape and territorial scale, defining it as an opportunity into integrate a multifunctional vision to the concept of heritage. Only in 2018, in relation to the *Leeuwarden Declaration on Adaptive Re-Use of the Built Heritage* (Architects' Council of Europe 2018), the use of the term "Adaptive Reuse" began to find a definition in the International Charters through specific characteristics that contemporary projects should satisfy, such as, for example, "flexibility, participatory approaches, innovation, quality-based procurement, multidisciplinary teams, financial viability and good story-telling"[12]. The proposal of a dynamic and constantly changing vision of Cultural Heritage is offered by the *Faro Convention* (COE 2005) or in the *Valletta Principles for the Safeguarding and Management of Historic Towns and Urban Areas* (ICOMOS CIVVIH 2011). These documents introduce a new focus on the communities of which Cultural Heritage is an expression and place cultural values and local cultural development at the centre as opportunities for transformation, revitalisation and rehabilitation.

Headword: "Resilience"

The research has verified that the term "Resilience" is used only occasionally in the Charters, starting from *The Florence Declaration on Heritage and Landscape as Human Values* of 2014 (ICOMOS 2014b). This result has shown that the term analysed is often related to environmental issues or to strategies that aim at the revaluation of Cultural Heritage from a perspective linked to communities and the environment, such as, for example, in the *European Framework for Action on Cultural Heritage* (COE 2018) and the *New Urban Agenda* (United Nations 2017). "Resilience" can be investigated through a number of sub-categories starting from the link-words "Urban Resilience", "Sustainability-Sustainable", "Sustainable Development", "Change", "Risk" or "Disaster". This meaning of Cultural Heritage begins to be analysed and relevant for International Charters starting from *Convention for the Protection of the Architectural Heritage in Europe* (COE 1985) in which the locution of "Risk" associated with Cultural Heritage appears for the first time. In the *Florence Declaration on Heritage and Landscape as Human Values* (ICOMOS 2014b), "Resilience" relates to the involvement of local communities in protection and transmission actions related to the heritage through Cultural Heritage recognition operations in a coexistence of

[12] *European framework for action on Cultural Heritage*. CoE 2019.
In *CLUSTER OF ACTIONS 5: supporting the adaptive re-use of heritage buildings* the text of the *European framework for action on Cultural Heritage* underlines "the possible benefits of adaptive reuse and lists a number of quality principles to reconcile heritage values with contemporary architecture".

innovative and traditional practices contributing to their adaptability[13]. The analysis of the link-words highlights the relationship between this aptitude for change and a constant trend to innovate the Cultural Heritage linked to the (quick and fast) time of natural disasters, climate changes and contemporary world. In 2018, the *European framework for action on Cultural Heritage* (COE 2018) recalls the need to develop a series of actions aimed at researching, developing and disseminating strategies and tools based on the efficiency of specific and effective case studies, in terms of costs, transformations and cultural transmission.

The analysis of the most frequent combinations and words related to "Resilience" in the context of Cultural Heritage has shown that this expression has been used in recent decades and refers to the capacity of sites, objects, traditions and communities of Cultural Heritage to withstand, recover from and adapt to changing circumstances, including natural disasters, conflicts and development pressures. International Charters, especially those aimed at safeguarding cities and landscapes, underline the need for common and shareable implementation measures, such as disaster risk reduction planning, conservation practices, community engagement and education, and the integration of Cultural Heritage into sustainable development as in the *Charter for the Conservation of Unprotected Architectural Heritage and Sites in India* ICOMOS 2004; *The Florence Declaration on Heritage and Landscape as Human Values* (ICOMOS 2014b) and the *Declaration of San Antonio* (ICOMOS 1996) with a focus on the idea of "Sustainable Development".

Graphic Rendering. The Use of AI

The critical analysis of the selected documents used well-known and already validated investigation systems such as the "Charta von Venedig", project developed by the team of Dr. Hanns Michael Küpper[14]. The objective of the project was to verify the validity of the principles and the semantic system contained therein through the identification of five key concepts (ECRAN)[15] and other link-words for the description of a strategic model of planning actions on Cultural Heritage.

Data analysis software enacted dynamic schemes, integrating the results with an AI model. This approach made it possible to understand the relationships between the terms "Reconstruction", "Reuse", "Resilience" and the related link-words in a dynamic mode.

---

[13] *Florence Declaration.* ICOMOS 2014b.

In Art. 2 in point 1 of the *Symposium of Florence Declaration*, declared that: "The involvement of local communities, the recognition of, and respect for, their Cultural Heritage, as well as innovative and traditional practices can promote more effective management and governance of multifunctional landscapes, contributing to their resilience and adaptability".

[14] "Charta von Venedig", project. 2014.
https://www.charta-von-venedig.de/conservation-restoration_strategic-model.html

[15] In the project, the acronym defined are: renew (E), conservation (C), restoration (R), add (A) and build (N).

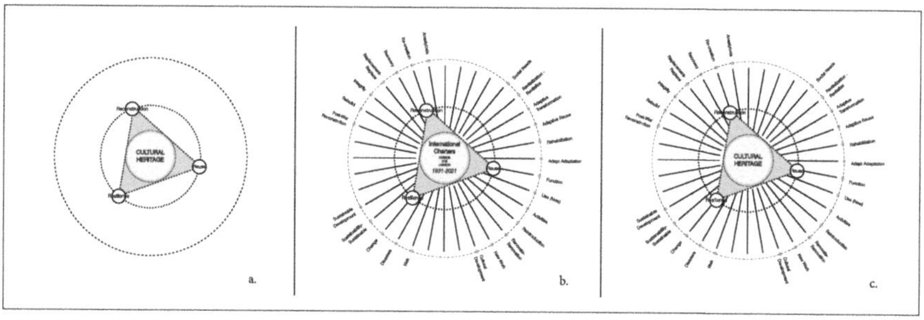

Fig. 11.2 The figure shows how the static graphical diagram was structured from the information acquired by relational software and artificial intelligence (Image by the authors)

The dynamic graphic representation of the concepts offered by the AI revealed the relationships between the topics, the link-words and the International Charters, visually restoring the interconnection of the temporal, terminological and thematic evolutions linked to the concepts expressed by the International Charters.

Starting from these diagrams, static graphic charters have been drawn up for each charter consisting of a radial system of elements—the link-words—with a central nucleus—the International Charter—intersected by the pyramid whose constituent elements are *Reconstruction, Reuse, Resilience.* This scheme can be used to understand two levels of further exploration of the International Charter investigations. The first level linked to the monitoring of the link-words and the second level linked to the interpretation that the three topics (*Reconstruction, Reuse, Resilience*) have had in the specific charter (Figs. 11.2 and 11.3).

The static scheme thus generated shows the relationships that the terms *Reconstruction, Reuse* and *Resilience* have with the link-words and with the International Charters over time.

This graphitisation confirms the possibility of considering the Charters as privileged observers for the relationship between Heritage Awareness, in terms of the evolution of practices and concepts, and Cultural Heritage.

Experimental Verification. Applied Research

On the basis of the research group's survey, an activity was proposed to a sample of subjects with geographical origin on a global scale, aimed at stimulating a reflection and a shared dialogue on the terms *Reconstruction, Reuse* and *Resilience.*

The reference sample for the survey was selected from the Master's Degree in Architecture in English at the Università Iuav di Venezia which has an international class. The choice is motivated by the global origin of the students. The homogeneity of the students' training typology (Bachelor in Architecture) was an opportunity to verify the multicultural and transgenerational approach that the International Charters can offer (Di Biase and Albani 2019, 17–20).

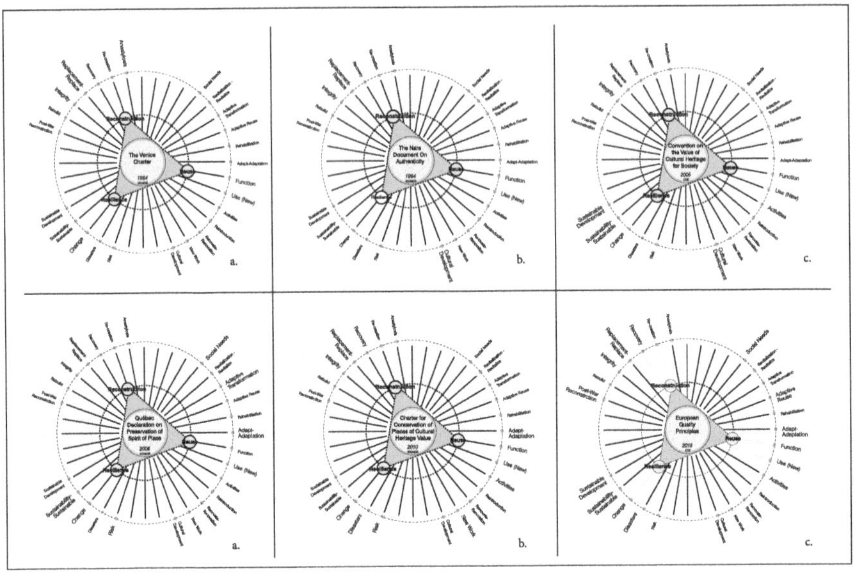

Fig. 11.3 The static graphic scheme of some of the most frequently cited International Charts in the paper to make understand the close relationship between link-words, headwords and their evolution over time (Image by the authors)

Starting from the proposal of a chronological, interdisciplinary and multicultural vision of International Charters, it was possible to activate a process of knowledge and sharing of concepts and practices related to the terms *Reconstruction*, *Reuse* and *Resilience.*

The verification of the method was based on three phases:

- Tutorial topic:
  - Analysing a case study of a project on Cultural Heritage, identifying a research term/topic among Reconstruction, Reuse and Resilience.
  - Understanding its evolution through the use of matrix and graphics of the AI to select the Charters and the link-words referring to the selected term.
- Type of subjects: working groups of max two or three persons from different geographical contexts.
- Collective presentations of the results and verification of the concept sharing process.

The selected sample consisted of approximately fifty subjects from more than twenty different countries located in Europe (Italy, Germany, Poland, Sweden, Greece, Cyprus, Kosovo), the Middle East (Egypt, Turkey, Iran, Armenia, Georgia), Asia (Russia, China, India, Thailand, Philippines) and the Americas (North America, Brazil, Mexico, Colombia). Twenty groups were created freely of which more than 50% brought together students from very close geographical and cultural areas with a number of exceptions such as: a group that was created with students from Italy and South America, a group that involved students from India, Germany and Turkey and a group with students from North America, South America and Eastern Europe.

The application of the research confirmed that the *Venice Charter* (ICOMOS 1964) is a common (and well-known) basis to understand the value and significance of interventions on the existing built heritage. In fact, a significant number of groups (roughly 30%, 6 out of 21) chose to investigate the topics suggested through the *Venice Charter* (ICOMOS 1964).

The analysis of case studies in cultural fields other than those in which the International Charters were drafted has shown that the principles enunciated in these documents can be applied and shared in multiple contexts (Fig. 11.4).

Most of the students who trained in Europe during their bachelor's degree (around 40%) showed a greater interest in issues related to "Reuse" and preferred to use International Charters and documents as a reference that define "Reuse" as an opportunity for "Adaptation" of the built heritage and which propose a vision of the heritage inherited from the past as a possibility of conserving places of historical and cultural value

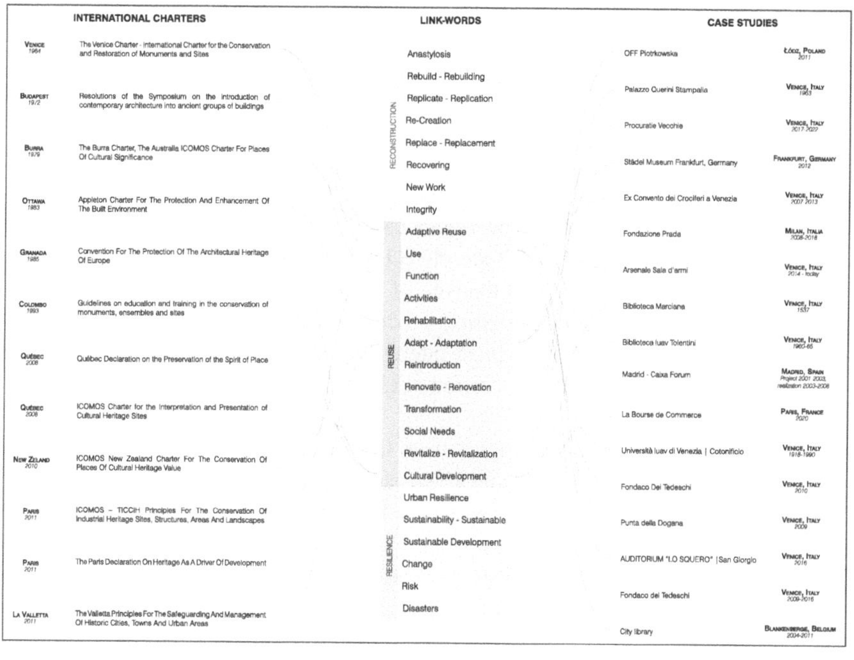

Fig. 11.4 Matrix that relates case studies, students' provenience and International Charter. For example, it was possible for German students to investigate the case study of the Städel Museum in Frankfurt, through an International Charter drawn up in New Zealand (*Charter for the Conservation of Places of Cultural Heritage Value*, ICOMOS 2010). In the same way, Polish students were able to investigate a building in their hometown—the OFF Piotrkowska Centre in Łódź—through an International Charter drawn up in Canada (*The Icomos Charter for the Interpretation and Presentation of Cultural Heritage Sites*, ICOMOS 2008b)

through a current purpose (as exemplified by the *Venice Charter* ICOMOS 1964, Art. 5 and by the *New Zeland Charter* ICOMOS 2010, Art. 8).

The students' investigations involved multiple types of buildings such as convents, markets, former industrial buildings, public buildings and private buildings. It has been highlighted that for some projects (Städel Museum, Frankfurt; Marciana Library, Venice; Palazzo Querini Stampalia, Venice; Monastery of San Servolo, Venice), reuse was a theme linked to continuous and relevant use for the communities, in line with the principles of accessibility, participation and sustainability set out in the International Charters analysed (such as *The Burra Charter* ICOMOS 1979–2013; *Convention for the Protection of the Architectural Heritage of Europe* COE 1985; *New Zeland Charter for the Conservation of Places of Cultural Heritage Value* ICOMOS 2010; *The Paris Declaration on Heritage as a Driver of Development* ICOMOS 2014a, b; *European Quality Principles* COE 2018). In the same way, it was recognised that Reuse can be viewed as valorisation of disused Cultural Heritage but able to adapt to the needs of local communities in representation of the values of the *Québec Declaration on the Preservation of the Spirit of Place* (ICOMOS 2008b), *The Paris Declaration on Heritage as a Driver of Development* (ICOMOS 2011) and of the *Principles For The Conservation of Industrial Heritage Sites, Structures, Areas and Landscapes* (TICCIH-ICOMOS 2011). Reuse can also be a guide for concrete action in the field of conservation, management and use of Cultural Heritage (referring to the Venetian case studies of the former convent dei Crociferi, Punta della Dogana and the Sale d'Armi dell'Arsenale).

The considerations on Reuse, developed by students of the Eastern continent (China, Philippines and Thailand), focused on the "Integrity" of the buildings analysed, with particular reference to the *Venice Charter* (ICOMOS 1964). The selected case studies were the Fondaco dei Tedeschi (OMA) and Caixa Forum (Herzog & De Meuron) and allowed to reflect on the role of critical evaluation of cultural, artistic and historical aspects in transformation processes (see *Venice Charter* ICOMOS 1964, Art. 4-Art. 11). Among the most interesting analyses from the point of view of multicultural dialogue are those carried out by groups with heterogeneous origins on a global scale.

A group with heterogeneous origins from Central, Eastern and Southern Europe focused its reflection on the concept of "Reuse" by analysing the value of heritage through immaterial aspects such as "Cultural Development" understood in the definition of *Guidelines on Education and Training in the Conservation of Monuments, Ensembles and Sites* (ICOMOS 1993). This concept was evaluated in relation to the architectural, functional and cultural proposal of the Marciana Library in Venice (Italy) as a model which, over the centuries, has dealt with the city and with the promoters of local and international culture, with preservation of its social identity.

The experimental verification proved to be effective with respect to the objectives of the survey: the groups involved had the opportunity to interact in a cultural sharing process in an international and intercultural context through mediation of the use of Charters.

International Charters as Multisemantic and Multicultural Reading

The research therefore returns a method of dialogue between different subjects which, through the reading of the International Charters, guarantees the understanding of the polygonality and multisemantic nature of the starting concepts and a sharing strategy for reading good practices on Cultural Heritage.

The research highlights that for a possible international and intercultural dialogue on the built Cultural Heritage, it is necessary to review the objectives of conservation and protection of the heritage and their transformations over time.

The Venice Charter underlines the importance of preserving monumental works for future generations, as a common heritage of humanity, and of guaranteeing access to cultural and historical values (Petzet 2004, 7–9; Stovel 2008, 9–17). This awareness of the deep connection between monuments, history and peoples (Occelli 2023) can be seen as the central point for the definition of Cultural Heritage as a common heritage, as underlined by the *Convention Concerning the Protection of the World Cultural and Natural Heritage* of 1972 (UNESCO 1972)[16]. The relevance of heritage in social life is underlined in many subsequent documents, for example, The *Amsterdam Declaration* (ICOMOS 1975) takes into account, as a central point for the transmission of heritage, in addition to the protection of monuments from destruction and deterioration, the conservation and protection from the disappearance of traditional and vernacular cultural assets[17]. The *Appleton Charter for the Protection and Enhancement of the Built Environment* (ICOMOS 1983) and the *Nara Document on Authenticity* (ICOMOS 1994) aim to propose an extension of the heritage by integrating intangible components to material assets, first linked to perception and subsequently to cultural factors such as "Cultural significance", "Condition and Integrity of the fabric", "Contextual value", "Diversity of cultures and heritage" and "Authenticity" (Fiorani 2023, 121–132; Jokkileto 2020, 55–72). In this respect, the development and implementation of policies and programmes for conservation of Cultural Heritage set out in the Venice Charter are reviewed and integrated into a number of documents, including those of a European nature, such as the *European Charter of the Architectural Heritage* (COE 1975a, b) and the *European Framework for Action on Cultural Heritage* (COE 2018) which propose the sustainability of development and protection plans in terms of inclusiveness, cultural diversity

[16] *Venice Charter.* ICOMOS 1964.

In Article 1 the concept of "historic monument" is extensively defined as "embraces not only the single architectural work but also the urban or rural setting in which is found the evidence of a particular civilization, a significant development or a historic event. This applies not only to great works of art but also to more modest works of the past which have acquired cultural significance with the passing of time".

[17] *Resolutions of the International Symposium on the Conservation of Smaller Historic Towns.* ICOMOS 1975.

In the preface of the declaration, it is stated that the objective of the document is to: "confirm the trend of abolishing a hierarchical difference between groups of buildings of outstanding artistic interest and those of lesser importance".

and local needs. In *European Quality Principles* (COE 2018), the essentiality of scientific, technical and historical research is reaffirmed to promote knowledge and understanding of Cultural Heritage, for its diffusion and dissemination (*Education for All Global Monitoring Report* UNESCO 2010). The conservation of Cultural Heritage is supported through programmes that aim to ensure balanced development while respecting the variety and uniqueness of national, regional and local cultures recognising as a necessary element the need to associate protection and enhancement policies with issues related to local communities. Relevant examples are those concerning regional and urban development, social cohesion, agriculture, maritime affairs, environment, tourism, transport, education, disaster risk management, digital agenda, research and innovation (*Treaty on the European Union* 2007, Art. 167).

The polysemantic nature associated with Cultural Heritage (Jokilehto 2007, 23–43) has activated a process by which the research pass from specific cases to universal conclusions. Each International Charter has contingent objectives which, through dialogue between local and international scholars, are provided with a lexicon and general theoretical concepts which can condense in a few words a synthesis between form, meaning and action that is sufficiently universal to preserve the double register of connecting an international community by enhancing local roots.

The terminological matter and evolution are addressed in most of the Charters with a combined use of the tool of definitions and linguistic pluralism. The *Venice Charter* (ICOMOS 1964), similar to subsequent International Charters and documents, has multiple official translations, shared and verified with the promoters. In addition to providing the sharing of text in different languages, this solution guarantees the unique transmission of concepts and meanings. Starting from the *Nara Document* (ICOMOS 1994), a linguistic downsizing takes place by reducing the drafting of the charters in English and French. Furthermore, "glossaries" and "definitions" become the only means of conveying, in a common solution, the concepts and meanings of the Charters. These tools allow to share homogeneous definitions when dealing with issues through translation into a bridge language (English or French) and whose meaning is characterised by the specific cultural, generational and geographical point of view (Fatica 2023).

The research and its application have shown that International Charters can be seen as general dialogue tools. This possibility is offered not only by the universally recognised character that some Charters have assumed over time (e.g. *Venice Charter* ICOMOS 1964, *Nara Document* ICOMOS 1994, *Faro Convention* COE 2005, *The Nara + 20* ICOMOS 2014a; cf. Petzet 2003; Jokilehto 1998 (2018), 453–458) but also by the presence of definitions and glossaries that allow the sharing of concepts and meanings.

This dialogue was conveyed during the research by a graphic visualisation through the use of AI which aims to restore the meaning of the evolution of terms and concepts in a dynamic time dimension.

These tools allow the research to find points of agreement on the meaning of complex truths, despite geographical and cultural origins, as they return multiple definitions and nuances of the terms investigated, therefore returns a method of dialogue between different subjects which, through the reading of the International

Charters, guarantees the understanding of the polygonality and multisemantic nature of the starting concepts and a sharing strategy for using good practices on Cultural Heritage.

References

Architects' Council of Europe. 2018. *Leeuwarden Declaration – Adaptive Re-use of the Built Heritage: Preserving and Enhancing the Values of our Built Heritage for Future Generations.* Leeuwarden.

Council of Europe. 1975a. *European Charter of the Architectural Heritage.* Bruxels.

Council of Europe. 1975b. *The Declaration of Amsterdam.* Amsterdam.

Council of Europe. 1985. *Convention for the Protection of the Architectural Heritage of Europe.* Granada.

Council of Europe. 1987. *Certification Cultural Routes of the Council of Europe.* Bruxels.

Council of Europe. 2005. *Convention on the Value of Cultural Heritage for Society (Faro Convention).* Faro.

Council of Europe. 2018. *European Framework for Action on Cultural Heritage.* Luxembourg.

Di Biase, Carolina and Albani, Francesca. 2019. The Teaching of Restoration at the Architecture School of the Politecnico di Milano. Traditions and Perspectives. In *The Teaching of Architectural Conservation in Europe*, ed. Carolina Di Biase and Francesca Albani, 17–75. Santarcangelo di Romagna: Maggioli Editori S.P.A.

European Union. 2007. *Treaty on the European Union.* Lisbon.

Fatica, Ottavio. 2023. *Lost in Translation.* Milan: Adelphi Edizioni s.p.a.

Fiorani, Donatella. 2023. Conservation beyond reuse and abuse? Notes on the current fate of deconsecrated churches. In *Handbook: International Handbook for Students on Research and Design for the Sustainability Heritage*, ed. Mar Loren-Méndez, Marta García- Casasola, Vladan Djokić, Ana Nikezić, Konstantinos Sakantamis, Maria Philokyprou and Emanuela Sorbo, 245–255. Belgrade: University of Belgrade, Faculty of Architecture. https://hersus.org/06-international-handbook-for-students-on-research-and-design-for-the-sustainability-heritage/.

ICOMOS. 1964. *The Venice Charter.* Venice.

ICOMOS. 1975. *Resolutions of the International Symposium on the Conservation of Smaller Historic Towns.* Bruges.

ICOMOS. 1979–2013. *The Burra Charter: The Australia ICOMOS Charter for Places of Cultural Significance.* Burra.

ICOMOS. 1982. *Declaration of Dresden on the Reconstruction of Monuments Destroyed by War.* Desden.

ICOMOS. 1983. *Appleton Charter—For Protection and Enhancement of the Built Environment.* Ottawa.

ICOMOS. 1993. *Guidelines on Education and Training in the Conservation of Monuments, Ensembles and Sites.* Colombo.

ICOMOS. 1994. *The NARA Document on Authenticity.* Nara.

ICOMOS. 1996. *The Declaration of San Antonio.* San Antonio

ICOMOS. 1998. *The Stockholm Declaration.* Stockholm

ICOMOS. 1999. *Charter on the Built Vernacular Heritage.* Mexico.

ICOMOS. 2000. *Riga Charter on Authenticity and Historical Reconstruction in Relationship to Cultural Heritage.* Riga.

ICOMOS. 2003. *Principles for the Analysis, Conservation and Structural Restoration of Architectural Heritage.* Vicoria Falls

ICOMOS. 2004. *Charter for the Conservation of Unprotected Architectural Heritage.* New Delhi.

ICOMOS. 2008a. *Québec Declaration on the Preservation of the Spirit of Place.* Québec.

ICOMOS. 2008b. *The Icomos Charter for the Interpretation and Presentation of Cultural Heritage Sites*. Québec.
ICOMOS. 2010. *New Zeland Charter for the Conservation of Places of Cultural Heritage Value (1993)*. Auckland.
ICOMOS. 2011. *The Paris Declaration on Heritage as a Driver of Development*. Paris.
ICOMOS. 2014a. *Nara + 20: On Heritage Practices, Cultural Values, and Authenticity*. Nara.
ICOMOS. 2014b. *The Florence Declaration on Heritage and Landscape as Human Values*. Florence.
ICOMOS. 2017. *Delhi Declaration on Heritage and Democracy*. Delhi.
ICOMOS. 2018 (2021). *European Quality Principles for EU-funded Intervention with Potential Impact Upon CulturalHeritage*. Charenton-le-Pont.
ICOMOS. 2021. *Guidelines on Fortifications and Military Heritage*.
ICOMOS-IFLA. 2017. *Principles Concerning Rural Landscapes as Heritage*.
ICOMOS CIVVIH. 2011. *The Valletta Principles for the Safeguarding and Management of Historic Cities, Townsand Urban Areas*. Valletta.
International Congress of Architects and Technicians of Historic Monuments. 1931. *The Athens Charter for the Restoration of Historic Monuments*. Athens.
Jokilehto, Jukka. 2000. A History of Architectural Conservation. *European Journal of Archaeology* 3 (1): 137–139.
Jokilehto, Jukka. 2007. International charters on urban conservation: Some thoughts on the principles expressed in current international doctrine. *City & Time* 3 (3): 23–43.
Jokilehto, Jukka. 1998 (now 2018). *A History of Architectural Conservation*, 393–458. London; New York: Routledge.
Jokilehto, Jukka. 2020. Questions of Authenticity. *Conversaciones* 8: 55–72.
Musso, Stefano Francesco. 2023. Re-constructions: Meanings and ways. In *Handbook: International Handbook for Students on Research and Design for the Sustainability Heritage*, ed. Mar Loren-Méndez, Marta García- Casasola, Vladan Djokić, Ana Nikezić, Konstantinos Sakantamis, Maria Philokyprou and Emanuela Sorbo, 138–145. Belgrade: University of Belgrade, Faculty of Architecture. https://hersus.org/06-international-handbook-for-studentson-research-and-design-for-the-sustainability-heritage/
Occelli, Chiara L. M. 2023. Between the world and the earth: A necessary reconciliation. The role of the restoration project. In *Handbook: International Handbook for Students on Research and Design for the Sustainability Heritage*, ed. Mar Loren-Méndez, Marta García- Casasola, Vladan Djokić, Ana Nikezić, Konstantinos Sakantamis, Maria Philokyprou and Emanuela Sorbo, 187–197. Belgrade: University of Belgrade, Faculty of Architecture. https://hersus.org/06-international-handbook-for-students-on-research-and-design-for-the-sustainability-heritage/
Petzet, Michael. 2003. Place, memory, meaning: Preserving intangible values in monuments and sites. In *14th ICOMOS General Assembly and International Symposium: 'Place, memory, meaning: preserving intangible values in monuments and sites'*. Icomos open archive. http://openarchive.icomos.org/id/eprint/470/1/4_-_Allocution_Petzet.pdf. Accessed 10 Feb 2023.
Petzet, Michael. 2004. Principles of preservation: An introduction to the International Charters for Conservation and Restoration 40 years after the Venice Charter. In: *International Charters for Conservation and Restoration. Monuments & Sites*, 7–29. München: ICOMOS.
Petzet, Michael. 2009. *International Principles of Preservation*. Berlin ICOMOS, Hendrik Bäßler verlag.
Stovel, Herb. 2000. Nara revisited: the impact of the Nara Document on understanding and use of the authenticity concept. In Il *restauro fra identità e autenticità*, Atti della tavola rotonda 'I principi fondativi del restauro architettonico', ed. Giuseppe Cristinelli and Vittorio Foramitti, 243–250. Venice: Marsilio.
Stovel, Herb. 2008. Origins and influence of the Nara document on authenticity. *APT Bulletin* 39 (2/3): 9–17.
TICCIH-ICOMOS. 2011. *Principles for the Conservation of Industrial Heritage Sites, Structures, Areas and Landscapes*. Dublin.

UNESCO. 1972. *Convention Concerning the Protection of the World Cultural and Natural Heritage*. Paris.
UNESCO. 2003. *Convention for the Safeguarding of the Intangible Cultural Heritage*. Bali.
UNESCO. 2010. *Education for All Global Monitoring Report*. Paris.
United Nations. 2017. *New Urban Agenda*. Quito.
World Commission on Environment and Development. 1987. *Report of the World Commission on Environment and Development: Our Common Future*.

Part IV
Design Strategies and Actions

Sustainable heritage management is not outside the debate on the role of creativity in conservation projects. In this sense, the first contribution of this fourth part puts forward the innovative idea of implementing coevolution in intervention processes (S. Della Torre).

Working with heritage from a sustainable perspective implies working with temporality, understood as the interpretation of the time passing: past, present, and mostly, future. Because architecture is the discipline that is able to think about the future, to project, and to manage change. Conservation heritage projects are working with the change of something valuable. That is the reason why all decisions based on the identification of values have to do with knowledge and significance. It is important to not lose the responsibility to think that this situation will change again and again because it is impossible to stop the passing of time.

Different experiences in heritage intervention show the heterogeneity of the situations, but all of them are inserted in a (cultural and historic) landscape and must consider the influence of the tourist enhancement. A different approach to sustainable heritage management is also tested through student work (M. Marzo, V. Bertini).

The refurbishment of social housing is one of the main tasks that characterize the architectural expertise is currently focused on. The practice mainly concerns mass housing complexes built in the fifties and sixties, which have a high degree of obsolescence (M. Dousi, D. Sakkos). Consideration of values, as these buildings are modern heritage, is also crucial to restoring and preserving this architecture following a rigorous process.

At the other extreme is the preservation of ruins and abandoned places, from which heritage action originated. Certain interventions in collaboration with the authorities characterize the approach to research and thinking in the results of the projects, which focus on service-learning (E. Sorbo, G. Spironelli).

Chapter 12
Implementing Coevolution in Design Methodology for the Conservation and Reuse of Architectural Heritage

Stefano Della Torre

Abstract In several scientific fields, the Coevolution concept proved to be inspiring for innovative attitudes. The idea of Coevolution is a relatively recent development of Darwinian evolution and adaptation models, based on the detection of species, whose evolution processes show some interdependencies with each other, allowing to describe evolutionary changes as based on a reciprocal interdependence. The example of plants and butterflies is the most popular, but interdependencies may be much more complex. Besides the field of Biology and Natural History, studies on Coevolution have been carried out in many other contexts, such as History and Archaeology, Humanities, Management, Sociology, Applied Economics, Organization Science, and even Aesthetics.

Keywords Built cultural heritage · Conservation design · Conservation planning · Relational notions

Introduction

In several scientific fields, the Coevolution concept proved to be inspiring for innovative attitudes. The idea of Coevolution is a relatively recent development of Darwinian evolution and adaptation models, based on the detection of species, whose evolution processes show some interdependencies with each other, allowing to describe evolutionary changes as based on a reciprocal interdependence. The example of plants and butterflies is the most popular, but interdependencies may be much more complex.

Besides the field of Biology and Natural History, studies on Coevolution have been carried out in many other contexts, such as History and Archaeology, Humanities, Management, Sociology, Applied Economics, Organization Science, and even Aesthetics. A synthetic review of this panorama has been carried out in Della Torre

S. Della Torre (✉)
Department of Architecture, Built Environment and Construction Engineering, Politecnico di Milano, Milano, Italy
e-mail: stefano.dellatorre@polimi.it

M. Loren-Méndez et al. (eds.), *Sustainable Heritage Management - Design: Notions, Methods and Techniques*, https://doi.org/10.1007/978-3-031-81970-4_12

(2023), arguing that coevolutionary thinking could be inspiring also for architectural design methodologies, at least as dealing with conservation and reuse of historical buildings, or for the revitalization of landscapes and urban settings.

For instance, Italian "territorialist" scholars deal with the territory not as with a simple system, but as with a complex living system (Magnaghi 2017): ecosystems are studied by the implementation of sophisticated tools, whose background is exactly the metaphor of territorial components as coevolving living beings, and the idea that cultural landscapes should be understood as complex adaptive systems. Several studies (as an example Rescia et al. 2012) included both the concepts of emergence and coevolution, while the concept of "extended evolution" has been used to understand traditional agriculture landscapes as "living knowledge systems", whose preservation becomes a resource for the future, promoting interactions with new players in the frame of new societal processes (Laubichler and Renn 2015; Niles and Roth 2016).

Moving from the fundamental attitudes to the management of territorial processes, it is worthy to observe that in recent years, coevolution has been increasingly applied in planning studies, to analyze and model development and resilience processes, identifying interdependencies that are the bulk of coevolutionary processes, as the key to manage the involved actions and processes (Brunetta et al. 2019; Haider et al. 2020).

Actually, heritage is obviously connected with historic evolution and time, and according to some important authors, such as Amedeo Bellini (Bellini 2000) and Bernard Fielden (Fielden 2003, p. 3), conservation should be understood as a careful management of change. On the other hand, an increasing awareness can be detected of the need of giving preservation activities a sound basis rooted in social and economic processes. Therefore, the implementation of coevolutionary models could be really useful to produce resilience in a becoming world and in front of the contemporary big challenges.

The ultimate purpose of historic preservation has been generally defined as giving a future to the past and a past to the future. At once it was a matter of historicism in European culture, but now this mindful attitude on Time and Becoming means to implement the ethics of responsibility. This attitude has also an evident relationship with the concept of strong sustainability (Moioli et al. 2014; Buriti 2019).

Nevertheless, heritage processes cannot avoid being connected with life: in particular, architectural conservation happens in the frame of the social, economic, and practical rules of the contemporary world. Therefore, there is a high risk that conservation activities on architectural heritage are thought of as ordinary building works, following the logic of a short-term service life. The aim risks being limited just to meet the needs, or maybe the appreciation of the present, instead of giving the future more opportunities, even still vague here and now.

In the field of conservation, thinking in terms of long-term responsibility means to open to different criteria for decision making: the set of values coming from the past should be appreciated not in the mirror of the present attitudes, but in terms of coevolutionary potentialities. Referring to the step "from being to becoming" proposed by Ilya Prigogine, some years ago I proposed that conservation should focus on preserving, above all, the potentialities for coevolution (Della Torre 1999). What matters in this proposal is the displacement of the focus, from the level of the facts to the level of potentialities. It would be impossible to imagine such a step, without the previous step to anthropological approaches that, few decades ago, produced a renovation of heritage understanding, moving from the traditional authorized heritage discourse to the interest for other and different contents and relationships, which build the territorial tissue and the involvement of broader communities (Smith 2006). In this perspective, "heritage studies are no longer a matter of chasing the significance of the best artworks, but of recognizing the significance of networks" (Della Torre 2023).

On the basis of previous papers (Della Torre 2019, 2020, 2023) proposes to structure the implementation of coevolutionary thinking in heritage studies on four lines, which could be developed and implemented in the practice of built heritage conservation, and therefore also in the consistent educational programs:

"Long term vision: in the step from conservation to planned preventive conservation, the whole process is re-designed, and the purpose of conservation gets redefined as well, conserving not just objects, but their the coevolutionary potential;

Evolving significance: a methodology based on the creativity of architectural design, which takes the significance as forever given, risks to miss the aim. Designers' attitude should include research on the values included in the significance concept: that entails a step from adaption to coevolutionary reuse;

Heritage as a world of interdependencies: heritage reuse programs are easily understood under the perspective of territorial planning; hereby heritage is recognized no longer on the basis of excellence, but of the density of relationships, and social engagement in heritage becomes a major factor for sustainability, implementing coevolutionary practices in heritage programs;

The relationship among program and project: design not understood as creation (from nothing) but as a problem-solving activity, implementing coevolution between the program and the project".

While the themes of long-term vision and evolving significance have been already introduced in other papers, pointing out the idea that conservation should aim at transmitting to the future the coevolutionary potential of architectural heritage; the other two lines are more directly related to design activities and may deserve a wider discussion in the context of this book.

A Design Methodology Implementing a Coevolutionary Attitude

The implementation of coevolutionary thinking in conservation design means to exploit the awareness of coevolution for a better design methodology, ready to be also a reference for architectural conservation education.

On this point, it is unavoidable to deal with the discourses, which advocate for the value of creativity for design in unstable environments. According to some authors, research-by-design methodology would be a suitable approach for highly complex problems, which require counterintuitive thinking and the development of new knowledge (Roggema 2017). Architectural conservation is a critical point in this picture, because it encompasses a difficult balance between knowledge and creativity. Put otherwise, the challenge is in the designer's attitude: architecture (architectural design) aims at being considered as a form of knowledge, but conservation of built cultural heritage applies several scientific analyses to build the knowledge that gives the foundations to operational decisions. In other words, the knowledge produced "by-design" tends to be quite free, while conservation uses to be linked to a set of strict conditions, beginning with the aim of passing to the future (as much as possible) what has been inherited from the past. Designers often tend to minimize that methodological knot, separating what has to be conserved and what can be removed. Conservation therefore is a boring duty, limited just to the protected elements, while creativity is architectural creativity, often oriented to form a new architectural realm. Under a coevolutionary perspective, this is undoubtedly a mistake: nothing can survive if not evolving, and nothing is worthy to be demolished without very good and multiple reasons. The last sentence comes from a strong sustainability perspective, in the frame of a long-term vision, and the imagination of coevolutionary potentialities. As said above, conservation is the management of change, not freezing something unchanged in a completely new frame. Creativity will help to find the best solutions for careful changes. Coevolutionary thinking opens to innovation and change and makes significance dynamic, but it aims at keeping for the future the inspiring role that heritage can play, as well as it implies a deep and never ceasing research on the significance and discussion of the significance-based program.

Such a set of bounds makes the role of creativity, in comparison with the customary practices of architectural design, definitely different and peculiar. In the above-described perspective, creativity is implemented to reset and solve problems, conserve buildings through innovative uses, introduce new devices while exploiting the native performances of historical solutions, and also engage users in more careful attitudes.

Some methodological proposals, some decades ago, went in the direction of coupling (material) conservation and (freely designed) additions to comply with functional requirements. Conservation would be not value-based but total and absolute, additions designed with a great freedom, even creating dramatic contrasts as concentrating the design bounds only in the material permanence of the built fabric. But in most cases, this method doesn't fit the multiple aims of a long-term sustainability, so that it is necessary to refer to a well described, and open to developments, program in order to manage requirements and performances of the building elements.

Other theories on design methodology are inspiring on that point, even if not recent, and referred to projects in fields other than architecture and more interested in computational models (see for instance Maher et al. 1996; Maher and Tang 2003), as they apply the concept of coevolution to creativity in the design process, in terms of a mutual and evolving interdependency between the solution and the problem (Dorst and Cross 2001). As Nathan Crilly says, "rather than just solving a problem, design involves developing both an understanding of the problem and the possible solutions until a good match is found. Ideas for solutions might drive a reinterpretation of the problem, which, in turn, prompts new ideas for solutions. These changes to the representation of problems and solutions might be relatively minor, such as a change in emphasis or a refinement. However, they can also be radical, opening up entirely new parts of the design space. Creativity can thus be seen not just in the solutions that are offered, but also in the problems that are posed, and in the dynamic interplay between these processes" (Crilly 2021).

In architectural preservation, such paradigms find very clear examples of implementation, also in the field of reuse. The emerging keyword here is the "performance-based approach". For example, the "save and preserve" researches on energy efficiency for historic buildings have been carried out mainly by exploring traditional solutions, the performances of traditional materials and habits, getting out of the unthinking implementation of standards. In seismic prevention, a performance-based approach had to be implemented in order to customize retrofitting measures for historic masonry buildings, taking into account the construction history, the embodied resources, and the kinematics of damage, instead of staying focused on numerical models and safety standards defined for limit states not realistic for historic structures. For accessibility as well, a performance-based approach opened to better, non-standard solutions (Treccani and Arenghi 2016: 112–113). The same happens with safety in case of fire, to be obtained with a comprehensive strategy, not just with some adaptations.

The conceptual basis of the performance-based approach should not be understood as chasing a customized solution, but exactly as the creative development of the problem, through the investigation of both the ultimate target (energy saving, accessibility, safety, seismic resistance…) and the "hidden" resources of the historic building: resources that are hidden because not recognized through the customary models.

The coevolutionary methodology, therefore, has to do with the program and a twofold work of understanding and developing: on one side, there is an investigation and evolution of the needs to meet, and on the other side, there is an investigation of the resources of the historic building. This investigation goes hand in hand with the production of knowledge, which is an idiosyncratic character of conservation as a discipline. The problem–solution coevolution happens through the development of knowledge and of the acknowledged significance (i.e., cultural values).

The significance of the building is therefore a crucial part of the design methodology, and requires special attention paid on the nature and the statute of the values taken into account.

The requirements involved in the use of the buildings are evolving along with societal needs and attitudes. That entails frequent misalignments with the actual performances of old buildings, adding obsolescence to the natural ageing of historic constructions. Such problems can be solved, managing the related changes in order to answer the needs while keeping the values, implementing the long-term vision depicted above. Put otherwise, in the frame of the built environment industry a conservation sector can be identified, which deals with the cultural significance and the enduring life of historic buildings as facilities for contemporary needs. The capability to make old buildings fit with new uses has been identified with a condition for architectural conservation, advocating adaptive reuse as an alternative to demolition and substitution of the built stock. Furthermore, reuse is seen as being part of broader urban regeneration strategies, thereby also promoting a more sustainable built environment, exploiting the energy embodied in existing buildings, and also thanks to community involvement and related social benefits.

In the conservation sector, adaptive reuse is part of a conservation strategy, with its methods and tools: adaptation is described in the Burra charter, for instance. Outside the specialized sector, in the built environment sector at large, adaptive reuse is often carried on in a short-term perspective, focusing on the use-values, and implementing a design-based methodology.

Adaptive reuse has also been proposed as an autonomous discipline (Plevoets and Van Cleempoel 2013), in which design should encompass the targets of producing something new as well as conserving and exhibiting the nonmaterial significance of the place (Plevoets and Prina 2017). In this perspective, significance used to be defined once forever, thought as a permanent character of the place itself: the task of the designer would be not to deepen understanding and to produce new knowledge (put otherwise: to develop the significance), but to enhance just that given character, by creativity. The word and the concept of "character" are here introduced citing a recent paper by Elwazani and Katara, who recall the ICOMOS 2008 Québec Declaration on the Preservation of the Spirit of Place to identify the "character" with concepts as the spirit of the place, often called the genius loci (Elwazani and Katara 2019). Actually the sense of place links with cultural values, building configurations, and building integrity, but to be productive and to become a methodological tool, the operational connection with time, actors, and attitude to investigation has to be explored, and somehow modeled. For that purpose, the coevolutionary perspective could help. Coevolutionary thinking helps to understand how and why the sense of a place can be understood. It could be meant as steady and permanent, maybe threatened by new events and trends, and therefore to be protected or restored; or it can be understood as highly dynamic, as it emerged and emerges at any time from the meeting of the place with the social actors and the percipient subjects, so that changes that altered the original character may be appreciated as additions, which shaped a new character, more complex but just for that more interesting, and open to different futures. Put otherwise, nonmaterial significance and architectural character require to be understood not as static attributes of things, but as the outputs and outcomes of curiosity, scientific research, and social engagement.

In other words, "creativity has to be found in the investigation attitude, and coevolution is not binary, between the solution and the problem, but involves the evolution of knowledge: a knowledge that has to be built investigating in many directions and implementing many tools, getting rid of the schemes, that is often unlearning the prejudices linked to contemporary mainstream attitudes" (Della Torre 2023).

In terms of methodology and education, therefore, the implementation of coevolutionary thinking leads to multidisciplinary investigation as the distinctive feature of a productive methodology for conservation design.

Real-Task Education Practices: The Link Among the Project and the Program

The implementation of coevolutionary thinking in conservation design requires, therefore, the development of a twofold reasoning: taking into account long-term effects, the sustainability of projects will depend both on the capability of the program to deal with coevolutionary processes, and on a positive response of the design to the evolution of the program itself, which is necessary to frame the project into the reality of the context.

Under the educational perspective, the experience of preservation (see Ottoni 2021 for reflections on the Italian academic environment) often takes the role of introducing students to the reality of materials, to the issues of durability and repair techniques, and to the real issues of finding reuse opportunities.

While education to creativity often aims at keeping creativity free from the weight and the limits of reality, the courses of Preservation (Conservation, Restoration) use to involve students in a real-task experience, leading them to a hands-on study of the historic building, reasoning on the reuse through contacts with the real territorial context, that is exploring the interdependencies (for instance see Della Torre et al. 2019).

The core of the real-task exercise is reasoning on the complexity of sustainability as a matter of long-term vision, taking into account an evolving significance, in the frame of participatory programs, which could produce more sustainable references just because they accept to be not definitely fixed.

Heritage conservation/reuse processes must include territorial links and take care of competent communities, as this is crucial to make the process sustainable. For that purpose, coevolutionary thinking is a mighty tool to improve the understanding and management of social dynamics.

Focusing on the program phase, programs of heritage reuse are easily understood as related to planning disciplines. That is the good reason why some programs for the reuse of historic properties have been developed implementing coevolution concepts, mainly dealing with the social implications of reuse processes. Among these studies, it is worthy citing the paper by van Knipperberg and Boostra, working on the outputs of the Horizon 2020 Openheritage project, which analyzed several cases of heritage reuse, spread across Europe (van Knippenberg and Boostra 2021; van Knippenberg et al. 2022). The interpretation of territory as a field of interdependent forces influencing each other produced successful models In planning studies, (see for instance Gerrits 2008; Wang et al. 2017), to deal with complexity referring to coevolution. It seems possible to detect a common will and a common background with the model of the "trading zone" developed by Christer Gustafsson as a theoretical foundation for comprehensive wide-area projects implementing an "upstream" perspective, based on the idea of a four-pillars sustainability (Gustafsson 2011; CHcfE Consortium 2015: 195–197).

The above quoted experiences put the emphasis on community involvement and social impact, to the capability of heritage reuse to answer needs. The emphasis seems to be on adaptation more than on the results in terms of heritage conservation. To be clear, coevolutionary heritage programs, as seen above, seem to be very interested in nonmaterial issues, but at the very end, they deal with the physical reality of heritage sites, and cannot avoid taking some more "brick and mortar" responsibilities for conservation and reuse. This is a perspective for the future: the lesson to be learnt is that there is a space for the implementation of the same coevolutionary attitude both in the program phase, fostering the involvement of people and communities, and in the design phase, fostering investigation and the development of tailored solutions.

Conclusions

To wrap-up, the definition of a design methodology based on coevolutionary thinking could be very consistent with the aims of architectural heritage preservation.

The reasoning path presented above leads to underscore the link between program and design, as a precondition to place the design exercise into the conditions of a real-task. This means that a more comprehensive meaning of planned conservation, oriented to program and mindful of community involvement, has been posted in the background of our reasoning (Van Balen and Vandesande 2015; Vandesande et al. 2018).

In reality, practices of programming based on coevolution are not only timely, but also very proficient to build sustainability and resilience.

Furthermore, the attitude to elaborate and develop the program enables to describe coevolutionary design methodology not only as a good way to problem solving, but also as a good way of setting, or to say better fine-tuning the problems.

The key activity of such fine-tuning of the program requirements is investigation and the production of knowledge.

In theoretical terms, the proposed methodology can be described as a ternary coevolution between problems, solutions, and knowledge (Fig. 12.1).

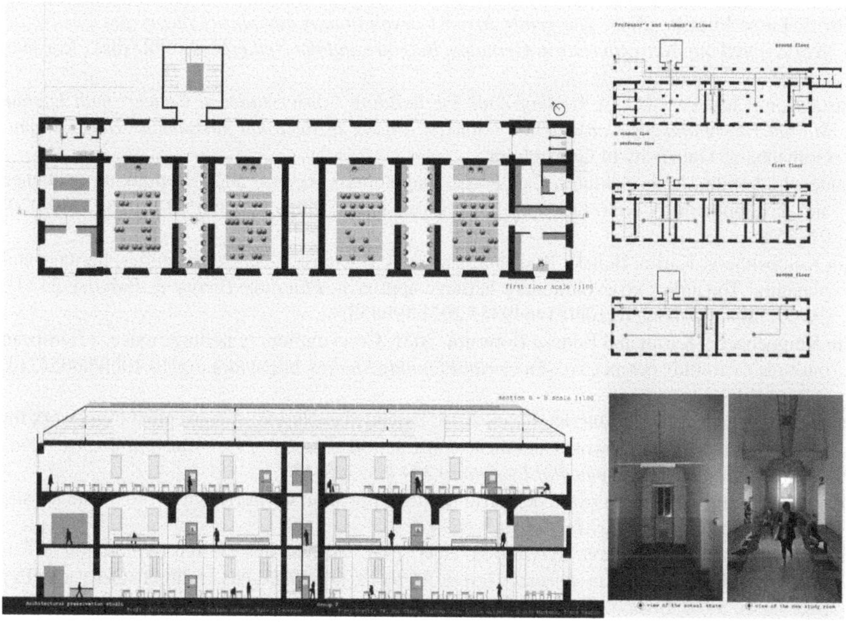

Fig. 12.1 S. Martino Psychiatric Hospital, Como. Architectural preservation studio. Professors. Stefano Della Torre, Giuliana Cardani, Sylvie Duvernoy. Students: Elena Ghetti, Filippo Giano, Giacomo Gola, Enrico Guidetti, Giulia Macheda, Irene Mastro. Politecnico di Milano, 2023

References

Bellini, Amedeo. 2000. De la Restauracion a la Conservacion: de la Estetica a la Etica. *Loggia* 9: 10–15 (English version: 105–106).

Brunetta, Grazia, Rosario Ceravolo, Carlo Alberto Barbieri, Alberto Borghini, Francesca De Carlo, Alfredo Mela, Silvia Beltramo, Andrea Longhi, Giulia De Lucia, Stefano Ferraris, Alessandro Pezzoli, Carlotta Quagliolo, Stefano Salata, and Angioletta Voghera. 2019. Territorial Resilience: Toward a Proactive Meaning for Spatial Planning. *Sustainability* 11: 2286. https://doi.org/10.3390/su11082286.

Buriti, Renata. 2019. "Deep" or "Strong" Sustainability. In *Encyclopedia of Sustainability in Higher Education*, ed. Leal Filho, 1–10. Cham: Springer. https://doi.org/10.1007/978-3-319-63951-2_503-1.

CHCfE Consortium. 2015. *Cultural Heritage Counts for Europe: Full Report*. Krakow: International Cultural Centre.

Crilly, Nathan. 2021. The Evolution of "Co-evolution" (Part I): Problem Solving, Problem Finding, and Their Interaction in Design and Other Creative Practices. *She ji: The Journal of Design, Economics, and Innovation* 7(3): 310–332. https://doi.org/10.1016/j.sheji.2021.07.00.

Niles Daniel and Roth Robin. 2016. Conservation of Traditional Agriculture as Living Knowledge Systems, not Cultural Relics. *Journal of Resources and Ecology* 7(3): 231–236.

Dorst, Kees, and Nigel Cross. 2001. Creativity in the design process: Co-evolution of problem–solution. *Design Studies* 22 (5): 425–437.

Elwazani, Salim and Katara, Pooja. 2019. Architectural character in conservation design Projects. *SHS Web of Conferences* 64, 03008. https://doi.org/10.1051/shsconf.

Fielden, Bernard. 2003. *Conservation of Historic Buildings*, 3rd ed. Oxford: Butterworth.

Gerrits, Lasse-Martijn. 2008. *The gentle Art of Coevolution. A complexity theory perspective on decision making over estuaries in Germany, Belgium and the Netherlands*. PhD diss., Erasmus University Rotterdam.

Gustafsson, Christer. 2011. *A Trading Zone for Building Conservation in Concert with Labour Market Policy and the Construction Industry, Aiming at Regional Sustainable Development*. Gothenburg: University of Gothenburg.

Haider, L. Jamila., Maja Schlüter, Carl Folke, and Belinda Reyers. 2020. Rethinking resilience and development: A coevolutionary perspective. *Ambio*. https://doi.org/10.1007/s13280-020-01485-8.

Van Knippenberg, Karim, Beitske Boonstra, and Luuk Boelens. 2022. Communities, heritage and planning: Towards a co-evolutionary heritage approach. *Planning Theory & Practice* 23 (1): 26–42. https://doi.org/10.1080/14649357.2021.199858.

Van Knippenberg, Karim and Beitske Boonstra. 2021. Co-evolutionary heritage reuse: a European multiple case study perspective. *European Planning Studies*. https://doi.org/10.1080/09654313.2021.2019684.

Laubichler, Manfred D., and Jürgen. Renn. 2015. Extended evolution: A conceptual framework for integrating regulatory networks and niche construction. *Journal of Experimental Zoology (Part B, Molecular and Developmental Evolution)* 324 (7): 565–577.

Magnaghi, Alberto. 2017. La storia del territorio nell'approccio territorialista all'urbanistica e alla pianificazione. *Scienze Del Territorio* 5: 32–41.

Maher, Mary Lou, Josiah Poon, Sylvie Boulanger. 1996. Formalising Design Exploration as Co-Evolution. In *Advances in Formal Design Methods for CAD*, eds. John S. Gero and Fay Sudweeks, 3–30, Boston, MA.: Springer. https://doi.org/10.1007/978-0-387-34925-1_1

Maher, Mary Lou, and Hsien-Hui. Tang. 2003. Co-evolution as a computational and cognitive model of design. *Research in Engineering Design* 14: 47–64. https://doi.org/10.1007/s00163-002-0016-y.

Moioli, Rossella, Koenraad Van Balen and Aziliz Vandesande. 2014. Costing the built environment: towards a policy of strong sustainable development. In *Quale sostenibilità per il restauro?* eds. Guido Biscontin and Guido Driussi, 457–467. Venice: Arcadia Ricerche.

Ottoni, Federica, ed., 2021. La didattica per il restauro. Strumenti, internazionalizzazione, competenze. *ArcHistoR* (Extra n. 9/2021). http://pkp.unirc.it/ojs/index.php/archistor/issue/view/57.

Plevoets, Bie and Daniela N. Prina. 2017. Introduction. In *Conservation-Adaptation. EAAE Transactions on Architectural Education n. 65*, eds. Donatella Fiorani, Loughlin Kealy and Stefano Francesco Musso, 1–8. Hasselt: EAAE.

Plevoets, Bie, and Koenraad Van Cleempoel. 2013. Adaptive reuse as an emerging discipline: An historic survey. In *Reinventing architecture and interiors: A socio-political view on building adaptation*, ed. Graham Cairns, 13–32. London: Libri Publishers.

Rescia, Alejandro, M. Esther Perez-Corona, Paula Arribas-Ureña and John W. Dover. 2012. Cultural landscapes as complex adaptive systems: the cases of northern Spain and northern Argentina. In *Resilience and the Cultural Landscape. Understanding and Managing Change in Human-Shaped Environments*, eds. Tobias Plieninger and Claudia Bieling, 126–145. Cambridge University Press.

Roggema, Rob. 2017. Research by Design: Proposition for a Methodological Approach. *Urban Science* 1(2). https://doi.org/10.3390/urbansci1010002.

Smith, Laurajane. 2006. *Uses of Heritage*. Abingdon-New York: Routledge.

Della Torre, Stefano, Rossella Moioli and Lorenzo Cantini. 2019. The historic centre of vimercate: Investigation, education, community involvement, in transdisciplinary multispectral modeling and cooperation for the preservation of cultural heritage (TMM_CH 2018). *CCIS* 961: 1–10. https://doi.org/10.1007/978-3-030-12957-6_22.

Della Torre, Stefano. 2023. Coevolutionary Thinking put into practice. *Intrecci*, 3.

Della Torre, Stefano. 1999. "Manutenzione o "Conservazione"? La sfida del passaggio dall'equilibrio al divenire. In *Ripensare alla manutenzione*, eds. Guido Biscontin and Guido Driussi, 71–80. Venice: Arcadia Ricerche.

Della Torre, Stefano. 2019. A Coevolutionary Approach to the Reuse of Built Cultural Heritage. In Il *Patrimonio Culturale in mutamento. Le sfide dell'uso*, eds. Guido Biscontin and Guido Driussi, 25–34. Venezia: Arcadia Ricerche.

Della Torre, Stefano. 2020. A coevolutionary approach as the theoretical foundation of planned conservation of built cultural heritage. In *Preventive Conservation—From Climate and Damage Monitoring to a Systemic and Integrated Approach*, eds. Aziliz Vandesande, Els Verstrynge, and Koenraad Van Balen, 11–18. London: CRC Press.

Treccani, Gian Paolo and Alberto Arenghi. 2016. Different Design Approaches to Accessibility to Cultural Heritage. A Decalogue. In *Accessibility as a Key enabling Knowledge for Enhancement of Cultural Heritage*, eds. Alberto Arenghi, Ilaria Garofolo and Oddbjørn Sørmoen, 105–114. Milano: Franco Angeli.

Van Balen, Koenraad and Aziliz Vandesande. 2015. *Community Involvement in Heritage*. Antwerp—Apeldoorn: Garant.

Vandesande, Aziliz, Koenraad Van Balen, Stefano Della Torre, and Fausto Cardoso. 2018. Preventive and planned conservation as a new management approach for built heritage: From a physical health check to empowering communities and activating (lost) traditions for local sustainable development. *Journal of Cultural Heritage Management and Sustainable Development* 8 (2): 78–81.

Wang, Jiang, Jilong Zhao, Wu. Tianyu, and Jin Li. 2017. A Co-Evolution Model of Planning Space and Self-Built Space for Compact Settlements in Rural China. *Nexus Network Journal* 19: 473–501. https://doi.org/10.1007/s00004-017-0334-z.

Chapter 13
Teaching Experiences in Designing for Heritage Places and Their Tourist Enhancement

Mauro Marzo and Viola Bertini

Abstract This paper presents some experiences of the international network of schools of architecture known as *Designing Heritage Tourism Landscapes*. The first part summarizes the objectives of the network and the main issues it addresses. The second focuses on teaching experiences promoted by the DHTL network and their experimental nature, which has influenced both the teaching and learning methods and the way in which a particular project was interpreted and tackled within this context. Finally, the third part provides examples of the international design workshops that have been conducted since 2015.

Keywords Heritage tourism · International network · Teaching methods

Introduction

This paper presents some experiences of the international network of schools of architecture known as *Designing Heritage Tourism Landscapes*. The first part summarizes the objectives of the network and the main issues it addresses. The second focuses on teaching experiences promoted by the DHTL network and their experimental nature, which has influenced both the teaching and learning methods and the way in which a particular project was interpreted and tackled within this context. Finally, the third part provides examples of the international design workshops that have been conducted since 2015.

M. Marzo (✉)
Dipartimento di Culture del progetto, Università Iuav di Venezia, Venezia, Italy
e-mail: mauro.marzo@iuav.it

V. Bertini
Dipartimento di Architettura e Progetto, Sapienza Università di Roma, Roma, Italy
e-mail: viola.bertini@uniroma1.it

M. Loren-Méndez et al. (eds.), *Sustainable Heritage Management - Design: Notions, Methods and Techniques*, https://doi.org/10.1007/978-3-031-81970-4_13

The Designing Heritage Tourism Landscapes Network

> [...] the heritage field is today the theatre of an unequal and dubious battle—in which the power of individuals remains great and in which the determination of a mayor, of an architect, an urban planner, or a heritage administrator can still change the fate of a monument or an ancient city (Choay 2001).

The *Designing Heritage Tourism Landscapes* network, founded in 2015 and headed by the Università Iuav di Venezia, brings together 2019 European and non-European architecture schools from six different countries: Argentina, Italy, France, Greece, Portugal and Spain[1] (Fig. 13.1). Through joint research programmes and educational experiments, which have taken the form of international design workshops in conjunction with field trips to heritage sites, the network's member schools seek to develop shared reflections on the relationship between architectural design and heritage tourism landscapes.

Widely debated in the relevant literature is the twofold nature of tourism in relation to the sense of places, or their structure, those material and immaterial elements which underlie them, their character and the way in which they are lived and perceived by those who visit or inhabit them. While on the one hand, tourism can contribute to the cultural, social and economic development of territories, promoting knowledge and development of their landscapes and heritage and ending up as a determinant factor in the processes of territorial regeneration; on the other, by consuming non-relocatable and non-renewable resources, it can also introduce changes in the sense and physical conformation of places, producing falsified and standardized images of territories and, not infrequently, alter their demographic, socio-economic and cultural dynamics (Marzo 2020).

The DHTL network develops its activities starting precisely from a reflection on this oppositional condition, so inherent to the tourist phenomenon. In particular, among the multiple heterogeneous forms into which tourism is divided and diversified, the network focuses its attention on heritage tourism, namely, that particular sector of cultural tourism which presupposes a contemporary 'use' of elements of

[1] The following schools belong to the DHTL network: Universidad de Alcalá, Escuela de Arquitectura, Departamento de Arquitectura—Università di Camerino, Scuola di Ateneo di Architettura e Design 'Eduardo Vittoria' Ascoli Piceno—Technical University of Crete, School of Architecture—Universidade de Évora, Departamento de Arquitectura—Università degli Studi di Firenze, Dipartimento di Architettura—Politecnico di Milano, Dipartimento di Architettura e Studi Urbani—Università degli Studi di Napoli Federico II—Università degli Studi di Palermo, Dipartimento di Architettura—École Nationale Supérieure d'Architecture de Paris-Malaquais—Università degli Studi Mediterranea di Reggio Calabria, Dipartimento dArTe—Sapienza Università di Roma—Università degli Studi Roma Tre—Universidad Nacional de Rosario, Facultad de Arquitectura, Planeamiento y Diseño—Universidad Nacional del Litoral, Facultad de Arquitectura, Diseño y Urbanismo, Santa Fe—Universidad de Sevilla, Escuela Técnica Superior de Arquitectura—Università degli Studi di Catania, Struttura didattica Speciale di Architettura, Sede di Siracusa—École Nationale Supérieure d'Architecture de Strasbourg—Universidad de Valladolid, Escuela Técnica Superior de Arquitectura—Università Iuav di Venezia (*network leader*)—École Nationale Supérieure d'Architecture de Versailles.

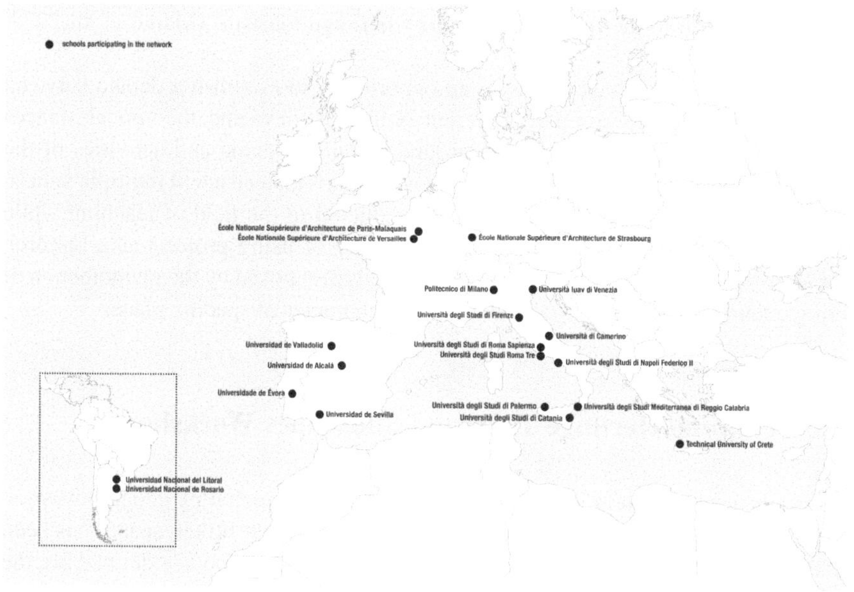

Fig. 13.1 Map of the schools which join the DHTL network. Drawing by V. Bertini

the past to which a condition of value has been recognized, and which a community has chosen to pass on to future generations (Tunbridge and Ashworth 1996; Timothy and Boyd 2003). Siding with this point of view, a two-pronged possibility is explored: on the one hand, using tourism in an instrumental sense as a lens to read heritage landscapes and, through the inventive action of the project at different scales, to recognize, enhance and valorize its qualities; on the other, using the eye of an architect to preserve and safeguard consolidated heritage places from those phenomena of consumption and standardization which are too frequently implicit in touristification processes (Amirante 2022).

Both possibilities, tied to each other in the form of a dialogue, lie at the centre of the reflections developed in the international research seminars promoted by the network, whose themes have ranged from 'the times of tourism' to the relationship between 'tourism and mobility', from 'cultural landscapes in rural environments' to the theme of 'abandoned villages', from access to 'iconic heritage places' to the relationship between 'tourism, sacred places and projects'.[2] These possibilities were also covered by the international conference *Sense of Past and Sense of Place. Designing Heritage Tourism*, whose results, published in a volume, have made it possible to highlight the vastness of the field of investigation and to assemble 'a multifaceted but coherent mosaic of research goals of contemporary architectural

[2] For the activities promoted by the DHTL international network of architecture schools see: https://www.iuav.it/it/strutture-didattiche-di-ricerca-e-di-servizio/designing-heritage-tourism-landscape.

design in relation to the questions posed by heritage tourism' (Marzo et al. 2022, p. 38).

If the research seminars represent an opportunity to establish a debate between DHTL schools and to compare different points of view and theoretical stances regarding that capital 'T' of tourism around which the name and activities of the network are based, the international design workshops instead adopt multiple values. Essentially, they involve experimentation conducted in the field of teaching while offering the opportunity to explore the possibility of using projects as a research tool to investigate the complexities and multiple issues posed by the phenomenon of heritage tourism in relation to the identity and character of specific places.

The Designing Heritage Tourism Landscapes Workshops

Tourism is a specific field of study for many disciplines, including geography, sociology and economics, to cite but a few examples. However, little research has been conducted by those who deal with architecture, and equally poorly defined are the tools and methodologies which projects have employed to deal with this phenomenon (Marzo 2016). It is on this basis that the educational activities promoted by the DHTL network have been structured.

The DHTL workshops, which take place annually and are organized and hosted from time to time by one of the network's member schools, focus on a specific design theme, which nonetheless always forms part of a wider field of investigation. The workshops organized to date have addressed the themes of the relationship between architectural design, tourism and art cities (Venice 2015), marginal areas (Évora 2016), the heritage of the modern (Roquebrune-Cap-Martin 2017), urban regeneration (Guanabacoa, Havana 2017), archaeological sites (Palazzolo Acreide 2017, Santa Severa 2022), rural landscapes and areas subject to processes of depopulation and abandonment (Guizhou 2019, Venice and Shanghai online 2021), monuments and contexts of belonging (Versailles 2020) (Fig. 13.2).

The Teaching Method

DHTL workshops are intensive educational experiences which take place over a period lasting from 8 to 15 days. Through dialogues within variable-geometry groups, but always composed of students and teachers from different universities and different nationalities, the area and the project theme are rapidly assimilated. This way of working allows a form of learning which is not only vertical, that is, from teacher to student, but also horizontal: between student and student and between teacher and teacher. In this way, the teachers also take part in a continuous learning process, based on dialogue and the hybridization of different points of view, traditions, and teaching methods. Although this approach is not unique in the national and international scene, it does present an innovative trait when likened with the more traditional forms of teaching architectural design in universities. Compared to the latter, it stands out for its brevity, for the ample room devoted to the project development phase, aimed

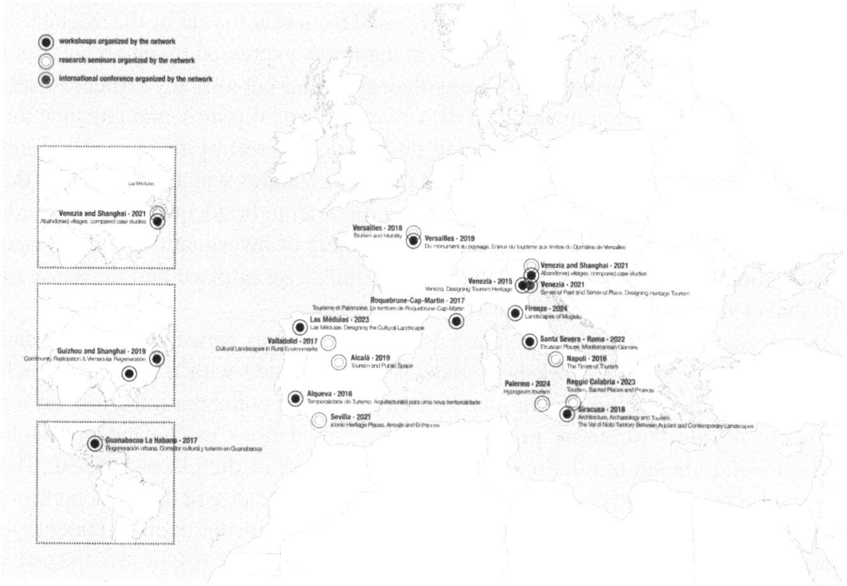

Fig. 13.2 Map of the activities (workshops, research seminars and international conferences) promoted by the DHTL network. Drawing by V. Bertini

at identifying in the short time available the most fitting answers to the requests posed by the project theme and the condition of exchange and dialogue between the participants. The project proposals elaborated at the workshops have an exploratory and open character rather than being definitive and settled; they propose visions and identify possible strategies which, while responding to the specific features of the individual projects, also seek to explore the more generic themes which the workshops focus on, each structured starting from one of the possible forms assumed by the relationship between architectural project and heritage tourism.

The Project as a Research Tool

The teaching experiences promoted by the DHTL network are based on the idea that a project can serve as a research tool, or rather, that it can contribute to the advancement of knowledge, as well as to the transmission of this knowledge. The focus is on the process and the possibility of exploring a working method, rather than on the actual results. The phase of getting to know the site and the intensive project elaboration that ensues is understood, as a whole, as a '*tool* for the teaching of design' (Amirante 2018, p. 18). Indeed, rather than successive phases, we can speak of a single process with continuous passages between the analytical-cognitive dimension of the places characterized by the heritage asset and the summary-design dimension. The site itself, along with the program proposed each year by the teachers at the organizing schools, raises questions, reveals the needs expressed by inhabitants and visitors of certain territories, and highlights both their potential and any critical issues. The site itself and the programme proposed from year to year by the teachers at the organizing schools pose questions reveal the needs expressed by inhabitants and visitors of certain territories, highlighting their potential but also any critical issues. To respond to the programme and needs, solve the critical issues, and enhance the potential identifiable at a site, the young designers, assisted by their teachers and tutors, must in a truly short time, pin down the main features which characterize the places identified for the projects. And the latter, in order to be adequately developed, must interrogate these places, take them as an object of investigation, and become a lens through which to progressively focus on the sense of place and the sense of time that characterize the site in question.

If the general field of investigation is the relationship between a project and heritage tourism, at each workshop students are presented with a question which concerns both the specific programme—often defined in collaboration with various local entities (administrations, protection bodies, foundations, etc.)—and a possible division of the theme in relation to the specific aspects of the places studied. The design responses formulated by the students, under the guidance of their teachers and tutors, define a range of hypotheses that, when compared with one another, encourage reflection on the most appropriate forms the project could take in response to the given research question. This comparison can also highlight what the common and recurring elements are, to which of them it is possible to attribute a general character, and which particularities pertain to the dimension of individual sensibilities, inherent to the nature of the project itself which is never 'logical-deductive, but *creative*, memorable, analogical, automatic, meta-metaphorical, metaphorical, random and even explicitly *irrational*' (Amirante 2018, p. 56). The objective is therefore not so much a deepening of theoretical knowledge as the development of operational reasoning on the design practice, as part of the overall teaching (Bianchetti 2018).

The design responses result from a non-linear process that unfolds through choices, reconsiderations, and a gradual refinement of possible solutions. These solutions emerge from discussions between students and teachers, interweaving into a network of relationships between problems and solutions, questions and answers. The teachers, despite significant differences in their approach determined by different educational backgrounds and origins, are asked to experiment with a working method which places at the centre of reflection the relationship between place and project, that is, the ways in which the project can get to know, interpret and rewrite the memory, structure, character and sense of the site in question. In the context of the workshops, it is the cognitive and interpretative capacity, sooner than the inventiveness of the project, that represents a further element which attributes the role of research tool to it. Students are thus offered the opportunity to tackle, albeit in a short time, a process which develops starting from a given research question, proceeds through an interpretative reading of certain places and, in a non-linear way, comes up with possible answers and solutions.

Several Teaching Experiences

Architecture, Tourism and Art Cities

Held in Venice in 2015 and organized by the Università Iuav di Venezia, the workshop *Venezia. Designing Heritage Tourism Landscapes*[3] was an opportunity to reflect on the relationship between architecture, tourism and art cities. With an annual footfall of around 8 million,[4] the historic centre of Venice offers the opportunity to use a project to reflect on the delicate balance between visitors and resident population, between the tourist enjoyment and valorization of a heritage asset and its protection and conservation. The objective was to design a City of Venice Museum, which could also function as a place to welcome visitors. In order to present the city and its territory, the imagined exhibition programme focused on both the history of Venice, starting from its origins, passing through its transformations over time, to arrive at its contemporary condition, deeply marked as it is by the processes of touristification, and on the relationship between the city and the lagoon, or rather, that particular ecosystem from which the city has drawn its lifeblood and without which it would not exist. Two project areas were chosen: one lying along the southern side of the Santa Lucia railway station, in a triangular space directly overlooking the Grand Canal; the other between the building of the former Venetian cotton mill, now home to the Università Iuav di Venezia, and the age-old cold stores. Together with the design of the museum, a redesign of its public space was imagined, conceived as a leisure spot for the resident population and as a transitional flexible meeting place for both residents and tourists. In fact, it was taken for granted that the entire museum, in addition to being designed as a tool to foster tourists' knowledge of the city, could also play a cultural role for citizens, by hosting activities and events addressed to the community at large.

Some of the questions posed by the relationship between tourism and art cities have been lucidly investigated by Marco D'Eramo in his essay *Urbanicidio a fin di bene* (D'Eramo 2014). In this text, which deals with those historic centres, like that of Venice, which are included in the UNESCO World Heritage List, the essayist highlighted the critical issues related to the complex 'balance' that exists between 'building and preserving', where the action of preserving, stopping time itself, can impede places 'from changing and becoming', thus depriving them of their very sense. Developing the project for a museum dedicated to the city thus meant trying to imagine, on the one hand, the designing of a space of mediation between those

---

[3] International workshop *Venezia. Designing Heritage Tourism Landscapes,* Venice, 18–25 September 2015. Organized by the Università Iuav di Venezia; and coordinated by Mauro Marzo from the same university—Dipartimento di Culture del progetto, along with João Rocha, Universidade de Évora, Departamento de Arquitectura. See: https://www.iuav.it/it/design-heritage-tourism-landscape/learning.

[4] In 2022, the footfall in the historic centre of Venice numbered 7,978,355; instead, in 2019, a figure of 8,817,623 was recorded. Data provided by the Veneto Region, Regional Statistical System, https://statistica.regione.veneto.it/jsp/venezia.jsp?comune=27042&periodo=202212&B1=Visualizza+il+report.

who permanently inhabit Venice and those who enjoy it temporarily and, on the other, the possibility of encouraging knowledge of the place by the latter, soliciting a reflection not only on the heritage value of the lagoon city but also on its specific characteristics and fragility.

Architecture, Tourism and Marginal Areas

In 2016, at the University of Évora, a workshop took place entitled *The Temporality of Tourism. Architecture for a New Territoriality,*[5] which focused on the relationship between architecture and tourism in relation to marginal areas. The theme was addressed through the case study of Alqueva. Located on the River Guadiana, on the border between Spain and Portugal, the Alqueva is marginal geographically, demographically, and economically speaking. In addition to its geographic marginality, quantifiable through its distance from other main centres, poor accessibility and the inadequacy of communication routes, there is a low employment rate, a demographic dynamic leaning towards depopulation and a lethargic economy mainly based on the primary sector. In order to contend with these critical issues, since 1976, the Portuguese Government has undertaken the construction of a dam—the *Barragem de Alqueva*—which, after coming into operation for the first time in 2002, led to the formation of one of the largest artificial lakes in Europe, determining an abrupt and radical transformation of the landscape. If, on the one hand, the creation of the water basin has offered the area new opportunities for economic and social repositioning—linked both to an increase in agricultural activities and a tourist reinvention of the territory focused on valorizing the lake as a resource—on the other, it has raised questions on the theme of preserving the memory of places, whose character and specific qualities, thanks to the area's condition of marginality, had long gone unchanged. The need to reconsider the territory in the light of its metamorphosis has also allowed a different interpretation of the concept of marginality, which, by assuming a different perspective, has in this case become an opportunity to promote slow and sustainable tourism (Bertini 2020). It was from this assumption that the activities of the workshop were developed, aimed at exploring how the marginality of an area could be read as an opportunity for development. The sites chosen for the project, located in the immediate vicinity of the *Barragem de Alqueva* and identified in collaboration with the dam management company EDIA, offered the opportunity to develop two different functional programmes: a territorial interpretation centre and a tourist reception facility. The former was conceived as a building in which to study the territory and to illustrate its transformations and peculiarities to visitors. Here, the conformation of the project area lying in an elevated position with respect to the lake's surface made it possible to work on open visuals which took the landscape as a design theme. The latter was imagined starting from the reuse of an ancient

[5] International workshop *Temporalidade do Turismo. Arquitectura(s) para uma nova territorialidade*, Évora, 09–16 September 2016. Organized by the Universidade de Évora, Departamento de Arquitectura; coordinated by João Rocha—Universidade de Évora, Departamento de Arquitectura, and Mauro Marzo—Università Iuav di Venezia, Dipartimento di Culture del progetto. See: https://www.iuav.it/it/design-heritage-tourism-landscape/learning.

monte[6] in a state of ruin. This inspired operational reflection on the theme of reusing heritage as a form of protection. In fact, in Alqueva, the rapid metamorphosis of the territory following the formation of the lake and a slower process of transformation in agricultural activities has led to the abandonment of many landed estates which, despite having lost their former function, still retain a strong representative capacity as a whole. Hypothesizing their possible transformation was a way to try out strategies aimed at ensuring that these patrimonial elements could retain their testimonial value while continuing to play an active role in the territory.

Overall, the design experiments developed during the workshop were aimed at reading and interpreting the places, valorizing their history and qualities, and envisioning slow and sustainable forms of tourist fruition that could encourage local development without compromising their specific characteristics.

Architecture, Tourism and Archaeology

The workshop Architecture, Archaeology and Tourism. The Val di Noto Territory Between Ancient and Contemporary Landscapes[7] (organized by the University of Catania, Special Teaching Structure of Architecture of Siracusa, and held in Palazzolo Acreide in 2018) and the workshop *Etruscan Places. Mediterranean Glances*[8] (organized by the Sapienza University of Rome and Roma Tre University and held at Santa Severa in 2022) were focused on the relationship between architecture, tourism and archaeological areas. In the first case, the subjects of investigation were two Sicilian UNESCO sites: the territory of the Val di Noto—with ancient Noto itself, the archaeological area of Akrai and the Castle of Palazzolo Acreide—and the city of Siracusa—with the Archaeological Park of Neapolis, the ruins of the Euryalus Fortress, and the Dionysian Walls. A consolidated destination for Grand Tour travellers, the territory studied has specific geographical peculiarities and a dense historical stratification which denotes it as a complex urban palimpsest in which classical remains coexist side by side with extraordinary Norman, Byzantine and late Baroque buildings as well as the contemporary city. Although differing in character, extension, history and the extent of their tourist flows, the areas of project experimentation inspired reflection on the role that archaeological sites might assume

---

[6] A *monte* is a traditional farm of the Portuguese Alentejo region whose architectural structures are associated with the production of a wider-ranging agricultural system.

[7] International workshop *Architecture, Archaeology and Tourism. The Val di Noto Territory Between Ancient and Contemporary Landscapes*, Palazzolo Acreide, 6–14 September 2018. Organized by the Università degli Studi di Catania, Special Teaching Structure of Architecture Siracusa; coordinated by Bruno Messina—Università degli Studi di Catania, Special Teaching Structure of Architecture Siracusa, Emanuele Fidone—Università degli Studi di Catania, Special Teaching Structure of Architecture Siracusa, Mauro Marzo—Università Iuav di Venezia, Dipartimento di Culture del progetto. See: https://www.iuav.it/it/design-heritage-tourism-landscape/learning.

[8] International workshop *Etruscan Places. Mediterranean Glances*, Santa Severa (Rome), 5–17 September 2022. Organized by the Sapienza Università di Roma and Università degli Studi Roma Tre; coordinated by Fabrizio Toppetti, Sapienza Università di Roma—Dipartimento di Architettura e Progetto, Luigi Franciosini—Università degli Studi Roma Tre, Dipartimento di Architettura. See: https://www.iuav.it/it/design-heritage-tourism-landscape/learning.

within a contemporary urban fabric and on the architectural operations to be implemented to patch up that condition of fracture between past time and present time which, often signalled by fences and abrupt changes in height, is particularly evident at the edges of the excavation areas (Marzo 2012). Working on the permeability of margins, on access systems and within sites so that they can be used and understood, some possible ways of enhancing what time has been handed down in the form of ruins were explored.

The second workshop focused on the archaeological sites of Veio, Cerveteri and Pyrgi as significant historical testimonies in the territory of southern Etruria. Also in this case, through project experimentation, conducted at both architectural and landscape scales, the theme addressed was the ways tourists can find pleasure in places in relation to their character (Fig. 13.3). Starting off from an examination of the specific features of the project areas, which share some common traits, the possibility was verified of reading these areas not as isolated spaces of an occasionally invisible archaeology, but as constituent parts of a territory in which historical traces, landscape values and outcomes of the transformation processes produced by modern and contemporary culture are concentrated (Marzo and Bertini 2022). Working on the confines, on the entrances, on the construction of a new museum and reception

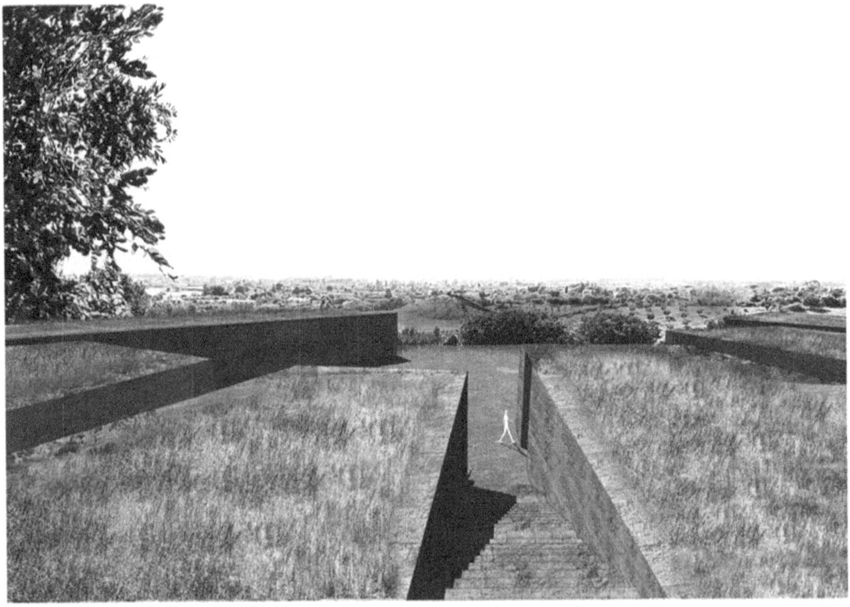

Fig. 13.3 *Cerveteri, city and form of the earth*, workshop *Etruscan Places. Mediterranean Glances*, 2022. Professors: R. Capozzi, E. Fidone, F. Foti, L. Franciosini, A. Gabbianelli, M. Marzo, M. Salerno, and F. Visconti. Tutors V. Bertini, N. Campanile, D. Careri, A. Carlini, C. Casadei, M. Erbani, G. Ferrarella, L. Flaviani, O. Lubrano, L. Pujia. Students: A. Allegra, F. Bertoncello, G. Di Biase, P. Fernández Ruiz, G. Galiero, D. Giersch, T. Hauswirth, D. Lais Zambiasio, E. Le Fauci, D. l Morales García, M. Pérez, M. Perrone, M. Rodríguez Rodríguez, E. Valori

services and on the designing of visitor routes and itineraries in the landscape, the projects developed within the context of the workshop were aimed both at a rearrangement of the areas studied in relation to their historical sense and at the entire territorial area of which they are a part.

Conclusions

The workshops presented briefly here were just some of the 9 which have been organized to date by the DHTL network. They can be taken as representative of the educational objectives which the network sets itself as well as the methods adopted to pursue them. By urging students to define a project response to a given theme, the intention is to encourage reflection on the role that architectural design can assume in reading and understanding heritage tourism landscapes before any actual transformation takes place. It goes without saying that the phenomenon of tourism could not exist without places. More in particular, it can be added that heritage tourism could not exist without signs and traces, be they self-evident or not, of what we have inherited from the generations that came before. Architects are and will increasingly be called upon to intervene on the conservation of signs, traces and artefacts, as well as on the transformation and valorization of places and landscapes, together, naturally, with holders of other disciplinary knowledge and professional skills. It is these issues which the workshops organized by the network tackle, allowing students to address in a form of intensive teaching a research theme that will go on to permeate their professional future: designing starting from the extant, valorizing those environmental pre-existences in which the communities of inhabitants recognize their identifying values and in which travellers find favourable opportunities to reflect on the sense of time and the sense of place.

References

Amirante, Roberta. 2018. Il *progetto come prodotto di ricerca. Un'ipotesi.* Siracusa: LetteraVentidue.

Amirante, Roberta. 2022. *I tempi del turismo (un'ipotesi tematica da condividere per una eventuale proposta editoriale)* (Unpublished essay).

Bertini, Viola. 2020. Alqueva, projects for a new territoriality. In *Architecture Tourism and Marginal Areas. Research and Design Proposals*, eds. João Rocha and Viola Bertini, 52–77. Siracusa: LetteraVentidue.

Bianchetti, Cristina (ed.). 2018. *La ricerca in architettura.* Siracusa: Lettera Ventidue.

Choay, Françoise. 2001. *The Invention of the Historic Monument.* Cambridge: Cambridge University Press.

D'Eramo, Marco. 2014. Urbanicidio a fin di bene. *Casabella* 982: 141–143.

Marzo, Mauro and Bertini, Viola. 2022. Heritage Tourism: luoghi e progetto. In *TRANSIZIONI. L'avvenire della didattica e della ricerca per il progetto di architettura*, eds. Samanta Bartocci, Giovanna Maria Biddau, Lino Cabras, Adriano Dessì and Laura Pujia. ProArch.

Marzo, Mauro, Ferrario Viviana and Bertini Viola. 2022. On the Sense of Time and a Sense of Place in Relation to Heritage Tourism. In *Sense of Time and Sense of Place. Designing Heritage Tourism*, eds. Mauro Marzo, Viviana Ferrario, and Viola Bertini, 26–41. Siracusa: LetteraVentidue.

Marzo, Mauro. 2012. Archeologia, Città, Turismo. Un possibile ruolo per il progetto di architettura. In *Architettura e Turismo*, ed. Luigi Coccia, 207–209. Milan: Franco Angeli.

Marzo, Mauro. 2016. Turismo e Architettura: una relazione disattesa. *Officina** Bimestrale on-line di architettura e tecnologia 11: 8–13.

Marzo, Mauro. 2020. A particular point of view. In *Architecture Tourism and Marginal Areas. Research and Design Proposals*, eds. João Rocha and Viola Bertini, 6–9. Siracusa: LetteraVentidue.

Timothy, Dallen J. and Boyd Stephen W. 2003. *Heritage Tourism*. New York: Prentice Hall.

Tunbridge, John E., and Gregory J. Ashworth. 1996. *Dissonant Heritage: The Management of the Past as a Resource in Conflict*. Chichester: J. Wiley.

Chapter 14
Social Housing Complexes Delivered by Workers' Housing Organization a Modern Building Stock to be Restored. Complexes of 50's–60's, The Axios Housing Complex: A Case Study

Maria Dousi and Dimosthenis Sakkos

Keywords OEK · Adaptive reuse · Sustainability · Greek modern heritage

Introduction

The building stock of the large urban centers within Greece includes several mass housing complexes, which have been built during the first two post-war decades. A 95% of the state-planned housing complexes are a work of the Workers' Housing Organization (Οργανισμός Εργατικής Κατοικίας, O.E.K.) (Gouvousi 2011). The designing principles of modernism are evident in the architecture of the buildings designed by the architecture department of O.E.K., in their typological and morphological characteristics, as well as in the overall urban planning of the settlements. In social housing complexes, built by O.E.K., thanks to the appropriate design, the housing needs were more effectively met than in the most common type of middle-class mass housing unit in Greece (polykatoikia). Large free spaces, low building density and the functional layouts of the apartments are to this day important features that improve accommodation in the dwellings, compensating for their limited area. The majority of mass housing complexes in Greece are poorly to moderately preserved. There is no organized preservation and maintenance or energy efficiency

M. Dousi (✉) · D. Sakkos
School of Architecture, Aristotle University of Thessaloniki, Thessaloniki, Greece
e-mail: mdoussi@arch.auth.gr

D. Sakkos
e-mail: dtsakkos@arch.auth.gr

M. Loren-Méndez et al. (eds.), *Sustainable Heritage Management - Design: Notions, Methods and Techniques*, https://doi.org/10.1007/978-3-031-81970-4_14

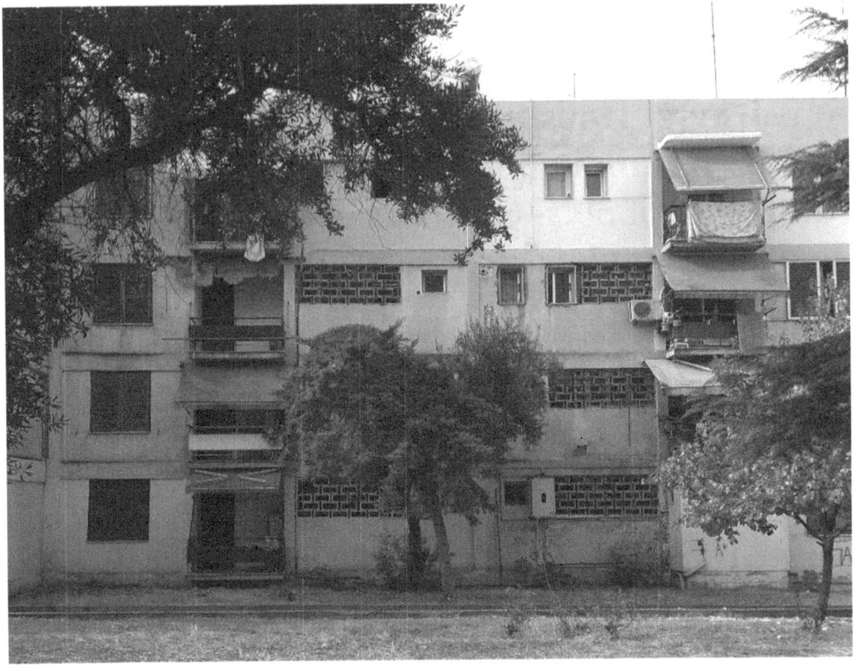

Fig. 14.1 Current condition of the majority of housing complexes in Greece, Axios Housing Complex (Sakkos D. 2018. Master Thesis *Restoration and rehabilitation of Axios Housing Complex*. Thessaloniki. A.U.Th.)

upgrading program of the buildings. Most of the maintenance work in the dwellings has been done by the owners-occupants and is partial and alters the character of the buildings (Fig. 14.1).

The mass housing complexes, built by O.E.K during the first two post-war decades are historical and architectural artifacts of the post-war modern architecture in Greece, they are of great importance and thus, their preservation is imperative, but in parallel their adaptation to contemporary functional and energy efficiency data is indispensable.

The Greek Workers' Housing Organization

After the end of World War II and the civil war that followed, during the two post-war decades, intense construction activity was documented in Greece. The need for rebuilding of all the infrastructures destroyed during the two wars and the following urbanization of a great part of population led to the construction of a large number of buildings within the main Greek urban centers.

The solution of the housing issue in post-war Greece was mainly based on private initiative, taking advantage of a land-for-flats practice called antiparochi (meaning in exchange) (Alifragkis and Kalfa 2020). The combination with the increase of the permitted building factors in the large urban centers, motivated private entrepreneurs to construct apartment buildings, in order to provide housing for the massively relocated population that had already accumulated in Athens and Thessaloniki since the 1950s.

There were four Greek ministries responsible for managing the housing issue, but only 2% of urban housing construction has been produced by state agencies (Gouvousi 2011). The main constructor of social housing units in Greece was the Worker's Housing Organization (O.E.K.) (Karanikas 1986). The aim of the organization was to construct and provide residences for workers who lacked private housing and lived mainly in large urban centers.

From 1954 until 2012, O.E.K. helped significantly in providing housing for many middle-class working families. The mission of O.E.K. was not only limited to the construction and distribution of apartments and houses to the beneficiaries, but also included a range of financing programs, either in the form of low-interest loans or in the form of rent subsidizing to the beneficiaries. In beneficiaries of O.E.K. assistance all employees who would pay monthly duties to the organization were included. Financial autonomy was desired since the foundation of O.E.K. and for this reason, the resources required for the construction of the complexes and for the provision of loans came from the revenues of the organization. In addition to the annual state funding, the funds from the monthly deductions made to the employees and from the interest on the loans which the organization had sponsored are added to the total revenues. However, resources were limited, and the desired financial autonomy was never achieved.

O.E.K. was one of the few public bodies in Greece that organized and supported the construction of mass housing complexes from the acquisition of the plot to the completion of the construction of the buildings and the final configuration of the surrounding area. After the plot purchase, its topographical mapping followed. Afterward, the design department proceeded to the urban planning of the wider area, followed by the architectural designing of both the buildings and the surrounding area. At the same time, the civil and mechanical engineering plans were being prepared. Finally, the landscaping of the surrounding area was completed by planning the tree planting project proposal. All stages of the studies were carried out in accordance with the current regulations in order to meet the required standards. The technical specifications and the construction quality were constantly improved in accordance with the average Greek urban residence.

During the first two post-war decades until 1972, the legal regime that applied to the operation of the O.E.K. assisted its constructing activity. During this time, studies for 152 settlements were prepared, but only 144 (Gouvousi 2011) of them were realized. Most settlements are located within the two largest urban areas of Greece, Athens and Thessaloniki, as well as in the largest cities-capitals of the prefectures. The construction activity of O.E.K. was interrupted from 1972 until 1975 as, according to law, the objective of the organization changed. The construction of new settlements was prohibited, except for special cases and the organization's role was exclusively lending in cooperation with commercial banks.

After 1975, construction of new settlements by O.E.K. was anew allowed. The need for housing was so imperative that in a period of only 7 years (1975–1982) 92 new settlements were designed, out of which 73 were finally constructed. Accordingly, during the next 12 years (1982–1993), 166 more settlements were designed and 156 were built. From 1993 until 2012, when O.E.K. discontinuing its operation, its role was different. Its mission is greatly limited to subsidizing rent and providing low-interest loans to beneficiaries, while the construction of mass housing complexes is ever more confined. In 2012, in the context of fiscal consolidation, the operation of O.E.K. was discontinued and the organization was merged with the Hellenic Manpower Employment Organization and serves exclusively as an archival database for transactions of O.E.K. apartments.

Mass Housing Complexes of the First Period of O.E.K. (1954–1972)

As previously mentioned, the first two post-war decades from 1954 to 1972 constituted the first period of operation of the Workers' Housing Organization. During this time, 152 settlements were designed and 144 of them were finally built. The largest concentration of mass housing settlements is found in Athens and Thessaloniki. In the wider area of Athens, 19 settlements were constructed, while in Thessaloniki 8 more were realized. A large number of mass housing complexes are located in Crete and within the largest capitals of the prefectures.

Most settlements constructed during this period were located on the outskirts of the cities. This option was compulsory, according to the decisions of the board of the organization, as the cost of purchasing land in areas within the city was increased and difficult to manage due to the limited financial resources of the first period. In addition, the issue of small ownership, a predominant characteristic of Greek urban areas, prevented the acquisition of a large area that would allow the construction of a settlement. In cases where settlements were built within the boundaries of large urban centers, the plot was usually located in more degraded and less developed urban areas where the cost of purchasing land was lower. In the smaller provincial towns, the new settlements were developed far from the urban center, usually on agricultural land plots. In several cases, the construction of the new settlements took place in areas far from the urban center, with uneven terrain and without easy access to it.

The architects of the organization, following the principles of post-war urban planning, sited the housing units, designed individual smaller outdoor public spaces for the residents and organized a distinct network of streets and sidewalks to serve transportation. According to modern plan-libre principles and with references from European case studies, in the complexes of the O.E.K., there was absolute separation of pedestrian and vehicular operations and movements. Provision of the necessary ventilation and optimal solar orientation for all the apartments were of great importance for urban planning. In most cases, the distance between the buildings was equal to their height, allowing sunlight even at the lowest levels. The different types of housing were repeated either in ribbon development or checkered columns, or they formed building blocks with internal courtyards, while sometimes the free standing of some types was chosen. A key axis in the design was the creation of a pleasant and healthy environment for the residents of all apartments. In the free spaces between the residential buildings, playgrounds, small squares and parks were planned for the recreation of the residents, as well as green spaces. In the complexes where the two-story residential units predominated, there was provision for small private gardens alongside the main facade of the residence. In the largest complexes of the O.E.K., there was a zone with uses other than residence. In this zone, buildings to house commercial uses were designed as well as spaces that served meeting and educational purposes, medical care and even religious spaces. In this way, the settlements were complete residential complexes offering their residents all the necessary services. However, this organization is less common and is found mainly in settlements far from the then urban centers. It has to be mentioned that during this first period of O.E.K. operation, due to the limited financial resources available, the pursuit was the maximum exploitation of the available space and the construction of the maximum possible number of dwellings, however, always ensuring a healthy living environment. For this reason, in case the settlement could be served by the area around it, the planning of the service zone was avoided. Since 1967, the quest for even greater exploitation of the land has become even more intense. Between 1967 and 1972, the residential complexes built by O.E.K. became more extended. In an effort to maintain adequate distances between buildings and as much free public and green space as possible, the construction of housing in multi-story towers of 6, 8 and 12 floors was investigated. These building typologies can only be found in large urban centers and were not repeated after 1975. In the smaller provincial towns, the building density was lower and two-story types of houses, either single-family houses or duplexes, were usually chosen. There, average gross densities range from 250 to 300 individuals/Ha, while the percentage of area built is between 20 and 25% and the building factor between 0.40 and 0.50. In the case of high building and the combination of high and low building typology the average mixed (gross) densities range from 300 to 550 people/Ha, while the percentage of area built reaches only 15–20% and the building factor reaches 0.60–0.80 (Gouvousi 2011).

Examining the different typologies of housing units, in this first period of operation of O.E.K. in the smaller provincial towns, ground and two-story single-family houses and two-story duplex types are mainly used. These types were more easily accepted by the beneficiaries, as they approach the type of single-family house with a yard, common in rural areas. In large urban areas, however, the limited space and the need for a greater number of houses impose the use of multi-story types of apartment buildings, combined with the above-mentioned types of single-family houses. During the first decade, the typologies of three-story apartment buildings were mainly used in the complexes built in Athens and Thessaloniki. From the 1960s onward, the typology

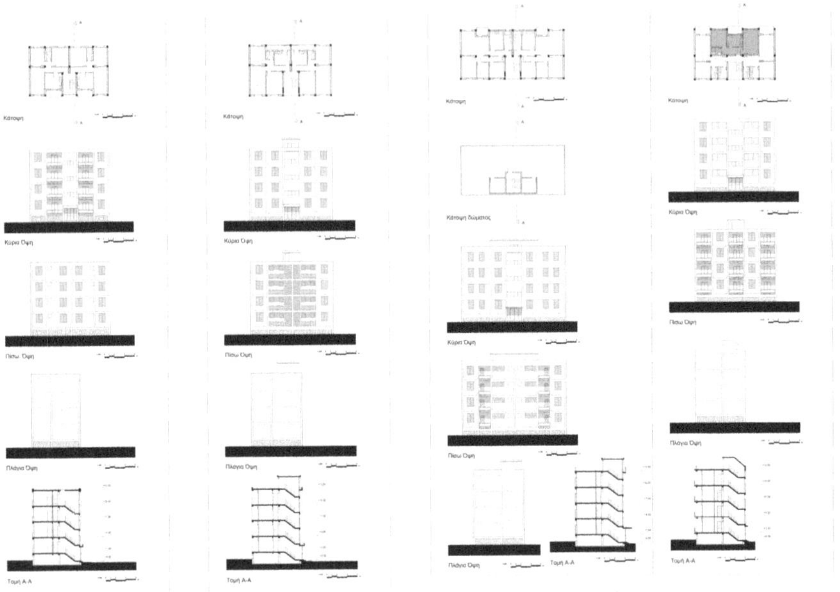

Fig. 14.2 Various O.E.K. housing typologies (Sakkos D. 2018. Master Thesis *Restoration and rehabilitation of Axios Housing Complex*. Thessaloniki. A.U.Th.)

of three- and four-story apartment buildings is also used in complexes of medium-sized provincial urban centers. Gradually, in the context of greater exploitation of the plot, but without the suffocating densification of the urban space, multi-story buildings of 8 floors also appear, as well as the unique typology of the twelve-story tower with a pilotis met only in the housing complex of the Axios in Thessaloniki (Figs. 14.2 and 14.3).

During the first period and especially in the first decade, from 1954 to 1965 the housing units were of limited area. The basic ground floor detached house unit was 55–65 m^2 large and consisted of two sleeping areas, a small living area, a kitchen, and a bathroom. In some variations of this basic typology, a small storage space is added to the main volume of the building. In the case of two-story detached houses and semi-detached houses, the area of the residence is larger and ranges from 60 to 85 m^2. The spaces included in the basic type of the two-story single-family house and the two-story duplex house are the same as the previous type and consist of two bedrooms, a living room, a kitchen, and a bathroom. In some larger variants, a W.C. under the internal staircase, as well as a small storage area were added. Spaces included in the apartments of the multi-story buildings are similar to these described for the two-story houses. The most common housing unit in the three- and four-story apartment buildings was the dwelling of 65 m^2, while there were also some typologies of buildings that included 3 apartments of 45 m^2 on each floor. Analysis of the basic apartment typologies built until 1965, renders the attempt to accommodate a large number of residents in the small area of the apartment evident. Since 1965, the

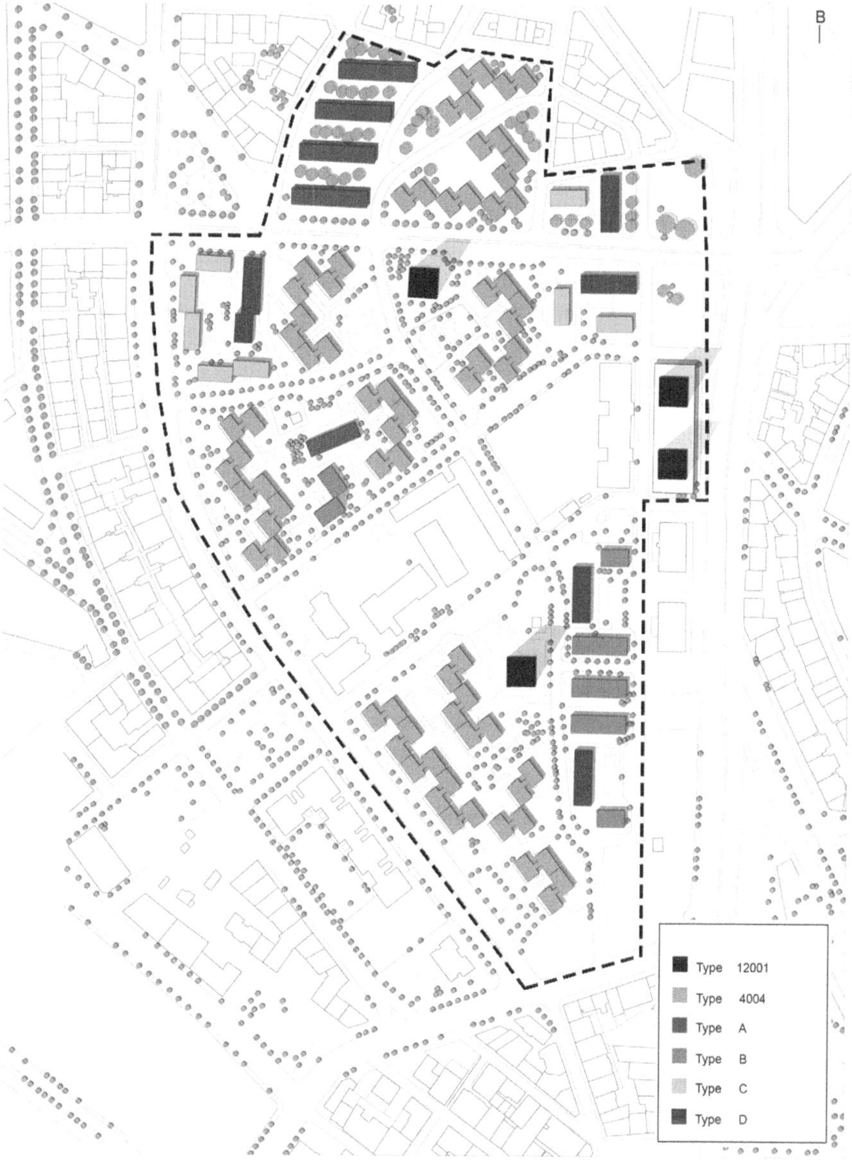

Fig. 14.3 Axios Housing Complex masterplan and buildings typologies (Sakkos D. 2018. Master Thesis *Restoration and rehabilitation of Axios Housing Complex*. Thessaloniki. A.U.Th.)

character of the apartments has changed, serving a more urban lifestyle. In this way, whereas initially emphasis is given on the existence of large sleeping areas and the living room area is limited to a small room that can potentially become a sleeping area at night, this situation changes in the units designed and built from 1965 onwards. In the new types of apartments, the living area is larger and far away from the sleeping rooms. At the same time, the area of residential units has been increasing. In the newest typologies of high-rise apartment buildings (eight- and twelve-story) there are apartments with an area of approximately 90–100 m^2.

What is also different in the middle of the first period is the relationship of the apartments with the public outdoor space. In the basic typologies constructed from 1954 to 1965, the apartments do not include large outdoor spaces. According to the single-family houses, the relaxation area of the residents is considered the nearby public free space. Small auxiliary balconies were designed in direct relation to the kitchen of the apartments, without, however, being able to cover the need for recreation into the public space. Extension of the apartment boundaries outside the shell enhances the sense of neighborhood and intimacy in public space. On the contrary, during the following period, the types of houses designed by O.E.K. include larger balconies, usually along the entire length of the building façade which serve as a continuity of the apartment in the open air. During this period, public green space within the complexes is perceived more as pedestrian and vehicular movement space and parking slots, than a recreational area. The sense of neighborhood fades and only the playgrounds continue to be used in the same way as previously.

Morphologically, the buildings of the first period (1954–1965) are characterized by the visible reinforced concrete structural system on the facades, the small balconies and the intense coloration of the coatings. Between 1955 and 1957, the design department was headed by Aris Konstantinidis, who designed several settlements during this time. His ideas influenced the building production of the organization in the 1960s. His main aim was to produce complete and comfortable living spaces. His design was in harmony with the Greek climate and whereas his references to modernism are evident, the final result reveals an adaptation to the place, its scale and its climate. The buildings designed after 1965 are more influenced by the form of typical urban polykatoikia with balconies predominating the whole facade.

The Case Study of Axios Housing Complex in Thessaloniki

A mass housing state-planned complex in Thessaloniki that is of particular interest is the Axios housing complex. In this settlement, one can discern the evolution of the issue of social housing in Greece during the first two post-war decades. In this case, older typologies designed between 1954 and 1965 are concomitantly used with the newer building types designed afterward. The twelve-story tower typology (12001) was uniquely used in this settlement and has not been ever repeated since.

The Axios housing complex is located in the Xirokrini area of Thessaloniki, near the passenger train station. It occupies an area of about 5.5 hectares and the percentage of area built and covered by residential buildings is about 31%. In total, in the complex, there are 58 building units that are arranged in parallel rows. In them, there are 722 apartments with an average area of 65 m^2. For the construction of the complex, the total area was divided into six sectors, five residential sectors and one sector where the services of the settlement are included. The whole area is divided by the service zone. Inside the complex, various green areas, small squares and playgrounds have been created among the buildings, whereas in the central area schools and a church were constructed.

In the remaining five sectors there is a combination of six different building typologies. These include four older types that are strongly influenced by the buildings that Aris Konstantinidis had designed during the time he was in charge of the design department of the organization. Added to the above is type 4004 which was designed after 1965 and is a variant of the typical urban apartment building (polykatoikia). This type is not of particular architectural interest. The last typology used in the settlement is type 12001, that of the twelve-story residential tower. Examining in detail the building typologies, the first four types include four-story apartment buildings. These types stand out morphologically as they have the reinforced concrete structural system visible on their facades. Type A and type B consist of two apartments of 65 m^2 per floor. Each residence includes two sleeping areas, a small living room, a kitchen and a bathroom. In the center of the floor plan lies the staircase. There is no elevator in these buildings, similarly to all buildings up to four floors. Type C has emerged as a variation of the two previous types in an attempt to add one more sleeping area. The layout of the standard floor plan consists in this case of two apartments per floor with an area of approximately 85 m^2. The fourth type of building, which also has the same morphological characteristics as the previous ones, includes three apartments per floor with an area of approximately 45 m^2. In the center of the typical floor plan lies the staircase. The basic apartment unit in this type consists of one sleeping area, a living room, a kitchen and a bathroom.

The two typologies of buildings that follow were designed after 1965 and have characteristics similar to the typical middle-class apartment building of Greek cities (polykatoikia). Type 4004 is also a four-story apartment building with two apartments per floor, one apartment of an area of 65 m^2 and one apartment of an area of 45 m^2. The rooms they include are similar to the previous types. The larger apartment has two bedrooms, a living room, a kitchen and a bathroom, while the smaller apartment has one bedroom, a living room, a kitchen and a bathroom. Characteristic of these buildings form is the continuous balcony on the main facade. The last type, which as mentioned above was built only in this settlement, is that of the twelve-story tower (type 12001). In total, four towers have been built in this complex. Two independent ones and two with a common base on the ground floor. The latter, in fact, due to uneven terrain, are thirteen stories high. The typical floor plan includes three apartments. Two apartments of approximately 85–90 m^2 and one apartment of approximately 100 m^2. These apartments, compared to the previously described, except for the larger area, have variations in the floor plan layout, which serves a more urban lifestyle. Contrary to the other forms, in this type, each residence includes an entrance hall, two bedrooms, a kitchen, a bathroom and a larger living area. Moreover, all rooms of the apartments have access to a balcony.

Examining the construction of the buildings, one can observe that their load-bearing system is made from reinforced concrete. The construction was done according to the Concrete Regulation at the time, and the first Greek Antiseismic Regulation was established in 1959. The structural elements made from reinforced concrete are small in the case of four-story apartment buildings, while the use of reinforced concrete antiseismic walls is limited. On the contrary, the structural system of the twelve-story towers is particularly dense and the square pillars of the ground floor have a side dimension of almost 45 cm. Moreover, in these buildings, there are corner antiseismic walls in the core of the staircase. For the brickwalls, different brick structures were used, a combination of headers and stretchers for the external masonry and stretchers for the internal partitions. There was no provision for thermal insulation as the thermal insulation regulation was instituted later. Regarding the electromechanical installations, all buildings had a complete electrical installation. In the lower buildings for the heating, there was provision for a chimney in each room as it was estimated that the tenants would use stoves. Only the twelve-story towers had a central heating installation with an oil burner from the beginning. Moreover, only in these buildings elevators were installed originally, two in each building. Of note, in the twelve-story towers, there is a second fire safety staircase, which was added later.

The frames in all the buildings were originally wooden with a single pane of glass, while on the outside there were protective shutters. Differences between the lower buildings and the twelve-story towers are also found at the level of final coatings. In the lower buildings, the use of terrazzo flooring in all areas was common, while, on the contrary, wooden glued floors were used in the sleeping and living rooms of the twelve-story apartments.

The Current Condition of the Settlements and Their Potential Prospects

The preservation state of the majority of mass housing complexes in Greece is poor to moderate. In the case of the Axios housing complex, the main problems at the level of urban space are related to the lack of accessibility. The fragmentation of the public space by various configurations, flowerbeds and paths, the encroachment of public area for the expansion of residential space, the creation of private courtyards, and the obstacles placed by the residents in their effort to strengthen their sense of security, have as a result the degradation of the settlement's image. The absence of use of certain spaces around the clock intensifies the feeling of insecurity for the pedestrian. This fact, combined with the insufficient lighting, degrades the image of the public space of the complex.

The condition of the buildings is moderate to poor for some building complexes. The preservation state of the two newest types (4004 and 12001) is better than the rest. The ownership status of the apartments plays an important role in this. Most of the apartments of these two types are owned by private individuals, while many of the apartments in the other building units are managed by the Ministry of Welfare and granted for use by financially weak people, immigrants and drug addicts. Most building preservation issues are related either to the absence of maintenance and repair or to incompatible arbitrary interventions by residents. Consequently, peeling of exterior coatings and moisture issues from roofs are very common. There is no organized program of restoration and energy efficiency upgrading for the buildings, while the procedures performed by the owners-occupants of the houses are fragmentary and alter the character of the buildings. The interventions are divided into two categories: those aiming at improving the energy behavior and comfort conditions of the apartments and those aiming at increasing the area of the residences at the expense of the exterior spaces. In order to improve the energy efficiency behavior of the apartments, the residents proceed mainly to change the external doors and windows, to installation of external thermal insulation and cooling-heating systems in the balcony. The absence of any planning and overall intervention program is a factor that degrades the image of the buildings and the complex as a whole.

The mass housing complexes constitute an important chapter of the modern Greek post-war architectural history. In several cases, they consist of the work of an important distinguished Greek architect. Furthermore, due to the urban sprawl of Greek cities, most complexes are now integrated into a very dense urban area with minimal green spaces. On the contrary, the building density of mass housing settlements is relatively small. In parallel, the correct orientation of the building units can provide a good living environment. The small area of the residential spaces and the accessibility problems due to the absence of an elevator, with the appropriate design can be resolved. Similarly, with appropriately organized interventions, the overall energy efficiency behavior of the complexes can be improved. In the context of the master thesis of the Interdepartmental master's Program "Protection, Conservation and Restoration of Cultural Monuments" of the Faculty of Engineering of Aristotle University of Thessaloniki, the prospects for the upgrading and rehabilitation of the Axios housing complex in Thessaloniki were explored. Sakkos (2018) As part of this work, the public space was redesigned in a way to limit its fragmentation. By designing two smaller squares and simultaneously adding commercial uses to them, an attempt is made to strengthen the sense of neighborhood and improve the feeling of safety for pedestrians. Finally, all residential units were redesigned according to the current functional standards (Figs. 14.4 and 14.5).

These proposals can be applied correspondingly to most mass housing complexes as the typologies of the buildings are repeated. However, small ownership makes any intervention difficult and renders an overall framework of action necessary, in order to achieve the restoration and adaptation of mass housing complexes to new functional data, without the risk of altering their character.

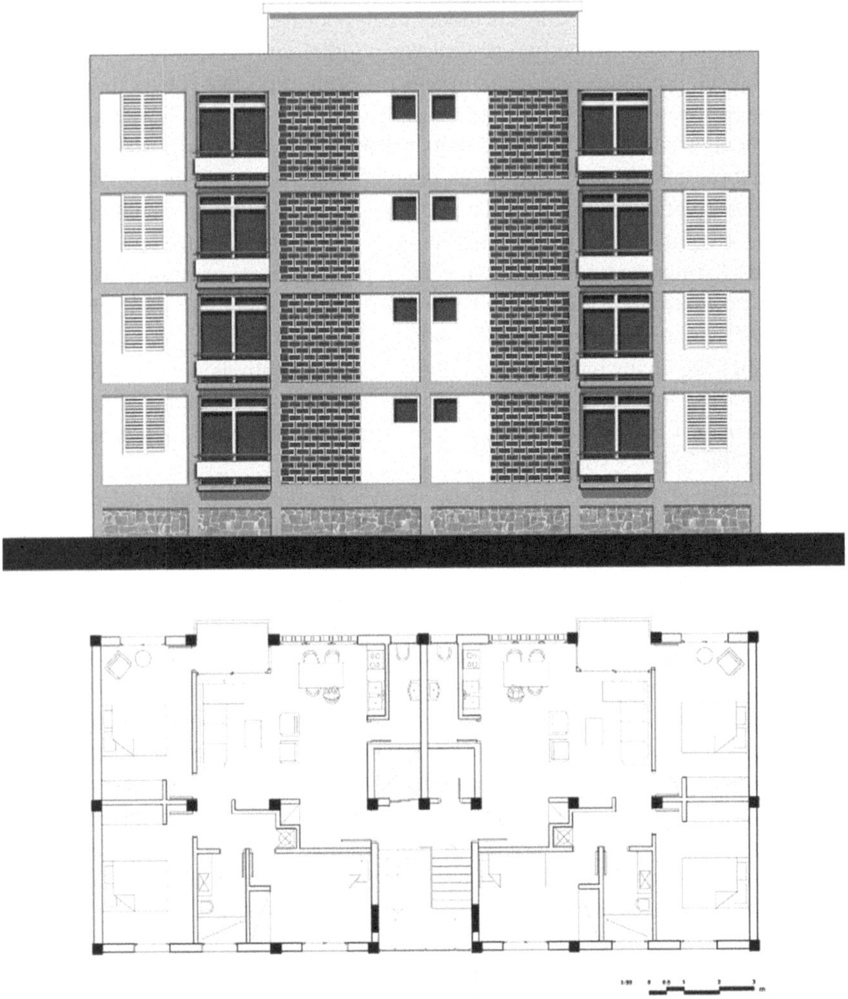

Fig. 14.4 Floor plan and façade of Type C, restoration proposal (Sakkos D. 2018. Master Thesis *Restoration and rehabilitation of Axios Housing Complex*. Thessaloniki. A.U.Th.)

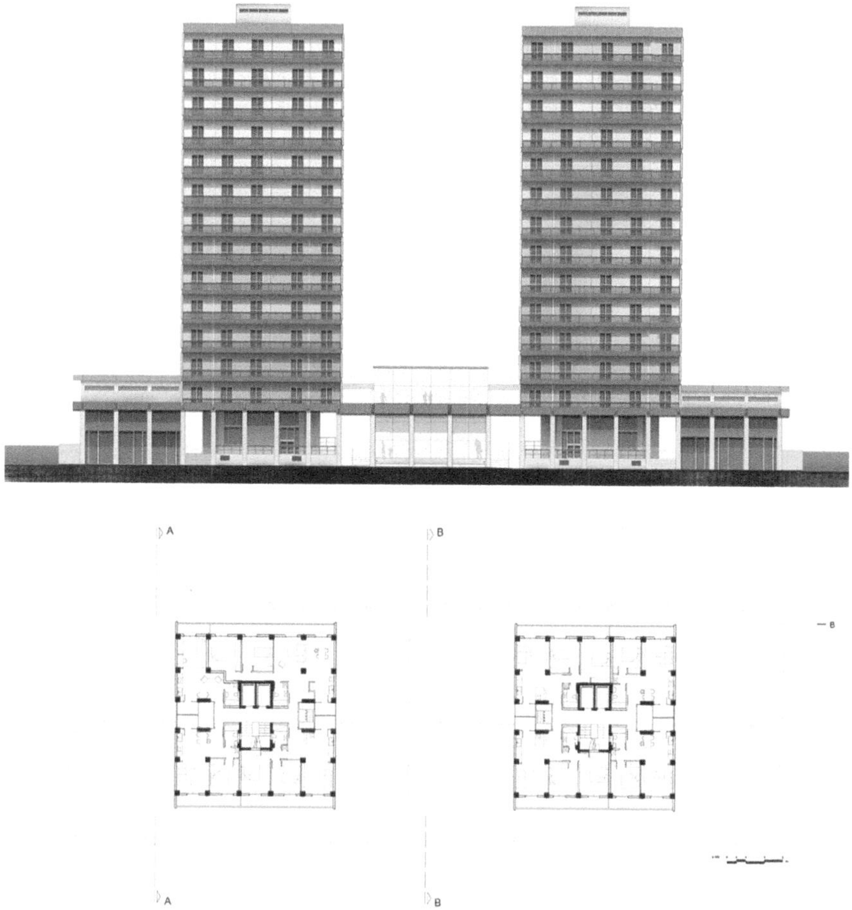

Fig. 14.5 Floor plan and façade of Type 12001, restoration proposal (Sakkos D. 2018. Master Thesis *Restoration and rehabilitation of Axios Housing Complex*. Thessaloniki. A.U.Th.)

References

Alifragkis, S. and Kalfa, K. 2020. *Antiparochi—A Short Introduction*. Athens: National Technical University of Athens.

Gouvousi S. 2011. *Housing Complexes in Greece and their prospects*, Phd thesis. Thessaloniki: A.U.Th. (in Greek).

Karanikas, I. 1986. *Worker's Housing Organization. The difficult road for the architecture of social housing*. Architectural Topics. Athens: Doumani publications (in Greek).

Sakkos D. 2018. *Restoration and rehabilitation of Axios Housing Complex*, Master thesis. Thessaloniki: A.U.Th. (in Greek).

Chapter 15
Ruins and Abandoned Places. Symbolic, Functional and Material Value. Research Methods and Conservation Strategies

Emanuela Sorbo and Gianluca Spironelli

Abstract Ruins and abandoned sites are a palimpsest of historical layers that embody material heritage, cultural identity, and collective memory. As John Ruskin pointed out, ruins function as mnemonic devices mediating historical consciousness transmission. At the same time, Georg Simmel theorised their capacity to expose the dialectical tension between human intervention and the inexorable passage of time. In contemporary society, abandoned places emerge as aesthetic relics and reflections of social, economic, and cultural transformations. The research conducted within the MATESCA Research Cluster at Università Iuav di Venezia explores such spaces' symbolic, functional, and material value, employing interdisciplinary methodologies that integrate historical analysis, digital tools, and conservation strategies. Through case studies—including the Guerra Gregorj Kiln, the unfinished Church of San Michele Arcangelo in Brendola, and the former Psychiatric Hospital in Rovigo—the study examines how public engagement and heritage awareness can bridge the gap between ruins and contemporary communities. The research contributes to the contemporary perspective on cultural heritage conservation studies, promoting the acknowledgement of abandoned spaces as a lens to investigate society as dynamic cultural vessels that shape contemporary socio-cultural narratives and collective memory.

This contribution was written by the authors with unity of purpose, however, the paragraph "Ruins" is to be ascribed to Emanuela Sorbo, the paragraph "The functional Ruin. Guerra Gregorj Kilns" to Gianluca Spironelli, "The unfinished. San Michele Arcangelo Church" to Emanuela Sorbo and Gianluca Spironelli, the paragraphs "Heritage Awareness and Right to Cultural Heritage", "The abandonment. Former Psychiatric hospital in Rovigo" and "Revealing Memories" to Emanuela Sorbo.

E. Sorbo (✉) · G. Spironelli
Department of Architecture and Arts, Università Iuav Di Venezia, Venice, Italy
e-mail: esorbo@iuav.it

G. Spironelli
e-mail: gspironelli@iuav.it

M. Loren-Méndez et al. (eds.), *Sustainable Heritage Management - Design: Notions, Methods and Techniques*, https://doi.org/10.1007/978-3-031-81970-4_15

Keywords Forgotten heritage · Cultural and collective memory · Tangible and intangible heritage · Public engagement · Heritage Building Information Modelling (HBIM)

Introduction. Ruins

A ruin is a device that can act as a filter of memory, as John Ruskin wrote (1889, p. 193). However, at the same time, the ruin reveals man's intense bond with nature—understood as the force of time (Simmel 1959)—overcoming a dimension dear to the picturesque (Dillon 2011). In the contemporary world, the creative action inspired by the dimension of time (Yourcenar 1992) is linked to the acceleration of social evolution, punctuated by the emergency impacts of economic crises, wartime conflict and by rapid changes in social customs. The theme of the ruin as an "image of time" is, therefore, added that of having the role of grasping dynamics and processes of social and economic development. In this perspective, contemporary society incessantly produces ruins/rubble, as Marc Augè (2003) described (prophetically). Abandoned places may (or not) have the value of a ruin (in the classical sense), but above all, they condense a different type of value, namely the need and/or the disappearance of a social role for some architectural typologies (such as ecclesiastical heritage, military forts, barracks, kilns, spinning mills—just to name a few). There is, therefore, a process of collecting images (even a dystopian one) of the city, to which the ruins, understood as consolidated places of an aesthetic experience, are added to other places, fragmented by society, in which it is not the pure time of Simmel's nature to act as much as social memory, in that they are places of collective memory (Nora 1984). Despite the social rejection that underlies them, abandoned places retain a patrimonial dimension linked to a cultural value that can be symbolic, functional or material (Nora 1989; Sorbo 2017a) in relation to the causes of abandonment (history), the dimension of memory (*forgotten heritage*) and the construction processes (matter).

The condition of abandonment projects them, in fact, in a peculiar research perspective because, despite being in many cases declared assets of considerable cultural interest, therefore, recognised in a collective (and normative) process, the abandonment to which they are subject inserts them, on a semantic level, in a different framework, linked to a distance from the communities in which they are inserted.

With these premises, the MATESCA Research Cluster of the Università Iuav di Venezia—coordinated by Emanuela Sorbo and Fabrizio Antonelli—conducts an experimental investigation on abandoned places of cultural interest that intersects research with the tools of the third university mission and has a systemic approach to the reconnection between ruin and society (therefore, between memory and matter) through research activities and actions of public engagement.

To date, the MATESCA Research Cluster has launched framework agreements and commissioned research (scientific responsibility of Emanuela Sorbo) with the Superintendence of Archaeology, Fine Arts and Landscape for the provinces of Verona,

Rovigo, Vicenza (from here on SABAP-Vr), the Municipality of Verona, the Municipality of Vicenza, the Municipality of Brendola, carrying out studies and research on important monuments such as the Valmarana Salvi Gardens and the annexed architectures, the Teatro Olimpico in Vicenza, the Arena di Verona, the Teatro Romano in Verona, the Castelvecchio Museum, the *Incompiuta (Unfinished)* church in Brendola, the Barchesse of the Villa da Porto in Montorso Vicentino, the church and buildings belonging to the psychiatric hospital in Rovigo, the church of San Pietro Martire in Verona, The Church of San Zeno Minore in Verona, the *Fabbrica Alta* building and the former *Asilo Rossi* in Schio. In each of the cases mentioned, the research projects carry out enhancement activities through research aimed at defining preliminary knowledge through Iuav scientific Laboratories, and the identification of conservation strategies, thanks to constant dialogue with the Superintendency, and dissemination and *public engagement* actions, involving communities through collaboration with public bodies, such as the Municipalities, and the media, both print and television.

The importance and topicality of the theme have become central and emerging in Italy also starting from the census promoted in 2019 by the Directorate General for Archaeology, Fine Arts and Landscape (DG ABAP), which launched a survey of the abandoned cultural heritage present on the Italian territory to valorise them and restore their accessibility to the public (Ministry of Culture 2023).

The Superintendencies implemented the recognition campaigns in the relevant territories of competence. SABAP-Vr, with the collaboration of the researchers of the MATESCA Cluster (Elisa Brusegan), participated in the programme with 72 sites, of which 33 in the province of Verona, 25 in the province of Vicenza and 14 in that of Rovigo.

The direct observation of the assets, assisted by bibliographic and archival research, has become the basis for developing a material and dimensional description of the main elements that characterise the architecture. In particular, it has been directed towards some fundamental parameters, grouped into three categories:

- the construction and material characteristics of the architectural envelope and its internal partitions, divided into foundations, walls, floors, roofs, connections, door and window fixtures;
- decorative apparatus referring to ceilings, walls and floors;
- any technological systems that may be present.

These parameters were described specifically through a photographic dossier, accompanied by brief descriptions, which could provide an instant photograph of the assessment of the conservation status of the surveyed asset.

The general outcome of the survey was merged, for each of the abandoned cultural assets, in a summarised sheet, structured in a database of information including:

- the historical name of the property;
- The historical and artistic description;
- The geographical position;
- The cadastral data;

- The owner;
- The protection regime;
- The chronological context;
- The state of conservation;
- The intended use;
- The type;
- The description of past intervention projects and past/ongoing enhancement activities;
- A photographic summary.

Looking at the abandoned cultural assets surveyed in the provinces of Verona, Rovigo and Vicenza (Municipality of Verona 2023), there are over 21 architectures with military function (Bozzetto 1990; Jacobacci 1981; Perbellini 1981), thus representing the most widespread architectural type (almost 30% of the total).

This picture represents a snapshot of the heterogeneity of abandoned places, which is why the research activities of the MATESCA Cluster have had as a guideline a process of identification of abandoned heritage sites in three thematic areas related to the identity-authenticity couple based on the theme of the place of memory by Nora (1989, p. 19):

- the functional identity;
- the symbolic identity;
- the material identity/authenticity.

To which correspond different analysis and research themes, such as:

- the ruin;
- the unfinished;
- the abandoned.

The matrix, thus composed, allows a frame of research scenarios and project horizons involving different theoretical tools to find a methodology of investigation preparatory to the conservation project for abandoned places of considerable cultural interest. It is accomplished through experimentation on case studies of interest of administrations, foundations, and public and private owners in which the central theme is the relationship between the social dimension of neglect and the collective dimension of conservation.

Heritage Awareness and Right to Cultural Heritage

In art. 5 of the Faro Convention (Council of Europe 2005), the *right to cultural heritage* for communities is one of the objectives of the protection strategies for Cultural Heritage (Manacorda 2018, p. 21). In particular, it defines enhancement activities in: *identification, study, analysis, interpretation, preservation* and *conservation*. The Convention delineates, therefore, both a right on the part of the communities to the use (and knowledge) of the Cultural Heritage, and a duty of dissemination. Only through an operation of enhancement of *heritage awareness* it is possible to promote the *public engagement* of the communities and, therefore, strengthen the right/duty process to the cultural heritage. On the other hand, the protection of Cultural Heritage is considered a determining factor for achieving some objectives, present in the UN 2030 agenda and referred to in the Faro Convention in art. 5, such as: sustainable development, cultural diversity and contemporary creativity. In particular, art. 9 defines sustainable use of Cultural Heritage as an action that promotes a clear understanding of the underlying cultural values, considering them as the basis for the evolution and change of the cultural assets. The abandoned heritage can be read as a resource of the territories, but also as a trace through which the identity of the places can be read. Therefore, the ability to understand and transmit cultural values is at the base of the action of *heritage awareness*. It seems more difficult to investigate what the tools for tracing these values may be, since they include some key concepts that embody two different dimensions: a tangible one, which dialogues with and analyses the cultural assets and the communities that produced them (their being places from an anthropological point of view), and an intangible one that includes themes such as collective cultural memory and the community value of abandoned places (Fiorani et al. 2017). The subjects of these actions: the communities and *heritage professionals* dialogue with tangible and intangible values, using as tools, at both extremes, the high quality of the project and the engagement process. (Dimitrova et al. 2019).

Such research is rooted in the concepts of identity and authenticity. The Nara Document on Authenticity (ICOMOS 1994) and Nara + 20: on heritage practices, cultural values, and authenticity, (ICOMOS 2014) highlight elements useful for densifying this notion in a dual form: tangible (referring to the design, materials, use, function, traditions, techniques, context) and intangible (such as reference to *spirit and feeling*).

The double polarity of the role of *cultural value* and *authenticity* (as a place of conservation in terms of *common cultural heritage*) places the abandoned places, in the perspective of a challenge for redesigning society.

Abandonment, therefore, exerts a reversal on the Cultural Heritage since the detachment from the daily life of communities underlies a process inverse to the recognition of value; it takes them outside the urban dimension of the city and excludes them from the horizon of society, as well as going beyond the dimension of authenticity.

Exercising the reading of the causes of the abandonment of the cultural heritage in civil society from recent economic (or pandemic) crises allows us to unveil the impacts and the anthropological dimension of the relationship with conservation, or in other words, with time. Discovering the tangible and intangible limits of this framework of relationships means following the boundary space between matter and history, between the dimension of abandonment and the dimension of the unfinished.

The Functional Ruin. Guerra Gregorj Kilns

Founded by Vincenzo Guerra in 1840 in the hamlet of Sant'Antonino di Treviso, Guerra Gregorj Kilns is among the first proto-industrial plants to be established in Veneto. At the territorial level, the plant was the protagonist of the process of unification and growth of the country, representing an element of conjunction between rural Italy and urban transformations resulting from the industrial development of the cities that took place on the basis of European experiences (Mancuso 1990).

As evidence of the central role the kilns played in building within this territory, with its production of bricks, is the vicissitude linked to the reconstruction following the collapse of the Campanile di San Marco (San Marco bell tower) in Venice on 14 July 1902.

In the technical and cultural debate established following the tragic event regarding the shapes, dimensions and colouring of the elements to be used, it became necessary—as Gregorio Gregorj himself explains (1904, p. 3) —to involve several kilns for the supply of the materials necessary for the validity of a philological and structural reconstruction (as it was and where it was) of the bell tower. In this context, part of the supply for the new brick elements would have been entrusted to the Guerra Gregorj Kilns.

This story testifies to the final role that the Guerra Gregorj Kilns assumed at the beginning of the twentieth century, associated with an experimental vision witnessed by the multifaceted composition of the complex that, in addition to being used as kilns for firing bricks and ceramics in the building and artistic field, seasonally hosted the important moments of agricultural activity, consolidating a link between work and society, of which Gregorj was an early believer.

In framing the role of the artistic production of the kilns, among the many artists emerges the figure of Arturo Martini, who, in Gregorio Gregorj, found the mecenate who supported him in the first steps of his production (Gregorj 2001).

The architectural character of the buildings is identified both by the use of traditional local construction techniques and exposed brick walls and by the introduction of experimental elements in the buildings, such as the extensive use of *mattoni smaltati a gran fuco* (glazed bricks), an original patent of the furnace of 1895, which combines the load-bearing capacity of the brick with ceramic decorations, through a surface treatment made with a coloured enamelled finish (Dal Pos and Spironelli 2018a).

In 1965, in the face of technical transformations in the production of bricks and ceramics, the kiln was definitively closed due to the absence of the capital necessary for updating the production processes.

From that moment on, the buildings of the complex have been subject to a process that would lead to the abandonment of the area. The absence of a maintenance plan, and the resources necessary for implementing it, led to the collapse of the roof structures of the kiln's building and of the horizontal planes, exposing the underlying structures to weathering.

On 10 June 1987, the kilns site was declared a place of cultural interest under Law no. 1089 of 1 June 1939 (Protection of Objects of Artistic or Historical Interest) as a "precious testimony of an old world in the process of disappearing that deserves to be preserved and protected".

Between 2017 and 2018, starting from the Master's degree thesis in Architecture "Un-finished. Former Guerra-Gregorj Kilns. The 'unfinished' as a limit in the project for the ruins" (by Laura Dal Pos and Gianluca Spironelli, supervisor Emanuela Sorbo), the tools for an approach to preliminary knowledge in the context of research on abandoned places had been investigated. This methodology included:

- the execution of an extensive survey campaign (both photogrammetric and topographical) supported by a direct observation activity;
- the cataloguing of construction techniques and materials used in the buildings;
- the elaboration of synoptic records aimed at identifying construction techniques and their conservation status, based on regulatory references and national guidelines such as the AeDES 2014 Manual and the MiBACT 2011 Synoptic Intervention Record.

This approach has made it possible to reach, through a diachronic reading of the complex and its buildings, an in-depth knowledge of their state of conservation, typological and constructive characteristics, and elements of vulnerability and damage (Dal Pos and Spironelli 2018b).

Concerning the condition of the place and its urban isolation, some actions to promote knowledge about it have been identified (Faro Convention, Council of Europe 2005, Art. 1) through *public engagement* actions that involve the Università Iuav di Venezia, the Superintendency of Archaeology, Fine Arts and Landscape for the Municipality of Venice and Laguna (SABAP-VE-MET), the Istituto Superiore per la Conservazione ed il Restauro (ICR), the Civic Museum of Treviso and the owners. This path of analysis and study of Cultural Heritage has focused on a process of *Heritage awareness* at various levels.

In the exhibition entitled: *"Il Giudizio di Salomone". Arturo Martini in Fornace negli spazi della Ex Fornace Guerra Gregorj,* (The Judgement of Solomon. Arturo Martini in the Kilns in the spaces of the former Guerra Gregorj Kilns) realised in collaboration with the openings of the FAI (Fondo Ambiente Italiano—Italian Environmental Fund) days, proposed in March 2019, promoting extensive involvement by the local community, with a total of 3000 visitors and a large press review on the media.

In the seminar entitled *Fornace Guerra Gregorj. Arte e Construction (Guerra Gregorj Kilns, Art and Construction),* the representatives of local and national administrations and the owner were involved in a *workshop* session (held as part of the educational activities that saw the kilns as an application case study of the Restoration course in the academic year 2018–2019) making it possible to explore visionary project scenarios, in a process of *cultural sharing* that otherwise would have been unthinkable. The process of knowledge transfer activated has made it possible to cross the limit of the link between architecture and community, using the project as a verification strategy (The Burra Charter, ICOMOS 1979; Québec Declaration,

ICOMOS, 2008) and has allowed to imagine the Kiln as a story in fragments of visions of reality that, working on the concept of utopia, reality and dystopia, have promoted the knowledge of the Kiln and developed possible scenarios of reuse (Fig. 15.1).

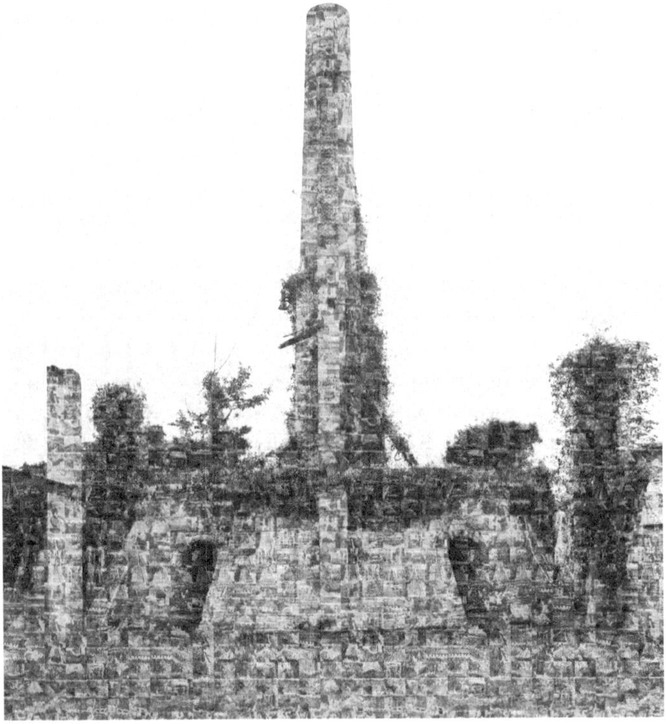

Fig. 15.1 Guerra Gregorj kilns. Reconstruction of the Hoffmann kiln's image obtained through a collage of the pictures taken during the survey on site. Edited by Laura Dal Pos and Gianluca Spironelli, 2018

The Unfinished. San Michele Arcangelo Church in Brendola

The process of building a new church for the community of Brendola dedicated to San Michele Arcangelo (Cecchin 1937) began in the 1930s, based on the project by the engineer-architect Fausto Franco (1899–1968) (Battista 2017; Liguori 2011; Spada 2017).

The construction site was interrupted in the 1940s due to the impossibility of finding the resources necessary for its completion, leaving the church as a landscape ruin nowadays recognised by the community as *Incompiuta (Unfinished)* (Fig. 15.2).

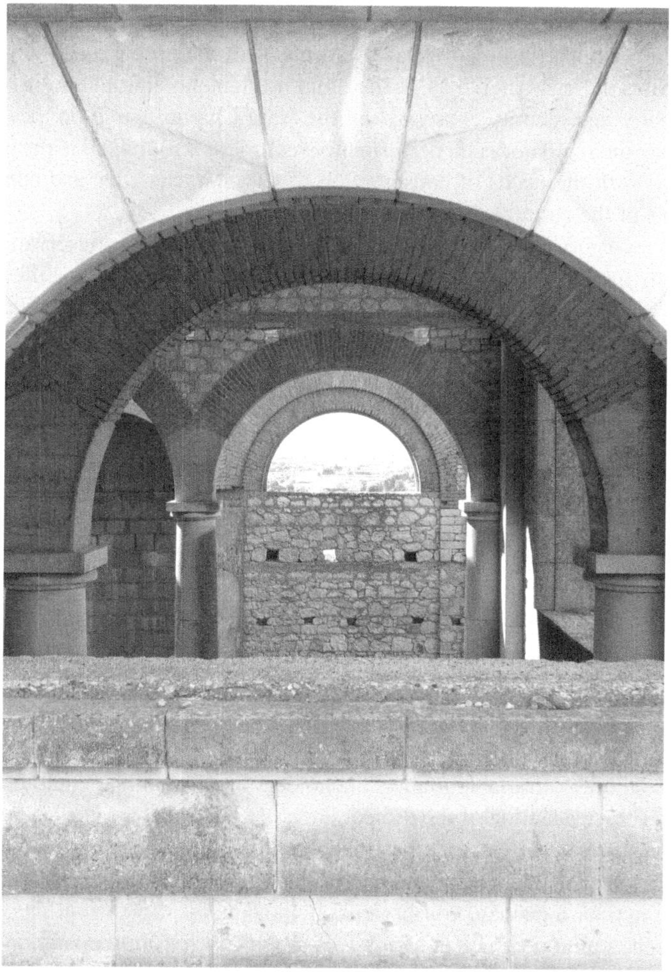

Fig. 15.2 The *Unfinished* church of San Michele Arcangelo in Brendola, west nave. Photo by Università Iuav di Venezia. Brendola, 2021

In 2021, following the report of the municipal administration regarding the precarious conservation status of the church, a multi-year research programme was promoted for the *Unfinished* church of Brendola between the Università Iuav di Venezia, the MATESCA Research Cluster, the Municipality of Brendola and the SABAP-Vr.

The research programme was based on the need to identify conservation and enhancement strategies through the construction of a knowledge system that concerned both the tangible aspects (linked to the architectural and material consistency of the church), as well as the intangible ones related to the social implications that the construction of the church has entailed for the community, which recognises in it a cultural value as "common property" (Settis 2014). The conservation needs, linked to the condition of abandonment and incompleteness, have conveyed the need to identify operational protocols that could measure themselves, on the one hand, by identifying the characteristics of the church (elements that define its character of authenticity and identity expressed by the Nara Charters), and on the other, the need to relate the regulatory and performance requirements—linked to the public use of the site—with the needs of conservation of the characteristics and construction singularities of the cultural property.

The research carried out a careful evaluation of the church's conservation status, supported by the development of a digital knowledge system based on the artefact's conditions (Sorbo and Spironelli 2021a) (Sorbo and Spironelli 2021b) (Sorbo 2023) through the creation of an informative model (HBim) (Fig. 15.3).

The model had the objective of structuring the knowledge information (unpublished)—acquired by the research group—within an integrated (and implementable) system that, in a logic of data systematisation, had built a digital tool for the management, protection and enhancement of the asset with a view to possible conservation strategies (Sorbo and Spironelli 2021c) (Sorbo 2023).

The archaeological knowledge has identified the main elements of vulnerability and the architectural themes to which the conservation project should have responded: the treatment of the wall ridges, the disintegration of the brick elements, the disposal of rainwater, the conservation and treatment of the surfaces, the expulsion of material from the wall coverings, the erosion of the joints, spontaneous vegetation and weeds, moderate architectural reintegrations such as flooring, roofing structures, horizontal and vertical accessibility and the urban dimension.

The research worked for the project through the definition of an abacus of application case studies—available to the stakeholders involved—that could provide the elements for the evaluation of the project interventions.

This approach to the research has made it possible to work on the (intangible) boundary between enhancement and the collective dimension of memory and, at the same time, to trace a guide to conservation in the tangible dimension. The operation resulted in the cultural choice to avoid demolitions by selecting operations of local reinforcement of the constructive elements.

At the end of the path of knowledge, the research activities led to the approval by the local authorities (SABAP-Vr) of a *Scheda Tecnica* (technical data sheet), preliminary elaboration of the final project (Fig. 15.4), presented to the local community

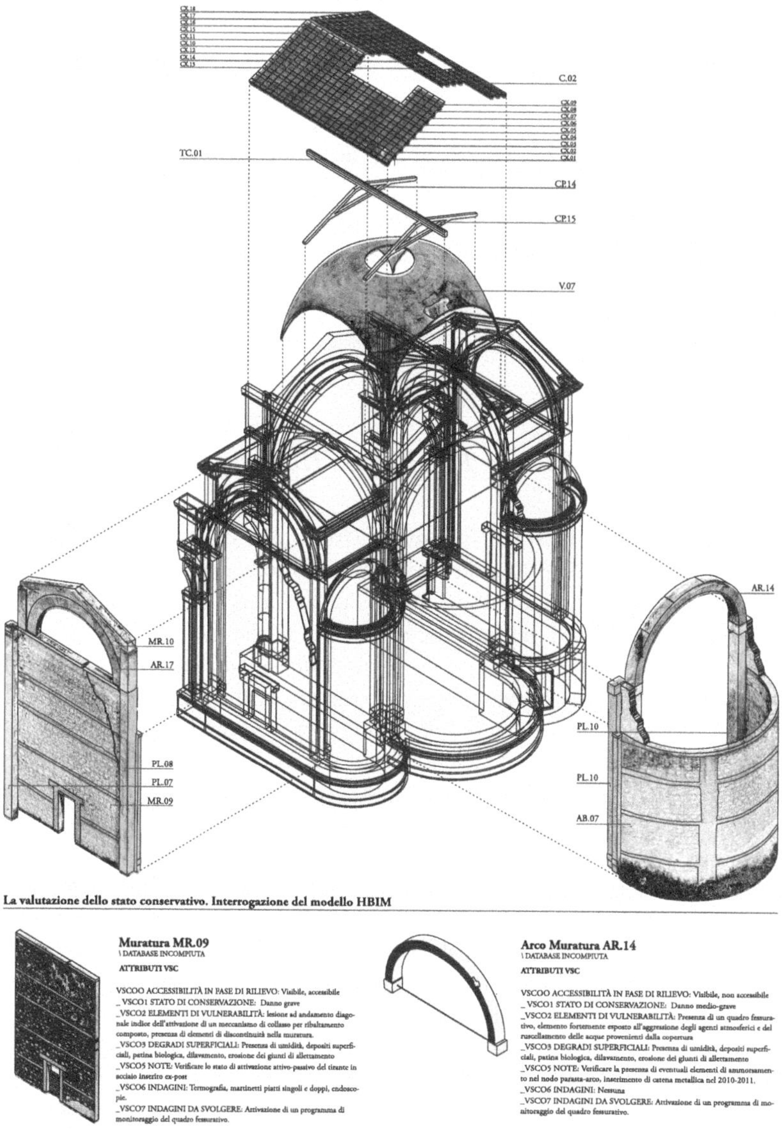

Fig. 15.3 The *Unfinished* church of San Michele Arcangelo in Brendola. Heritage Building Information Modelling (HBIM). Edited by Università Iuav di Venezia, 2023

within the seminar *Il limite della rovina. Processi metodologici di condivisione per la conservazione e valorizzazione della Chiesa 'Incompiuta' di Brendola* (The limit of the ruin. Methodological processes of sharing for the conservation and enhancement of the 'Unfinished' Church of Brendola) which involved: the municipal administration, the SABAP-Vr, the Veneto Region, the team of technicians appointed by the Municipality, experts in the field of European funding, the press and local broadcasters.

A knowledge path based on an applicative and design nature, within which it was possible to outline a preparatory *forma mentis* for the project through a "monographic study of the work" and the "construction of a systematic inventory of the materials and knowledge concerning the built environment as an object of protection" (Reichlin 2011, pp. 11–29).

The Abandonment. Former Psychiatric Hospital in Rovigo

The former Psychiatric Hospital of Rovigo (from here on OPP) (Fig. 15.5) was built from a project of 1906 as a structure intended to accommodate the poor and mentally ill in implementation of Law no. 36 "on asylums and psychiatric patients" of 1904, which determines the forced confinement of the mentally ill. The project involved the use of an architectural type—*no restraint*—that Augusto Tamburini had experimented with in the "observations" department of the Reggio Emilia asylum. The project plan is organised according to a horseshoe scheme to allow the visual control of the pavilions (24) and the absence of fences (Sorbo 2017b). It includes an extension of 130,000 square metres and 84,000 square metres of agricultural area.

In 1978, the nationwide implementation of Law no. 180 (Basaglia Law) ordered the closure of asylums and the regulation of compulsory health treatment with the establishment of mental hygiene services (Sorbo 2019). Part of the complex remained psychiatric hospital residue until 1995, when the "Transitional Residential Facilities to Territorial Structures" was established, only to be definitively abandoned in 1997. In 2004, the OPP was declared of considerable cultural interest pursuant to art. 12 of Legislative Decree 22 January 2004, no. 42 (Code of Cultural Heritage) "as it constitutes precious historical and architectural evidence of an asylum model typologically defined according to the reformist criteria that guided the technicians—based on psychiatric doctors—in the design of nursing homes at the beginning of the twentieth century".

Since 1997, the complex has been a place where time is suspended in its abandonment; it is a *lieu de mémoire*, because it condenses all the expressions of architecture as a narrative source: the immobile memories of the pavilions, the gardens, the greenery; the archival memories of the medical records, examination results, tags with patients' names; the mobile memories of the furniture, beds, typewriters, medical equipment and the invisible memories, the experience behind the walls, the unspeakable of human experience (Sorbo 2016b).

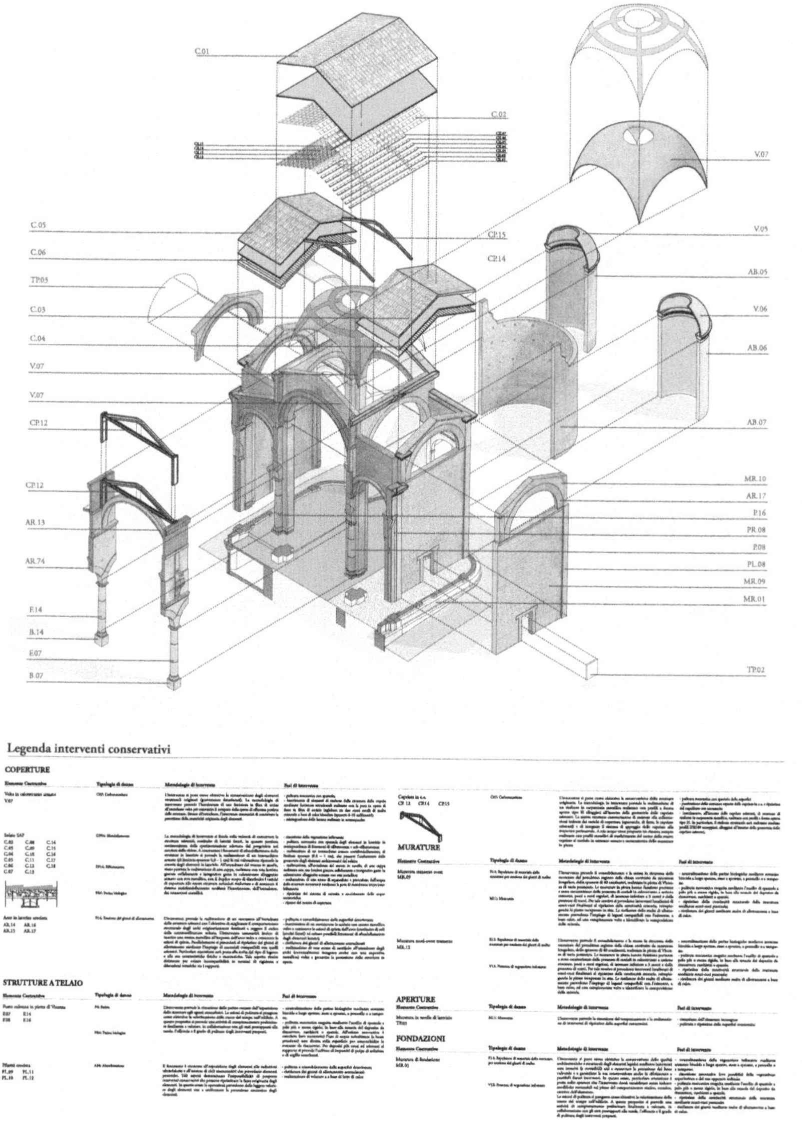

Fig. 15.4 The *Unfinished* church of San Michele Arcangelo in Brendola. Conservation project—transept. Edited by Università Iuav di Venezia, 2023

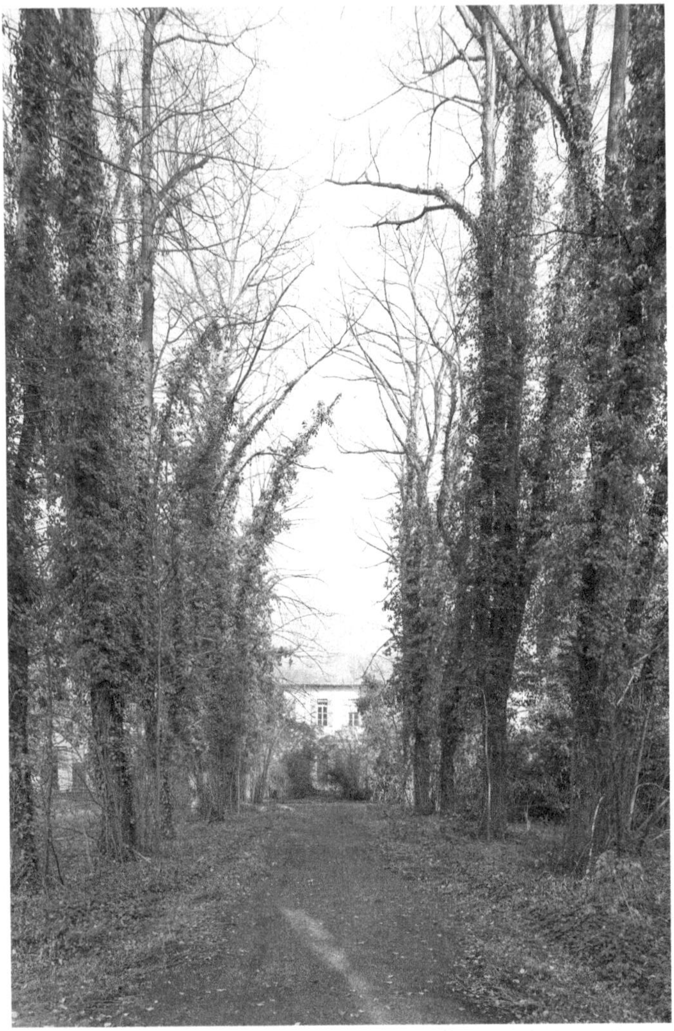

Fig. 15.5 The abandonement. Former psychiatric hospital in Rovigo. Entrance. Photo by Università Iuav di Venezia. Rovigo 2015

Remembering Simmel (1959), the OPP today is a sublime urban ruin where greenery has inserted itself into the pavilions, into things, which everywhere bring to mind a limit, the artifice. This limit, the preservation of the time of waiting, has become a new place to experiment with a research project. Here, the method—that is what could be the cognitive tools to investigate and define the identity of this temporality, which kneads through and transcends matter and memories of things—has constituted a new horizon. This approach has provided for institutional agreements with the Protection Bodies, the SABAP-Vr, public bodies such as the Rovigo Local

Unit no. 18 of the National Health Service, and the Province of Rovigo, the body that is the producer and conservator of the Historical Archive of the former OPP.

The documents were processed as *memories* of the place since the evocative capacity of *memory* could only be *revealed* through the detailed study of the *material* grafts, where a thorough study of every relationship between the construction techniques and the symbolic value hidden behind each element was carried out (Per un atlante degli ospedali psichiatrici (…) 1999).

In this relationship between collective memory (past and present) and history, the role that material plays becomes central to the conservation of the former OPP: so that it can continue to have that capacity for inspiration in the *Proustian* sense that grafts the process of recognition and/or reactivation of memory in the relationship that is established between material, construction and image. An ideal axis in which the poles of conservation between the invisible action of consciousness and the physical presence of architecture have (or lack) a form of correlation (Sorbo 2017c).

The laceration of material is the vehicle of this experience of oblivion, which must find a meeting point between *spectacularisation*, the artifice of reconstructing an experience, and the sense of conservation of the remaining traces (Musso 2014, pp. 95–100). The intangible value to which the former OPP refers is, therefore, different from the notion of monument, authenticity and tradition (The Nara Document on Authenticity, ICOMOS 1994; Nara + 20: on heritage practices, cultural values, and authenticity, ICOMOS 2014) because its cultural dimension lies in the sphere of memory as a reminder of an act of voluntary neglect. Its immaterial conservation is in this fluid absence that defines oblivion as an experience of recognition.

This condition is the immaterial value of the place, the epiphany that opens the possibility of witnessing a process, overcoming the concept of the monument to introduce it into the dimension of the story of the memorial.

A project for the former Psychiatric Hospital of Rovigo, which has been outlined experimentally by Università Iuav di Venezia and Sabap-Vr during 2025 as a case-study of Pavillon n. 4, summarises the material memories of the pavilions, the material elements of the boundary between the hospital and the city, the finishes, and the architectural lexicon of the minimal elements. They become the place of the project because it is in them, rather than in the monumental value of the elements, that the project's narrative lies.

Some elements of this lexicon could be listed as important elements of a design process, conceived as an architectural memorial of the places:

- the fixtures, the doors, acquire the value of the gaze between the patient and the outside world and as such, their conservation is the first moment of an action of recognition;
- the interior plasters where the frescoes of the patients are engraved;
- the decorative elements, indicators of a timid attempt to soften the buildings;
- the identity of the plans that define, through the use of C, H, L, T typologies, the role of architecture in a system of visual control, avoiding fragmentations that

would prevent the reading of this experience that is immaterial and material at the same time;
- the height of the rooms, because it is an expression of the hygienic reasoning of the early twentieth century (Morachiello 1982, p. 174).

In the same way, this strategy allows for the definition of other permanences in the recognition of the complex on a landscape scale:

- the semicircle type, among the few (if not the only one in Italy), that has not undergone transformations;
- the green system that defines in some places the light spot of the fences (where they were metal and where they were replaced with rows of tall trees), of the avenues for walks, of the presence of linden trees, of fruit trees, of this memory of construction of a landscape;
- the structure and axes of the relationships between the pavilions, the presence of the surrounding green areas as a memory of the agricultural colonies;
- the access along the river, a sign of the relationship with the surroundings, the material boundary of the immaterial meaning of isolation from the urban context.

This approach to material preservation allows the architectural discourse around mental illness to be retraced and recognised in places.

The process of recognition is thus the construction of a memorial project, to the extent that a conservation project is the expression of a story that is an outcome of history (Teyssot 1977, p. 31). In its being an architectural memorial of places, it can have a generative action that reveals the elements of history (Rella 1977a, b, p. 47).

Revealing Memories

In the examples shown, the centrality of the preliminary knowledge process for the restoration project emerges because, unlike the nineteenth-century experience of the approach to ruins, where the aesthetic value (connected to the presence of the nature of ancient ruin) is central, the perception of contemporary ruins, after the Second World War, is intimately linked to the evolution of society, in the balance between an action capable of revealing memory, enhancing history and preserving/reintegrating material (Sorbo 2016a).

The research and design strategies for ruins as a device for reading (also cultural) about abandonment are therefore inevitably linked to the interpretation of the architectural place in its being the outcome of a historical place "because, as Nietzsche had already said, there are no facts, but only interpretations (Focault: "everything is interpretation") that attempt to 'capture' silence in knowledge" (Rella 1977b, p. 8). This condition reveals the emotional and experiential charge of the encounter with different dimensions of knowledge: the "archaeological" dimension of ruins and the "social" dimension of abandonment.

In the realm of abandoned heritage, to consider the process of knowledge from an archaeological perspective - as revealed by the typologies described here - is, therefore, a guide to interpretation. It represents the connection to the conservation project that makes it possible to frame the strategies within a system of memory cultivation, reading the heterogeneity of the abandoned place as a form of heterotopia. And it is in this *heterotopia of interpretation* that we can push ourselves towards new kinds of imagination for abandoned places.

References

Augé, Marc. 2003. *Le Temps en ruines*. Paris: Galilée.

Battista, Giovanna. 2017. "La prima regola dell'arte" Il pensiero di Fausto attraverso alcuni suoi scritti. *In Tiepolo segreto*, ed. Guido Beltramini, and Fabrizio Magani, 31–36. Vicenza: Palladio museum.

Bozzetto, Lino Vittorio. 1990. L'architettura militare asburgica a Verona. In *Il Veneto e l'Austria. Vita e cultura artistica nelle città venete (1814–1866). Catalogo della mostra*, ed. Sergio Marinelli, Giuseppe Mazzariol and Fernando Mazzocca. Milan: Electa.

Cecchin, Francesco. 1937. *Il giubileo arcipretale nella parrocchia di San Michele arcangelo in Brendola: 19 Maggio 1912–19 Maggio1937*. Vicenza: Tipografia commerciale.

Council of Europe. 2005. Convention on the Value of Cultural Heritage for Society (Faro Convention). Faro.

Dal Pos, Laura and Gianluca Spironelli. 2018a. La Fornace Guerra-Gregori di Treviso: l'arte applicata all'industria. *CIL Costruire in Laterizio* 177: 66–74.

Dal Pos, Laura and Gianluca Spironelli. 2018b. Il ruolo dell'Archeologia Industriale nella Società Contemporanea: Metodologie di Indagine e Strategie di Riuso. In *ReUSO 2018: L'intreccio dei saperi per rispettare il passato interpretare il presente salvaguardare il futuro*, ed. Fabio Minutoli, 1433–1442. Rome: Gangemi Editore.

Dillon, Brian. 2011. *Ruins*. Cambridge: Whitechapel Gallery and MIT press.

Dimitrova, Elena, Marie-Laure Lavenir, Paul McMahon, Baiba Mūrniece, Stefano Francesco Musso, Gergely Nagy, Christoph Rauhut, Grellan D. Rourke, Erminia Sciacchitano and Bénédicte Selfslagh. 2019. *European quality principles for EU-funded interventions with potential impact upon cultural heritage*. Charenton-le-Pont: ICOMOS.

Fiorani, Donatella, Loughlin Kealy and Stefano Francesco Musso. 2017. *Conservation-Adaptation. Keeping alive the spirit of the place. Adaptive reuse of heritage with symbolic value*. Hasselt: EAAE European Association for Architectural Education.

Gregorj, Gregorio. 1904. *A proposito dei mattoni del Campanile di San Marco*. Treviso: Ed. Turazza.

Gregorj, Luisa. 2001. *Dietro Le quinte: artisti nella fornace Guerra-Gregorj a Treviso*. Susegana: Giacobino.

ICOMOS. 1979. The Burra Charter: The Australia ICOMOS Charter for Places of Cultural Significance. Burra.

ICOMOS. 1994. The Nara Document on Authenticity. Nara.

ICOMOS. 2008. The Québec Declaration on the Preservation of the Spirit of Place. Québec.

ICOMOS. 2014. Nara +20: On Heritage Practices, Cultural Values, and the Concept of Authenticity. Nara.

Jacobacci, Vittorio. 1981. *La piazzaforte di Verona sotto la dominazione austriaca 1814–1866*. Verona: Cassa di Risparmio di Verona, Vicenza e Belluno.

Liguori, Francesca Romana. 2011. Fausto Franco. In *Dizionario biografico dei soprintendenti architetti, 1904–1974*, ed. Ministero per i beni e le attività culturali, Direzione generale per

il paesaggio, le belle arti, l'architettura e l'arte contemporanee and Centro studi per la storia del lavoro e delle comunità territoriali, 275–284. Bologna: Bononia University Press.
Manacorda, Daniele. 2018. Il patrimonio culturale tra politica e società. *DigitCult|Scientific Journal on Digital Cultures* 3 (3): 21–40.
Mancuso, Franco. 1990. *Archeologia Industriale nel Veneto*. Cinisello Balsamo: Amilcare Pizzi Editore.
Ministry of Culture. 2023. *Census of Abandoned Cultural Heritage*.
Morachiello, Paolo. 1982. *I congegni delle istituzioni: ospedali, manicomi e carceri*. In *Italia moderna. Immagini e storia di un'identità nazionale*: *Dall'Unità al nuovo secolo: 1860–1900*, ed. Carlo Bertelli, and Omar Calabrese, 169–194. Milan: Electa.
Musso, Stefano Francesco. 2014. Conservazione, restauro e Patrimonio dell'Umanità. *Materiali e Strutture* 7: 95–100.
Municipality of Verona. 2023. *Verona fortificata (Fortified Verona)*. Municipality of Verona Archive.
Nora, Pierre. 1984. *Les Lieux de Mémoire*. Paris: Gallimard.
Nora, Pierre. 1989. Between Memory and History: Les Lieux de Mémoire. *Representations*. 26 Special Issue: Memory and Counter-Memory: 7–24.
Per un atlante degli ospedali psichiatrici pubblici in Italia. Censimento geografico, cronologico e tipologico degli "asili" pubblici italiani al 31 dicembre 1996, con aggiornamento al 31 ottobre 1998. 1999, ed. Domenico Luciani, Maria Federica Palestino, Francesca Rossi, Ida Frigo, Massimo Rossi and Teresa Marson. Treviso: Fondazione Benetton Studi Ricerche.
Perbellini, Gianni. 1981. Le fortificazioni austriache nel veronese. In *Vita veronese*, 1–2.
Reichlin, Bruno. 2011. Riflessioni sulla conservazione del patrimonio architettonico del XX secolo. Tra fare storia e fare progetto. In *Riuso del patrimonio architettonico*, ed. Bruno Reichlin, and Bruno Pedretti, 11–29. Milan: Silvana Editoriale.
Rella, Franco. 1977a. Un'economia politica del corpo. In *Il Dispositivo Foucault*, ed. Massimo Cacciari, Franco Rella, Manfredo Tafuri and Georges Teyssot, 47–56. Venice: Cluva Libreria Editrice.
Rella, Franco. 1977b. Introduzione. In *Il Dispositivo Foucault*, ed. Massimo Cacciari, Franco Rella, Manfredo Tafuri and Georges Teyssot, 7–22. Venice: Cluva Libreria Editrice.
Ruskin, John. 1889. *The Seven Lamps of Architecture*, 6th ed. Sunnyside: George Allen.
Settis, Salvatore. 2014. *Azione popolare. Cittadini per il bene comune*. Turin: Einaudi.
Simmel, Georg. 1959. The Ruin. In *Georg Simmel, 1858–1919: A Collection of Essays*, ed. Kurt H. Wolff, trans. David Ketler, 259–266. Columbus: Ohio State University Press. First edition: Simmel, Georg. 1911. Die Ruine. Philosophische Kultur. Berlin: Gesammelte Essays.
Sorbo, Emanuela. 2016a. Restauración como imagen de la memoria = Restoration as an image of memory. *LOGGIA* 29: 28–43. https://doi.org/10.4995/loggia.2016.5145.
Sorbo, Emanuela. 2016b. Ruins of Memory: A Sustainable Conservation for the Material and Immaterial Values of the Former Psychiatric Hospitals in Italy. *Procedia Engineering* 161: 2198–2202.
Sorbo, Emanuela. 2017a. The project as re-signification between 'lieux de mémoire' and 'lieux d'histoire'. In *Conservation adaptation. Keeping alive the spirit of the place adaptive reuse of heritage with symbolic value*, ed. Donatella Fiorani, Loughlin Kealy and Stefano Francesco Musso, 283–292. Hasselt: EAAE European Association for Architectural Education.
Sorbo, Emanuela. 2017b. *La memoria dell'oblio. Ex Ospedale Psichiatrico di Rovigo*. Venice: Marsilio.
Sorbo, Emanuela. 2017c. Un Memoriale (Im)maginario e (im)materiale per l'ex complesso psichiatrico di Rovigo. In *RiCerca/ REStauro—Sessione Ricerca e Didattica*, ed. Donatella Fiorani (coordination), 958–968. Rome: Ouasar.
Sorbo, Emanuela. 2019. Into the De/Construction of the Psychiatric Space. In *The Years of Alienation in Italy: Factory and Asylum Between the Economic Miracle and the Years of Lead*, ed. Alessandra Diazzi, and Alvise Sforza Tarabochia, 43–59. Cham: Palgrave Macmillan.

Sorbo, Emanuela, and Gianluca Spironelli. 2021a. Estetica e conservazione: Strumenti metodologici e teoretici nel rapporto tra rovina e paesaggio per la chiesa 'Incompiuta' di Brendola. *Materiali e Strutture X* 20: 67–80.

Sorbo, Emanuela and Gianluca Spironelli. 2021b. Digitalization strategies as a methodology for knowledge and management of cultural heritage. The "Unfinished" church of Brendola as a reference case study. In World Heritage and Design for Health, international congress Le Vie dei Mercanti XIX International Forum, ed. Carmine Gambardella, 328–337. Rome: Gangemi.

Sorbo, Emanuela and Gianluca Spironelli. 2021c. Informative models of cultural heritage. The "Unfinished" church of Brendola. In *ARQUEOLÓGICA 2.0 & GEORES, international congress ARQUEOLÓGICA 2.0—9th International Congress & 3rd GEORES—GEOmatics and pREServation*, ed. Polytechnic University of Valencia Congress, 127–133. Valencia: Gangemi.

Sorbo, Emanuela. 2023. *Rovina. Incompiuta. Chiesa di San Michele Arcangelo a Brendola. Strumenti di conoscenza, valorizzazione e conservazione.* Venice: Marsilio.

Spada, Irene. 2017. *L'Italia in Istria: Tutela, Conservazione E Restauro Dei Beni Culturali Tra Le Due Guerre Mondiali.* Venice: Marsilio.

Teyssot, Georges. 1977. Eterotopie e Storia degli spazi. In *Il Dispositivo Foucault*, ed. Massimo Cacciari, Franco Rella, Manfredo Tafuri and Georges Teyssot, 23–26. Venice: Cluva Libreria Editrice.

Yourcenar, Marguerite. 1992 (reprint). *That Mighty Sculptor, Time.* Farrar: Straus and Giroux.

Part V
Tools

The sustainable process must transfer the time of the monument and the time of its design, which is done through the architectural project. The project is the tool to intervene in heritage, transforming it based on a model of work that involves knowledge, values, and working with communities. The methodology described is applied through the development of a project understood as a tool to manage sustainable heritage preservation. The whole process is shown, highlighting the main steps: *(1) Tracing values and planning the conservation and consolidation works. Critical issues and dilemmas; (2) Restoration and reinterpretations. Designing the new* (K. Theologidou, M. Papanikolaou).

The different processes and tools are also applied in vernacular architecture, as a significant part of the built cultural heritage, trying to introduce best practices for environmental sustainability in different contexts (M. Philokyprou, S. Thravalou, A. Savvides). MOOCs (Massive Open Online Courses) are one way in which many experiences developed in heritage are transferred. It is indeed very important to share the experience and knowledge, and to consider educational responsibility, being accessible to students, practitioners, and researchers involved in this field of work.

Building technologies are also ways to be more effective in design approaches, especially those related to energy-efficient buildings. Technological advances and use of new materials introduce parameters conditioning the resulting environment, which qualifies the atmosphere created.

Design strategies for new and existing buildings are classified into passive and active design. The first group has a long path from an ecological and bioclimatic perspective, from environmental management, working with thermal conditions, passive solar measures, and daylighting. The second group has been increasing in the last few years, introducing tools and methodologies, such as building life cycle analysis, rating systems for green building assessment, building energy performance simulation, and in situ measurements of building performance. All of them aim to improve comfort conditions and quality of life, now also in heritage interventions where the requirements are made more complex in professional practice through the need to guarantee the preservation or generation of cultural values (A. Radivojević, D. Ignjatović, N. Ćuković Ignjatović, B. Zeković, N. Lukić).

Technologies have also allowed for the development of tools to improve graphic modeling through creative and accessible cartography. This incorporates a line of research to characterize the cultural landscape, involving mapping the territory, adding the identity as an attribute and thinking spatially (E. Larive-López, J. Peral-López). A new way of dealing with cultural heritage is understood as a good sustainable practice.

Chapter 16
The Regeneration of Agora Modiano: Regaining Its Role as the Central Food Market of Thessaloniki and a Landmark in the Area

Kleopatra Theologidou and Morpho Papanikolaou

Keywords History · Architecture · Values · Conservation · Designing the new

Introduction

Agora Modiano is located in the historic center of Thessaloniki in the heart of the commercial center. It was built in 1925 by Eli Modiano according to the architectural plans by Eli Modiano and Jacques Oliphant. Its architecture and construction were innovative and impressive, expressing the spirit of the time. Agora Modiano experienced many decades of splendor as the Central Food Market of Thessaloniki; however, in the last years it began to decline which brought degradation and decay.

Almost a hundred years after its construction, the building has been given a new chance to regain its grandeur and role in the area, with its conservation and reuse. In its function as a food market, a new role was added, that of a place of meeting and culture. A platform above the shop roofs was added to accommodate these new uses in harmony with the historic structure.

K. Theologidou (✉)
Consultant, Architect, Restorer, Thessaloniki, Greece
e-mail: patrathe5@gmail.com

M. Papanikolaou
Project Architect, Thessaloniki, Greece
e-mail: mail@sparch.gr

M. Loren-Méndez et al. (eds.), *Sustainable Heritage Management - Design: Notions, Methods and Techniques*, https://doi.org/10.1007/978-3-031-81970-4_16

The paper focuses on determining the value of the historic building and its role in the development of the area, through a detailed documentation. These values formed the framework for the maintenance and consolidation project, the design approaches with respect to its character and current life demands and the whole process of their implementation.[1]

Tracing Values and Planning the Conservation and Consolidation Works. Critical Issues and Dilemmas

Kleopatra Theologidou

The Location of Agora Modiano and the Urban Regeneration Plan of Thessaloniki in 1920s

Agora Modiano is located in the building block no 9, which is surrounded by Ermou, Komninon, Vas. Herakliou and Aristotelous streets. In 1917 a major event happened in Thessaloniki, a big fire that burned and destroyed the biggest part of its historic center, including the area where Agora Modiano would be built a few years later. A team of planners and architects from Greece and abroad, with Ernest Hébrard, as the head of the team, took over the new urban plan of the city that changed the urban landscape (Karadimou-Yerolimpou 2002).

Before the fire, the Jewish quarter of Kadi was located in this area. It was characterized by an irregular road network with narrow roads and small size building plots. Its population was mainly low-income Jewish and its character commercial with an emphasis on retail trade. The central synagogue of Thessaloniki, the Talmud Torah Agadol, was placed in the building block of Agora Modiano. It was built in 1520 and was associated with the exodus of the Jews from the Iberian peninsula and their settlement in Thessaloniki. The synagogue was destroyed by fire and rebuilt several

[1] Owner: F.L.S. PRIVATE COMPANY-FAIS GROUP.
Construction management: EMMETRON PRIVATE COMPANY,
Dr. Triantos Dimitrios, Doctor of mechanical engineering, NTUA.
Architects: SPARCH SAKELLARIDOU/PAPANIKOLAOU ARCHITECTS, project architect: MORPHO PAPANIKOLAOU.
Associate Architects: G. Papanikolaou, P. Liakou, G. Papageorgiou, K. Moustakas, P. Hrakleous, I. Tzampazi, M. Stauroulaki, E. Foutsitzoglou, I. Leivadioti, S. Georgou, A. Xari, G. Malliaropoulos, A. Avgerou, D. Daniil, F. Nteka, D. Nikandrou, Z. Chalkia.
Consultant Architect (first stage of the project): KLEOPATRA THEOLOGIDOU, Architect Engineer, Restorer.
Structural design: ATHANASIOS I. KONTIZAS, Bsc in Civil Engineering, NTUA, Msc.
E/M Design: SAMARAS & ASSOCIATES S.A., MEP design.
Lighting Design: ELEFTHERIA DEKO & ASSOCIATES LIGHTING + DESIGN.
Special advisor: J.E.P.A. E.I. PAPAGRIGORAKIS & ASSOCIATES.
Subcontractor in structural reinforcement, CONIC P.C.

times, most recently in 1899, after another great fire in Thessaloniki in 1890[2] (Zafiris 2016, p. 102). Its surrounding buildings housed religious, educational and administrative uses, while, at a short distance away was the first "Allianz" school, founded in 1873 by the World Israelite Union, as one of a series of schools established to upgrade the religious and European education of Jews.

The political decision for the regeneration of the entire historic center of Thessaloniki following the European trends and principles of the time, and the assignment of the urban planning to Ernest Hébrard and his team were critical for the city's future (Karadimou-Yerolimpou 1995, p. 128 and MIET 2017, p. 3). The so-called "Hébrard Plan", among others, included:

- A new road network with roads parallel and vertical to the sea front and diagonal to it with points of reference to important monuments of the city, thus emphasizing its byzantine past.
- The increase of the size of the land blocks.
- The maintenance of the existing bazaars and traditional open-air food markets located between Egnatia and Ermou streets.
- The positioning of luxury trade, administration, banks and leisure in the area defined by the sea front, and Venizelou, Aristotelous and Ermou streets, including the Aristotelous square (Karadimou-Yerolimpou 1995, p. 149–159).

The increase of the size of building blocks and the release of the land value in the auctions favored the land purchase by entrepreneurs with high incomes, thus changing the population composition in the area (MIET 2017, p. 12).

Regarding the architectural styles, eclecticism was the most widespread style in Thessaloniki, as early as the nineteenth century. Particularly, in the case of public buildings, banks, shopping centers, etc., monumentality was used to achieve impressiveness and prestige. At the same time, new building materials were introduced, such as reinforced concrete, metal and glass. Hébrard, a deep connoisseur of the monuments of Thessaloniki, which he highlighted in the urban planning of the city, adopted an architectural type closer to its Byzantine past, introducing the basilica type and elements of byzantine architecture, such as arched porticos on the ground floor and arched openings on the upper floors, which he combined with elements of European architecture. He adopted the absolute symmetry on the facades of the buildings with emphasis on a central vertical axis, the elevation of their central part and their organization into three parts: base, body and crown, like many of his contemporaries (Colonas 2012, p. 19–23).

Eli Modiano

Eli Modiano (1881–1968) was born in Thessaloniki. His family, the Modigliano family, came to Thessaloniki from Livorno Italy in 1817. His grandfather Saul Isaac

[2] More about the fire of 1890 and the changes that followed in the urban landscape in Karadimou-Yerolimpou 2004, p. 225–232).

Modiano (1816–1883), was considered the richest man in Thessaloniki. He founded the Real Estate Bank of Modiano and built Stoa Saul in the 1880s, the largest commercial center of the city at the time, possibly designed by Vitaliano Pozelli (Zafiris 2016, p. 117–118). He was one of the founding members of the "Alliance" school, as well as the founder of the Talmud Torah community schools. His three sons, Jacob, Samuel and Levi, further expanded his financial empire. Jacob, the father of Eli, took over the management of Saul's bank, as the main owner (Zafiris 2016, p. 80). Modiano brothers had a large estate that included many urban properties and peri-urban land. "Half of Thessaloniki belonged to them" was a well-known popular expression, while Saul Modiano bank was the largest local credit organization in Thessaloniki (Chekimoglou 1996, p. 230–232).

Eli studied architecture in Paris at the Ecole Centrale. He designed several public and private buildings in Thessaloniki; many of them are still landmarks in the city. It is worth mentioning:

- The old customs house and passenger station today (1911) in the port of the city. Built in a Renaissance style in its facades, with symmetry, tower-shaped ends, a glass roof and a higher central part. The old customs house was the first reinforced concrete building built in Thessaloniki.
- Villa Giaco Modiano (1906). It was the family's seaside holiday home. It was built in a Renaissance style with French art nouveau additions. The building currently houses the Folklore and Ethnological Museum of Macedonia and Thrace.
- The building at I. Dragoumi Street (1925) (Zafiris 2016, p. 133–134),
- The tobacco warehouse of Salonica Company on Ptolemaion Street (1924)
- The "Luxembourg" hotel, at the junction of Komninon and Kalapothaki Streets (1924).

Eli's origin from a family of bankers and businessmen and his long contact with European trends, where indoor food markets were built in all major urban centers, were critical to his decision to build the first indoor food market in Thessaloniki, "the Central Food Market", mostly known as "Agora Modiano". Taking advantage of the urban regeneration of Thessaloniki, after the fire of 1917 and the whole context which favored the development of entrepreneurship, he proceeded to the purchase of the appropriate plot, in the heart of the luxury trade area, where he built the Central Food Market, the CENTRAL STOA.

Agora Modiano. The Building, its Architecture and Construction

Eli was aware of the contemporary architectural movements abroad due to his studies in Paris and contact with Europe. Moreover, the planning of Agora Modiano coincided with a period of strong public discourse about the identity of Thessaloniki, as it could be expressed in its urban regeneration and architecture after the fire of 1917.

The use of closed arcade was already widespread in Thessaloniki in commercial centers, one of the oldest being Stoa Saul, built in the early 1880s by his grandfather. In this context, Eli Modiano together with Jacques Oliphant would use the type

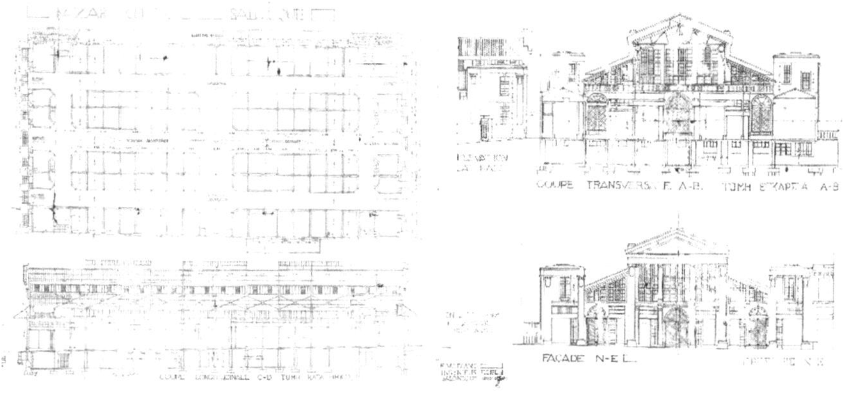

Fig. 16.1 Agora Modiano. Drawings of the architectural study by E. Modiano and J. Oliphant (Βλαστάρης et al. 2009)

of closed arcade in their architectural planning of the Central Food Market. At the beginning, in the architectural proposal the building was two-storied, with food stores on the ground floor and shops and offices on the upper.

In the final proposal that was submitted, approved and implemented, the building included food shops on the ground floor, a basement where the warehouses of the stores were located, while on one floor, inside the building in contact with its narrow sides, open balconies were placed overlooking the stores below. The four tower-shaped corners of the building were also two-storied; the offices of the company were placed there with autonomy to their operation in relation to the rest of the market. The number of architectural drawings submitted for approval was impressive, including, in addition to basic plans, many drawings with architectural and construction details and facilities (Fig. 16.1).

The architectural composition is characterized by symmetry in horizontal and vertical sections and in the facades. The end result is also characterized by monumentality and quality in both design and construction. The architectural type is mixed, combining the local trends of the time, that is local tradition neoclassicism and eclecticism. Elements from the local tradition are adopted in the configuration of the building's volumes and in the floor plan, by applying the type of three-aisled basilica with a "skylight". This type of basilica has a raised "nave", relatively to the sides, and thus light is given to the interior by side skylights in its elevated part. The choice of this type leads to an impressive effect inside, dominated by the empty space above the shops, which is bounded upwards by the impressive metal construction of the roof and the skylights. Neoclassical standards are mainly adopted on the two facades, on the Vas. Herakliou and Ermou streets. The interior decoration of the stores could also be described as neoclassical and particularly simple, with meanders and simple waves.

The total area of the building is 2.71179 m^2. The floor plan is organized by a 3×3 m grid of concrete columns, with fine proportions. Inside, two zones are distinguished: the zone of shops, with dense constructions and the zone above the shops, a free, open, essentially "empty", space, extremely impressive, where the roofs dominate, impress and seem to float thanks to the side skylights.

The floor plan is an oblong rectangle, with an average length of about 70 m and an average width of about 37 m. The ground floor is organized by 3 longitudinal and 3 transverse corridors. The two corridors located on the longitudinal and transverse axis of symmetry are wider than the lateral ones, thus acquiring the characteristics of central corridors, while at their intersection a relatively spacious space is created, a crossroads that facilitates the movement of customers. In the space delimited by the corridors, four "islands" of shops are created, the central shops. In vertical cross-section, the shops facing the central longitudinal corridor have a greater height and are additionally ventilated by skylights closed by metal mesh. The shops in the side aisles are lower and are additionally ventilated by air vents in their roofs, which are also covered with mesh. It is worth mentioning that the roof of these shops was not accessible. Side rows of shops were also arranged on all four sides of the building.

Access to the building is from the two main streets, Ermou and Vas. Herakliou and from the side street Komninon. Due to the height difference of the two streets, approximately 1.50 m, the floor of the ground floor has a slight slope, which, due to the long length of the longitudinal corridors, is not discernible.

Initially, there were 144 shops in total on the ground floor, and the uses were allocated by category, depending on the products: areas with fishmongers, butchers, greengrocers, cheese shops, etc.

The natural lighting and ventilation of the interior space was provided by skylights, by a part of the central roof that was covered with glass and by the opening windows, on the open balconies, on the two narrow sides of the building. The mechanism which helped the opening of the skylights was of interest based on a chain that reached the ground floor.

The roofs were treated differently in the central area, in the side shops and in the tower-shaped corners, both in terms of form and construction materials. The central area, which includes the four islands of shops and the balconies, is covered with a central double-gabled roof and lateral single-gabled ones, in the logic of a three-aisled basilica with a skylight. They are constructed by a metal frame, with interesting sections and details, which contribute significantly to the overall aesthetics of the space. The trusses are covered with wooden planks. The roofs of the side shops and the towers are of wooden construction, double-gabled above the shops and four-gabled above the towers. The whole was covered with French-style tiles.

The main facades on Ermou and Vas. Herakliou streets are characterized by monumentality, absolute symmetry and a tripartite arrangement of base, body and crown. They have many openings and are almost identical, with differences arising from their different heights. Neoclassical elements were mainly adopted in their decoration, i.e. pseudo-parastades with capitals, waves, toothed bands, meanders and a pediment ending with a circular skylight. The whole was completed by the elaborate metal rails on the external balconies, the metal entrance doors to the building and the skylights of the shops and windows, under the pediment, made of metal bars, in a rhomboid arrangement.

The construction is characterized by innovation and variety in materials and construction techniques. Agora Modiano is one of the first examples in Thessaloniki where reinforced concrete was used, the first being the customs office, also designed by Eli Modiano. In addition, the use of iron is extensive as a load-bearing element of the central roof and the central pillars that support it, from one height and above; iron was also used in most of the windows, in the rails, in the ventilation grilles, etc., with interesting cross-sections that upgrade the image of the whole.

The main load-bearing structure of the building is made of reinforced concrete, which consists of columns, beams and slabs. The columns are placed on a 3.00 × 3.00 m grid, which is maintained on all levels of the building. The small cross-sections of all the load-bearing elements are characteristic and contribute to the aesthetics of the space, which, it is worth mentioning, behaved very well over time. The masonry in the perimeter of the building is of stonework, in the basement, and of solid brickwork, on the ground floor and the first floor of the corner towers.

The interior decoration of the building was simple; it was based on neoclassical relief elements on the facades of the shops and on the visible part of the staircases. The decoration is complemented by the metal constructions and the elaborate rails of the interior balconies that gave a special aesthetic to the space, consistent with its use as a food market. Small details, which the visitor gradually could discover, gave the whole a particularity, such as the lamps in the central corridors, and the gutters, on the inner side of the walls of the side shops, which collected the rainwater and channeled it into the drainage system, and also the cut corners, curved triangles, the marble counters from the fish shops, their ventilation and security system, etc.

Of particular interest were the facilities of the building, such as the lighting system with lamps suspended from the metal trusses of the roofs, made of cast iron, in original shape and construction. Of particular interest was also the system for the runoff of rain waters from the roofs, which were collected externally and ended up in a system of vertical gutters, placed internally, in contact with the wall of the side shops. Through the gutters the water was channeled into the sewage system. The drainage system itself was well studied; parts of it were still in use, such as the surface small channels, built in the floor along the shop fronts and in contact with them, to collect water from their washing. These waters were carried through pipes located on top of the basement ceiling to the sewage system. The exact construction can be seen in the building plans.

The State of Preservation before the Conservation Works

Agora Modiano enjoyed many decades of glory, as it was the market where rare kinds of food and raw produce of a high quality could be found. In addition it gathered well-known customers, who frequented the taverns it hosted. Its operation characterized an entire area that became a destination for many citizens and visitors. In recent years, a gradual decline began, the famous taverns were replaced by places of mass consumption, food shops gradually began to close, and degradation, desolation and decay followed.

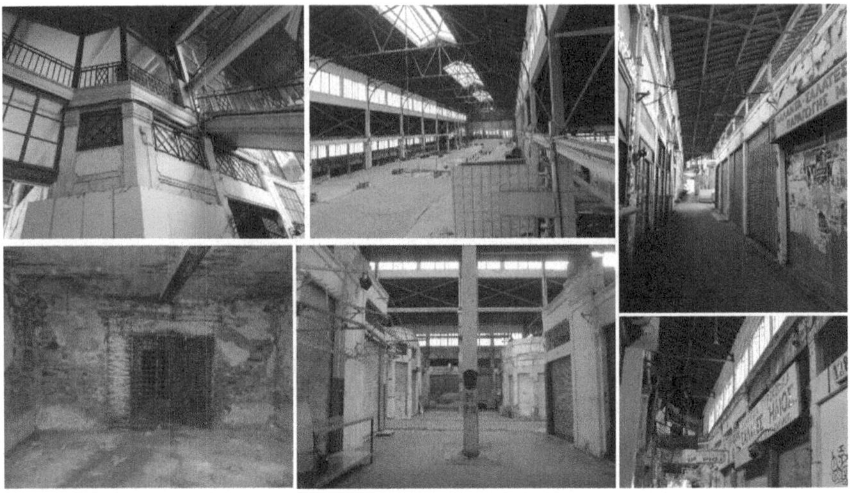

Fig. 16.2 Agora Modiano before the conservation works (Theologidou, K.)

The building, extremely dilapidated, presented an image of abandonment and decay. The image was burdened by the many closed shops, the coating of shop fronts with cheap materials and paints, the destruction of decorative parts of shop fronts by the various cables hanging everywhere, etc. (Fig. 16.2) However, one could argue that, behind this negative image, the building was kept in relatively good condition. An exception was the load-bearing system made of reinforced concrete and especially its section at the basement level, which however created a series of problems in the upper part of the building, as well. The main causes of its problems were the complete lack of maintenance and the intense humidity that prevailed, especially in the basement.[3]

Humidity was mainly due to the lack of maintenance, which caused extended leaks from everywhere including the roof, floors and drainage pipes. Humidity was also caused by underground water, which probably existed due to the building's proximity to the sea. It caused very serious problems related to the oxidation of the metal elements of the construction, mainly at the basement level, but partly also on the ground floor, to the creation of salts in extensive areas, which caused corrosion of the construction materials and flora in a primary form, i.e. mosses and lichens. Especially, the oxidation and consequently the swelling of the reinforcement in the majority of the concrete slabs and beams in the basement, combined with insufficient ventilation, non-existent maintenance and natural aging, resulted to the disintegration and pulverization of concrete in many areas.

All the other elements of the building, such as roofs, stonework, brickwork, window frames and shutters, metal elements in the upper parts of the building and coatings, were in a relatively good condition. Regarding its facades, although altered

[3] Limited repair works were carried out in 2001, mainly to parts of the roof, including the replacement of the tiles at the central part of the building with metal panels. (ΔΕΠΟΣ, 2001).

at shop level by later additions, it was clear that behind these additions they retained all the elements of the original construction.

Values and Protection Status

As presented in detail above, the Modiano Market has significant tangible and intangible values linked to its importance for Thessaloniki and the recent memory of its inhabitants, to the owners, their history and role in the city and to the building itself, its architecture and construction.

In brief, Agora Modiano was the first indoor retail food market built in Thessaloniki, according to European standards. It introduced a new concept for the way food markets should operate in the city. In the memory of the inhabitants and visitors, it was associated with the finest and highest quality food that could be found in the city and remained as a place of colors, food smells and music, especially from itinerant musicians. For a long time it was a meeting place for scholars and politicians and well-known people from all over Greece. Moreover, it was the meeting place of many Thessalonians, while, during specific celebrations and festivals the events that used to take place there gathered a large number of people.

The operation of Agora Modiano gave character to the entire area, where similar businesses related to retail trade and especially gastronomy gathered, enlivening it and making it a destination. It is worth mentioning that the decline of Agora Modiano affected the entire area, which, combined with the recent economic crisis, had become desolate.

The Central Food Market is associated with the Modiano family who had an important role in the economic development of the city in the late 19th and early twentieth centuries, due to the economic activities they developed. Its initiator and founder, Eli Modiano, architect and engineer, connoisseur of the European currents of the time, was a well-known architect and builder who also built important public and private buildings in Thessaloniki, which remain still landmarks.

The building with its characteristic typology, which follows the standards of the time both at European and local level, expresses the spirit and architecture of the time, and is also connected to a period of Thessaloniki, when the urban landscape changed dramatically, after the fire of 1917 and the plans of its regeneration. In addition, the great variety in construction materials and building techniques expresses the gradual transition to a new way of construction that will be consolidated in the second half of the twentieth century, where reinforced concrete is the dominant material.

The use of reinforced concrete in a 3 × 3 m grid gave alternatives and further possibilities to the internal organization of the building, while the metal structures to support the central roofs allowed the covering of large openings without the use of bulky pillars and columns. The adoption of the typology of the three-aisled basilica with a skylight allowed the best natural lighting of the space and, at the same time, resulted in the ceiling appearing as if it were hanging. The decoration adopted, on the facades of the building and the interior shops, is also characteristic. Especially in the facades, the Modiano market belongs to the few buildings in the city that adopted extensive neoclassical elements, in the spirit of the time, when Thessaloniki had to acquire a "new Greek identity" (Colonas 2012, p. 40).

Agora Modiano was listed as a historic building by the Ministry of Macedonia and Thrace in 1983 and by the Ministry of Culture in 1995. In particular, Ministry of Culture declared Agora Modiano as a historical listed building, because it is a typical example of the market buildings with an arcade, because it is a valuable representative example of architecture in a building ensemble, important for the study of the history of architecture and because it is connected to the memories of the residents of Thessaloniki. The Ministry of Culture also classified as preserved the use of the building as a food market.

Planning of Conservation and Consolidation Works. Critical Issues and Dilemmas

The values of Agora Modiano building, as have been documented, are linked not only to its architecture and decoration, but also to its construction methods and materials. It was therefore particularly important to preserve as much as possible of the authentic construction and materials. However, the load-bearing capacity of the reinforced concrete structure was reduced, as the building was essentially abandoned for years without any maintenance. Its condition was further accelerated due to the ephemeral nature of the material itself, the concrete and its aging problems. The situation was critical in the concrete slabs and beams of the basement, where most of the superstructure is supported.

This situation created concern and strong dilemmas regarding the conservation and consolidation methods to be applied to the building for its restoration and reuse. In this particular case there were two options.

1. The use of heavy reinforcements with concrete mantles.
2. The removal of the central part, where the situation was critical, and its reconstruction in the same dimensions, with the same materials, but with modern specifications.

The first case involved extreme alteration of the cross-sections resulting in the disfiguration of all elements of the reinforced concrete load-bearing structure. In addition, despite the treatment that should have been applied to stop the deterioration of the materials, the positive result was not a given, resulting in the risk that the original structure, encased in reinforced concrete coatings, would continue to deteriorate and even without the possibility of monitoring it, as it would no longer be visible.

In the second case, there would be a loss of the authentic materials and construction; however, the final result would preserve the building concept with the thin sections in the various elements of reinforced concrete, highlighting the technology of the time. In addition, the image of the building and its architecture with the delicate proportions of the whole structure would be maintained.

Between the two options, the second was chosen, which was considered to better preserve the values of the historic building and was in line with conservation theory and guidelines.

All the other elements of the building were planned to be maintained and conserved, including local replacements in the same form and materials. In the case of the facades, all additions were removed so that the original elements were revealed, which were maintained and highlighted (Fig. 16.5).

Restoration and Reinterpretations. Designing the New

Morpho Papanikolaou

Introduction

The following text refers to the basic principles of approach and the manipulation strategies that characterize the architectural proposal. The intention was to develop the architectural proposal within the framework of the general restoration principles, as they were formulated in detail in the earliest study phase. In this context, we acted with the intention to highlight the existing spatial and architectural qualities of the emblematic Agora Modiano building, as well as to add new.

The Time of the Monument and the Time of Its Design

A monument is inextricably linked to the concept of time. Agora Modiano is a building that speaks loudly of time. It is a space that narrates important passages in the city's history. It is the space that, through its morphological, material, and functional structure, weaves stories that connect it with its locale, time, and people.

With a synchronous and at the same time intertemporal process of approach, we asked ourselves:

What is the role and which are the marks of time on the body of the monument itself?

How can we perceive the concept of time and the catalytic impression of a century of life on the structure, the geometry, the form, and the materiality of the building, in a way that incorporates them as integral elements in the design process of the Market?

How does one approach the design of a modern food market in a building with strong signs of abandonment and wear, where each and every part has a familiar and active past?

What is the new element of intervention that will satisfy the modern framework of static reinforcement, security, and operational specifications, in a place where

the many needs at the level of the business enterprise and the multiple desires and expectations of integration in the city life converge?
How can we ensure an 'ethical' interpretation of the building in our architectural approach?

We formulated these questions to be answered within the course of our design process.

The central challenge of the approach lay in acquiring a deep understanding of the monument and actually learning from it. Our goal was to distinguish the building's values in terms of material and immaterial qualities, while negotiating its feasibility in every phase, and remaining unaffected by global trends.

The Architectural Approach

The conservation and re-interpretation of the Agora Modiano into a modern food market focused on preserving the building's authentic character, while highlighting the qualities it acquired over its former life. The overall approach of the project was characterized by examining the building's many important elements, in order to discern those that are potentially active and timeless: the Modiano Market, as a vessel of time, strongly retains material and immaterial qualities in its structure, as part of the history of the city and its people.

We focused on its atmosphere as the quality dominating its void yet engaging interior space. At the same time and throughout the design process, we explored the boundaries between the private use of the space and the public character of the monument. In doing so, we considered the private and public spaces as potential spaces of interaction for the people of the city.

The design process of the regeneration of the Central Food Market

During the four years of studies, we approached the project as a process of alternating between the research phase (Βλαστάρης et al. 2009) the evaluation phase, and the design phase, focusing on the interpretation of the predominant concepts of the monument. We identified those concepts as: Simplicity, Rhythm, Light, Lightness, Materiality, People.

We wanted to redefine and reinforce the qualities that the building had in its original form and life, while highlighting its atmosphere as the most powerful quality that the monument possessed. We realized that the atmosphere of the interior space, namely the void space, constitutes the dynamic element that contains and condenses history, construction methods, collective memory, materials, typology and rhythm. We felt the need to reinforce this atmospheric quality in our design of the new: a floating, square platform above the shops, that serves as a place of experience and connection and, at the same time, a place that once again belongs to the city.

The Agora is characterized by the "stoa" typology, with a linear arrangement of three parallel, lengthy streets. The small unit-shops adopt a repetitive form, lined up in a strictly functional arrangement, with a low height compared to the overall size and height of the interior space. In an evocative mix, the Agora combines the atmosphere of oriental bazaars with the planning and dimensions of western covered markets.

We studied the proportions of shops and corridors, factoring in the concentration of people and shops, which is exhausted at the ground floor level. At the same time, we recalled the emotional experience and the sense of tension evoked when viewing the space from the gallery. This space, while now serving as a dining area, was, in its original form, designed to be the market's observation area.

Since classical times, the market is the place identified with the heart of α city's public life, ever-present to observe its evolution and ceaseless course. The creation of a new and reinforced space to serve the public role of Agora Modiano became one of the decisive elements of our investigation. We were searching for a place where people would be able to stop and relax, where they could 'supervise' commercial transactions, and where social and cultural activities would be able to flourish, with the latter being a complementary yet necessary function which would accompany the monument in its new life cycle.

The Dual Character of the Monument

As we have thus far established, the main idea of the architectural approach was to highlight the duality of the monument, which emerged in every attempt to analyze its history, form, construction technique, and materials, in a constant balance of opposites:

- the simultaneous existence of an important past in combination with the creation of a modern identity
- its character as an autonomous building-gallery, in combination with its dynamic re-integration into the city street network, where it belongs
- the aura of 'exoticism' of the east, which is uniquely balanced with its strict 'west-oriented' structure
- the preservation of the traditional structure of a food market, in harmony with the requirements of the modern way of selling food, within the context of the growth of food markets at European level[4].

[4] Markthal Rotterdam. https://www.archdaily.com/553933/markthal-rotterdam-mvrdv
Mercado de San Miguel-Madrid. https://www.theguardian.com/travel/2015/jan/29/top-10-food-markets-madrid-spain
Mercado de San Miguel-Madrid. https://spainattractions.es/mercado-de-san-miguel/
Mercat de Sant Antoni. http://www.ginabarcelona.com/en/project/rehabilitation-of-sant-antoni-market#.W2leqNUzbcs
Mercat del Ninot. http://thebcnartproject2015.blogspot.com/2017/03/mercat-de-sants.html
https://www.archdaily.com/769005/remodelling-of-the-ninot-market-mateo-arquitectura
santa caterina

The Architectural Manipulations

We recognized the timeless structure and typology of the building and prioritized its basic properties as the starting point of the overall architectural approach.

We restored and highlighted the strict symmetry and linear repetition in both the organization of the floor plan and the views of the internal streets of the Agora.

We highlighted the transparent and extroverted nature of the Market. The stoa building connects two parallel streets, allowing the individual to "cross through and visit" (Papadopoulou 2003). At the same time, the robust character of the symmetrically arranged vertical openings really intensifies the concept of transparency. We took advantage of this spatial quality by providing viewing, moving, and standing possibilities on the roofs of the central islets.

We identified the power of empty space and utilized it by highlighting its enchanting effect. According to the standards of covered markets of the nineteenth century, the Agora is designed with an elaborate metal crown, according to the "type of the basilica".

We identified the intensity and density in the linear arrangement of the shops at street level, and, at the same time, the absence of a central space for standing, social interaction, and congregating.

Under the rhythmic lightness of the metal construction of the roof we proposed the creation of a new level of use, which would allow longitudinal movement at the level of the roofs of the central islets. This level wouldn't have a commercial character, but it would rather be designed as a place to stop and gather, adding the specific communing character that has over time been recorded in the collective memory and experience of the city.

Furthermore, a clean-cut and reversible construction was proposed, designed with geometric manipulations of symmetry, and resolved in a uniform silhouette. Its carving follows the geometry of the cells of the islets, and attends to their 'recessed' outline, so as not to add 'crowning' elements to the upgraded facades of the shops. In doing so, the construction is transformed into a discreet connecting bridge with curved cuts, which extends between the roofs and serves to emphasize the geometric and functional center of the Agora with its form (Fig. 16.3).

In this way and through an anthropocentric approach that seeks to reinterpret the classic perception of the Agora, the market can become a place of work, a modern place for trade, but also a place of exploration, communication and relaxation, a meeting place for both planned and spontaneous activities (Fig. 16.4).

---

https://en.wikiarquitectura.com/building/santa-catarina-market/
http://miesarch.com/work/682
Mercat Koszyki, Poland. https://www.koszyki.com/news/hala-koszyki-juz-otwarta-en-us/
Boqueria Barcelona. http://www.barcelonaconnect.com/la-boqueria/.

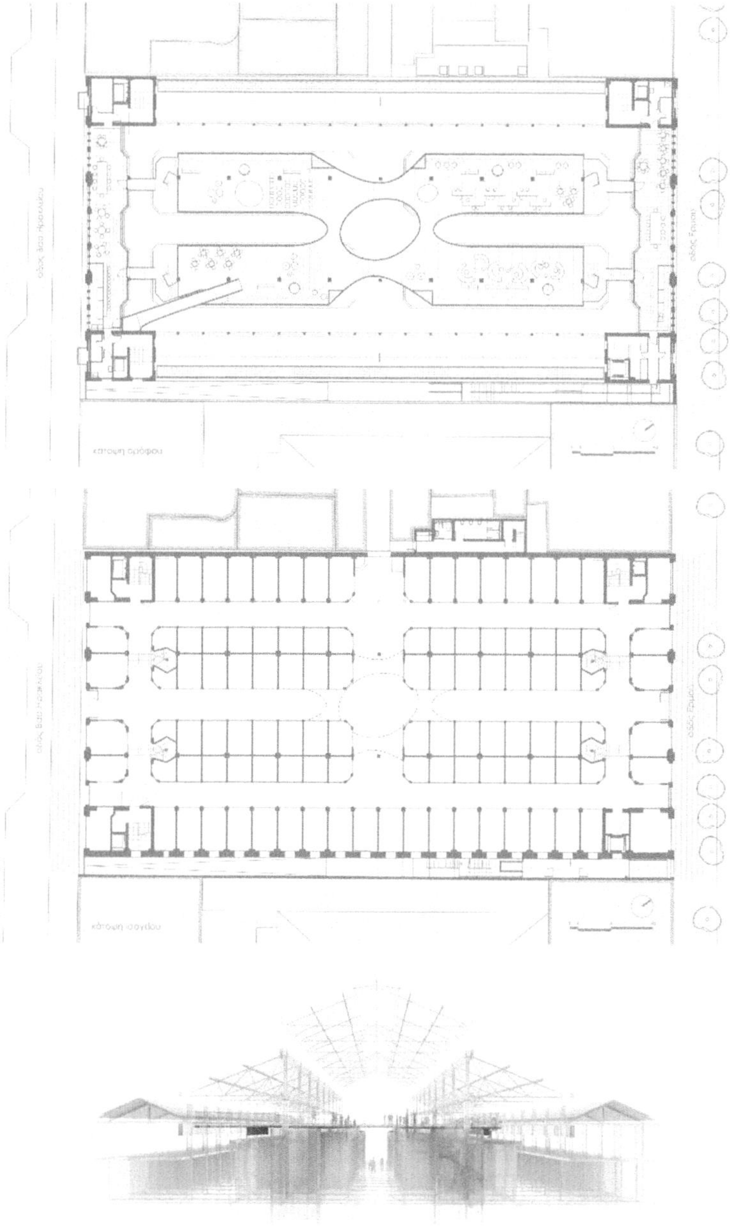

Fig. 16.3 Designing the new: simplicity, rhythm, light, lightness, memory (SPARCH)

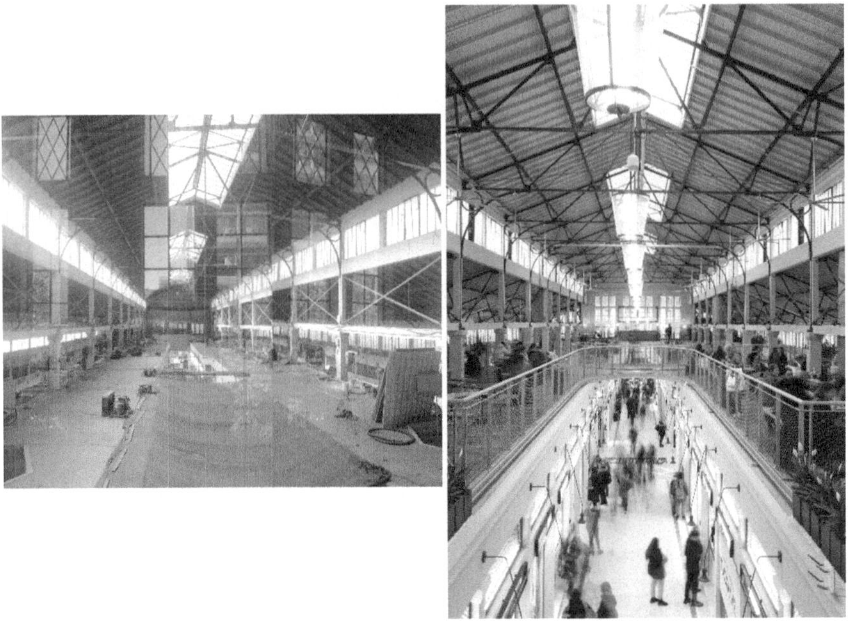

Fig. 16.4 Agora Modiano before and after the works. (Left: SPARCH, right: Deikou Olga)

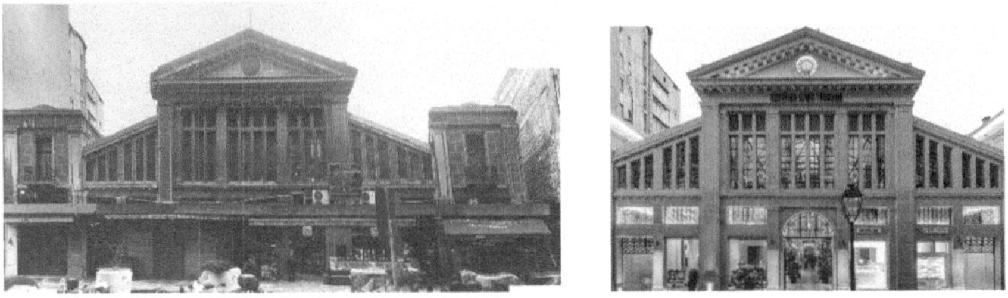

Fig. 16.5 Vas. Irakliou façade. Before and after interventions. (Left: SPARCH, right: Deikou Olga)

Brief Description

The entirety of the intervention included the preservation, reinforcement, and reconstruction of the monument. The central idea focused on the rendition of the modest appearance of the internal faces and the restoration of the clean and firm expression of the external faces of the Agora.

The basement level houses the storage areas necessary for the proper functioning of the Market, as well as the MEP facilities of the building. The spaces are organized into small repeating units, which follow the original corridor system in both the transverse and longitudinal direction. According to the original design, this level has direct access to the ground level via the external feeder ramp, as well as via two vertical communication cores. Adjacent to one of the vertical communication cores are the staff restrooms and the changing rooms.

On the ground floor, the primary level of the Agora shops is developed. The plans reflect the current image of the stores, as it has resulted from the merging of neighboring units. The restoration design takes into account a functional diagram for a flexible food market that organizes the basic infrastructure, so as to enable the new intervention to respond adequately and gradually to the more modern requirements for the organization and provision of infrastructure for a covered food market.

At the same time, each unit maintains its independence in appearance, in accordance with the structure and morphology of the restoration proposal. In plan view, it showcases the indicative and flexible types of organization of the stores according to their respective uses, while it simultaneously designates the enclosure of the external boundaries of each unit with strict limitations.

At the level of the galleries, the two dining areas located on the two main sides of the Agora were renovated. At this level, the new platform extending above the shops is configured with modest equipment, in order to be able to respond to the concepts of relaxation and gathering for the visitors of Agora. At the same time, it still allows for hosting small cultural events, as well as workshops or educational programs.

References

Boqueria Barcelona. http://www.barcelonaconnect.com/la-boqueria/

Chekimoglou, Evangelos. 1996. *Θεσσαλονίκη Τουρκοκρατία & Μεσοπόλεμος*. Thessaloniki: University Studio Press.

Colonas, Vasilios. 2016. *Η Θεσσαλονίκη εκτός των τειχών*. Thessaloniki: University Studio Press.

Colonas, Vasilios. 2012. *Θεσσαλονίκη 1912–2012, Η Αρχιτεκτονική μιας εκατονταετίας*. Thessaloniki: University Studio Press.

http://miesarch.com/work/682

https://en.wikiarquitectura.com/building/santa-catarina-market/

https://www.archdaily.com/769005/remodelling-of-the-ninot-market-mateo-arquitectura

Karadimou-Yerolimpou, Aleka. 1995. *Η ανοικοδόμηση της Θεσσαλονίκης μετά την πυρκαγιά του 1917*. Thessaloniki: University Studio Press.

Karadimou-Yerolimpou, Aleka. 2002. *Το χρονικό της μεγάλης πυρκαγιάς, Θεσσαλονίκη, Αύγουστος 1917*. Thessaloniki: University Studio Press.

Karadimou-Yerolimpou, Aleka. 2004. *Μεταξύ Ανατολής και Δύσης. Θεσσαλονίκη και βορειοελλαδίτικες πόλεις στο τέλος του 19ου αιώνα*. Thessaloniki: University Studio Press.

Mercat Koszyki, Poland. https://www.koszyki.com/news/hala-koszyki-juz-otwarta-en-us/

Mercat de Sant Antoni. http://www.ginabarcelona.com/en/project/rehabilitation-of-sant-antoni-market#.W2leqNUzbcs

Mercat del Ninot. http://thebcnartproject2015.blogspot.com/2017/03/mercat-de-sants.html

Papadopoulou, Titi. 2003. Το αίθριο, In *Επιστημονική επετηρίδα Πολυτεχνικής Σχολής, Αριστοτέλειο Πανεπιστήμιο Θεσσαλονίκης, Τμήμα Αρχιτεκτόνων*, Thessaloniki.

Markthal Rotterdam. https://www.archdaily.com/553933/markthal-rotterdam-mvrdv

Mercado de San Miguel-Madrid. https://spainattractions.es/mercado-de-san-miguel/

Mercado de San Miguel-Madrid. https://www.theguardian.com/travel/2015/jan/29/top-10-food-markets-madrid-spain

Santa Caterina. https://www.archdaily.com/769005/remodelling-of-the-ninot-market-mateo-arquitecturasantacaterina

Zafiris, Christos. 2016. *Η Θεσσαλονίκη των Εβραίων*. Thessaloniki: Epikentro.

Α. Βλαστάρης, Γ. Κακές, Φ. Καραγιώργος, Ν. Κορώνη, Κ. Μπόλη, Κ. Σταματοπούλου, Μ. Χατζή - επιβλέποντες καθηγητές: Μ. Νομικός – Α. Αβδελάς, Εργασίας μεταπτυχιακή, Προστασία και ανάδειξη, στοά Μοδιάνο, Αριστοτέλειο Πανεπιστήμιο Θεσσαλονίκης, Πολυτεχνική Σχολή – Τμήμα Αρχιτεκτόνων, Διατμηματικό Πρόγραμμα Μεταπτυχιακών Σπουδών, Προστασία, Συντήρηση και Αποκατάσταση Μνημείων Πολιτισμού, Θεσσαλονίκη 2009.

ΔΕΠΟΣ. 2001. Μελέτη: Επισκευή – Αποκατάσταση – Συντήρηση Αγοράς Μοδιάνο, Thessaloniki.

Σύντομος οδηγός έκθεσης ΜΙΕΤ. 2017. *Το τέλος της παλαιάς μας πόλης. Θεσσαλονίκη 1870–1917*, Συλλογικό έργο, Thessaloniki.

Chapter 17
Conservation and Environmental Sustainability of Vernacular Heritage. Lessons Derived from the Development of an Open Online Course on the Environmental Aspects of Vernacular Architecture

Maria Philokyprou, Stavroula Thravalou, and Andreas Savvides

Abstract Vernacular architecture, as a significant part of the built cultural heritage, reflects the identity of a place. This is because, over time it has developed through a continuous process of adaptation, to meet the social, economic, and environmental needs and requirements of the society. Furthermore, our understanding of sustainability can be enriched through a holistic and interdisciplinary view of both the tangible and intangible values embedded in these vernacular dwellings. However, heritage and sustainability are usually studied and taught as separate fields, not necessarily connected to each other. Meanwhile, it is noted that despite the recent wide adoption of digital tools in universities, their use in the field of cultural heritage conservation is still deemed poor. In response to this need, this article describes the design of an online course on the topic of environmental conservation of vernacular architecture, developed in the framework of the Erasmus Plus programme 'Smart Rehabilitation 3.0'. This Erasmus Plus programme aims to create new digital educational tools for cultural heritage conservation in order to address different types of challenges, as well as to present a new curriculum that takes a holistic approach to the field of conservation. The development of the aforementioned online course is divided into five modules, including an introduction and overview of the sustainable aspects of vernacular architecture (module I), a brief reference to environmental assessment tools (module II), an insight into traditional building materials (module III), as well as passive heating and cooling strategies (module V), and finally, an overview of the principles of environmental conservation through case studies (module V). The creation of this online course highlights the great opportunities for online learning and continued education in this field, in turn underlining the need to apply these tools to improve the sustainable features of vernacular architectural heritage.

M. Philokyprou (✉) · S. Thravalou · A. Savvides
Department of Architecture, University of Cyprus, Nicosia, Cyprus
e-mail: philokyprou.maria@ucy.ac.cy

M. Loren-Méndez et al. (eds.), *Sustainable Heritage Management - Design: Notions, Methods and Techniques*, https://doi.org/10.1007/978-3-031-81970-4_17

Keywords Vernacular architecture · Environmental aspects · MOOCs · Sustainability · Environmental conservation

Introduction

Teaching conservation and introducing best practices for the environmental sustainability of vernacular architecture constitutes an important part of architectural education. Today, in the face of an ageing housing stock across Europe that has to mitigate the effects of climate change, sustainable development and conservation of the built environment—and especially of architectural heritage—is becoming more and more necessary. Although many initiatives and agreements promote and encourage activities related to cultural heritage conservation and energy savings in Europe, most European universities still do not sufficiently address education on the sustainable aspects of heritage conservation (Djordjevic et al. 2022). Recent experience has shown that the preservation and reuse of heritage buildings, and especially vernacular dwellings, can have a positive impact on the regeneration of urban and rural areas (Akande et al. 2016). Consequently, good knowledge of the principles of sustainable conservation is an important professional skill, especially in view of the large number of existing buildings that can be reused). Yet only a small number of courses addressing the connection between conservation and the environmental sustainability of vernacular architecture have been integrated into university curricula in European universities. Moreover, although digital education tools seem to be quite effective, they are not widely used in universities in the area of sustainable development of cultural heritage (Senevirathne et al. 2022; Otto et al. 2019).

The adoption of the European Green Deal (European Commission 2019), the Renovation Wave Strategy (European Commission 2020) and the European Directives on the energy performance of buildings (European Parliament 2010, 2018) underline Europe's shift of attention towards the energy performance of existing building stock. The most challenging building category to upgrade is the heritage type. This is due to the difficulty in implementing conservation and energy upgrade practices, without affecting their architectural character.

It should be noted that architectural heritage, and especially vernacular dwellings, incorporate a large number of environmental features in their design (Coch 1998; Philokyprou et al. 2013, 2018; Desogus et al. 2016). At the same time, they provide important lessons on sustainable development that are friendly to both the people and the environment (Correia et al. 2014). Buildings that fall under the designation of vernacular architecture tend to be constructed with local materials that are more sustainable as they also tend to avoid high transportation costs and CO_2 emissions (European Commission 2022). The continuous use and reuse of vernacular dwellings has several benefits related to the social and economic life of their inhabitants. The values of vernacular dwellings depend on both the material fabric of the buildings, structures and spaces, but also on the intangible associations attached to them

(Philokyprou 2011). Thus, during the conservation of vernacular dwellings, internationally accepted conservation charters should be respected, taking into serious consideration both the environmental and also the intangible social values of these structures.

Many governmental and intergovernmental organisations—such as the International Centre for the Study of the Preservation and Restoration of Cultural Property (ICCROM)—emphasise the importance of creating courses on sustainable development and integrating them into university programmes via physical and/or online learning platforms (Philokyprou 2011). The Declaration of Amsterdam has promoted a large number of educational programmes, while the American National Council on Conservation Education has asserted the importance of coordinating educational programmes to further support the conservation of heritage buildings.

At the same time, Massive Open Online Courses (MOOCs) on conservation, sustainable development and heritage, may enrich university education offerings (Onecha et al. 2021, 2022). In this context, Smart Rehabilitation 3.0—an Erasmus Plus research project has involved four universities (University of Cyprus, Polytechnic University of Catalonia, University of Palermo and Kaunas University of Technology), two other related organizations (Cesie, Italy and AEEBC Association of European Experts in Building & Construction, Ireland) and the research institute RehabiMed–Rehabimed (Spain) for address this purpose. The partners of the programme have proposed the creation of a series of online courses on the sustainable aspects of heritage. An added value is that the four MOOCs developed by the collaborating universities are interlinked, combining both theoretical knowledge and practical examples. In this way, students and professionals are trained using an interdisciplinary and holistic approach (Philokyprou and Thravalou 2022). In this context, the course prepared by the University of Cyprus (UCY) and which focuses on the topic of environmental conservation of vernacular architecture, is presented in this paper, through the lens of a methodological and intellectual context.

Methodology

As a first step, to prepare an online course on the environmental conservation of vernacular architecture, extensive online research was carried out, to find, review and evaluate similar courses in terms of content and structure. The search came up with only a limited number of online courses on the subject in question. Furthermore, various MOOCs currently on offer were studied in detail, so as to proceed with final decisions concerning the structure of the new MOOC to be developed.

After selecting the final title and the main teaching objectives for the UCY MOOC, the next step entailed the gathering of all relevant material. At this stage, meetings and discussions were carried out with stakeholders, students and experts, to get feedback and address current problems in the field. Additional material was also gathered from relevant ongoing activities and research conducted by members of UCY. A significant contribution to selecting the teaching material was made by collaborating experts in the field. More specifically, the content of the three main modules of the MOOC is derived mainly from the deliverables of two multidisciplinary research programmes that dealt with the prevalent environmental features found in vernacular architecture of Cyprus (BioVernacular 2012; BioCultural 2013). At the same time, material from relevant European-funded research programmes and studies was also incorporated (Potts 2022; HERSUS 2020; BIPV Meets History 2018; BEEP 2019; 3D PAST 2016; RIBuild 2015; HiBERatlas n.d.; VERSUS 2012). Many links and external references were included in the MOOC, giving prospective students the opportunity to see useful relevant material and to broaden their knowledge in the field. Finally, a large part of the last module of the proposed MOOC was based on the material given by professionals in the field of conservation, as previously mentioned.

The main teaching objectives of the UCY MOOC are the following: To provide students with the fundamental knowledge and tools that allow them to recognise the environmental aspects pertinent to these dwellings; to cultivate in them a consciousness with regards to the environmentally appropriate reuse of such structures; to help them develop critical thinking in analysing such structures and to embrace the level of complexity involved in contemporary theories of conservation and sustainability; and finally, to encourage them to acquire cultural awareness highlighting the intangible and tangible values of heritage.

Regarding the hosting platform and domain of the UCY MOOC, as well as the MOOCs prepared by the other universities, the interactive online platform of the Polytechnic University of Catalonia was used (https://mooc.upc.edu/). Once all the MOOCs were prepared, a pilot was released on a trial basis for five weeks and a number of participating students were asked to send feedback for improvement.

Results

Structure of the UCY MOOC

In terms of structure, the UCY MOOC is composed of five different modules (Fig. 17.1). Each module incorporates, (a) an introductory video (trailer); (b) a PowerPoint presentation with theoretical material, as well as examples of real case studies; (c) links to other publications and websites, as well as data from two digital databases prepared within the framework of the Smart Rehabilitation research programme, including innovative strategies on conservation; and finally, (d) a multiple-choice questionnaire for the students' self-evaluation. The content of the MOOC is in English, while the trailers are in Greek with English subtitles, in order to be accessible to a diverse group of potential participants. The MOOC is free of charge and asynchronous, so students can follow it at any time.

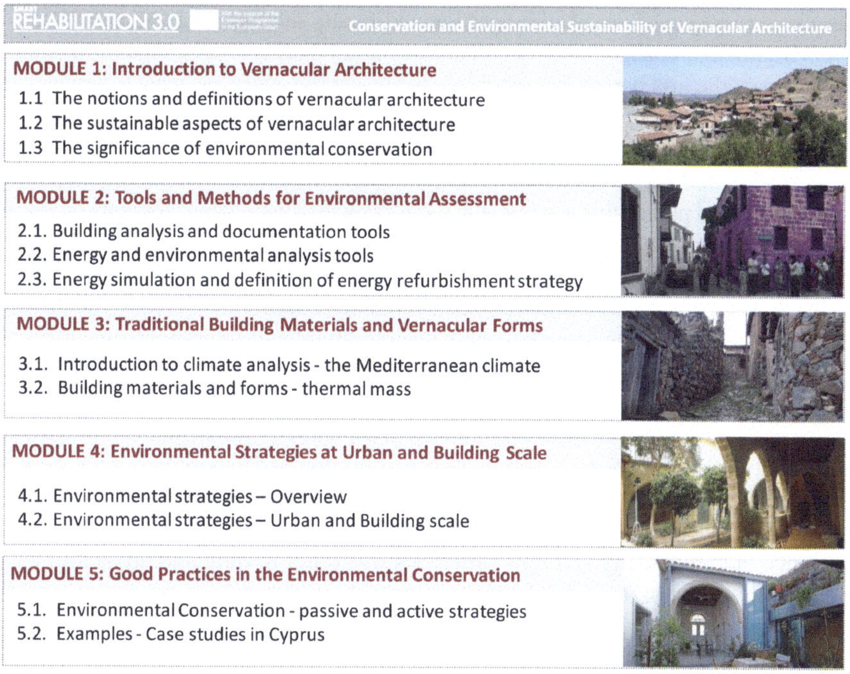

Fig. 17.1 Table showing the five modules and the main divisions of each module. *Source* Authors

Course Context

The five modules of the MOOC developed by the University of Cyprus team cover various theoretical subjects relating to vernacular architecture and sustainable conservation (Fig. 17.1). The first module introduces vernacular heritage, while the main environmental aspects of vernacular architecture are explored in modules II, III and IV. Finally, best practices for implementing sustainable and environmentally aware conservation are introduced in module V (Fig. 17.1). The MOOC includes not only theoretical issues, notions and discussions, but also practical tools and real case studies that aim to cover a variety of aspects of vernacular architecture. It deals mainly with the Mediterranean area, considering Cyprus a characteristic case study. However, examples of vernacular dwellings from other areas with distinct climatic conditions are also included.

Introduction to Vernacular Heritage

Module I, is entitled '*Introduction to vernacular architecture*'. It is divided into three parts and includes introductory notions and definitions of vernacular heritage, the sustainable aspects of vernacular heritage and finally the meaning and significance of environmental conservation of vernacular buildings (Fig. 17.1). One of the main objectives of the first part of this module is to answer the question "what is vernacular architecture?", as many different terms have been used to describe it. The various values encompassed in vernacular architecture are also discussed in detail (both tangible and intangible ones).

In the second part of this module, the focus is on the sustainable elements inherent in vernacular architecture. Lessons derived from various relevant international research programmes are briefly presented. The thematic subject regarding sustainability includes the pillars of environment, economy, culture, and society. The principles and strategies of the environmental, socio-cultural and socio-economic aspects of vernacular heritage are explored by an overview of international bibliography, emphasising the climatic and geomorphological adaptability of such dwellings and settlements. Furthermore, various tools used in the study of vernacular dwellings and settlements in Cyprus are demonstrated, such as the digital database VernArch, developed by the Department of Architecture at UCY. Other local research programmes (e.g., BioVernacular and BioCultural) focus on monitoring temperature, relative humidity and air velocity levels in vernacular dwellings in rural and urban areas.

In the third part of this module, international charters and declarations on conservation are presented and discussed. The focus here is on the significance of all aspects of sustainability and authenticity in heritage in general, and vernacular architecture in particular. Finally, through this module, the positive impact of the adaptive reuse of vernacular architecture—in the sustainability context—is explored. By the end of this module, one is expected to be able to identify vernacular dwellings and to appreciate their component characteristics relative to sustainable development, as well as the significance of reusing such structures in an environmentally sustainable way.

Environmental Aspects of Vernacular Architecture (Tools and Strategies)

Module II, entitled '*Tools & methods for environmental assessment*', examines the methodology and tools for assessing the environmental performance of heritage buildings, while also aiming to assist in the decision-making process for their energy consumption improvement (Fig. 17.1). Specifically, in this module emphasis is placed on digital tools used in, (a) the architectural and historical documentation of a vernacular dwelling, (b) the climatic analysis of the site, and (c) the evaluation of a heritage building's environmental performance. The main features of Building Information Modelling (BIM) tools, together with energy simulations, are also shown, as they significantly contribute to shaping the final strategy for environmental and energy upgrading measures, as well as monitoring all phases of project implementation (Thravalou et al. 2023; Deligiorgi et al. 2021).

For climate analysis, useful and easy-to-use tools are presented, to help interpret climatic constraints, and to give information about the potential development of a site. Among the most basic features mentioned, are, humidity charts and shading analysis, as well as a psychrometric chart that depicts thermal comfort zones for users, thus suggesting strategies for achieving comfort conditions in such spaces. The bioclimatic strategies implemented in the building envelope for heating and cooling are tabulated into data sheets. Laboratory and/or on-site measurements can determine the thermo-physical characteristics of the building shell. Non-destructive methods, such as thermography, can be used to identify masonry stratigraphy, as well as points of heat loss and rising damp. The use of simple temperature and humidity data loggers, light meters and anemometers can provide important information about indoor and outdoor environmental conditions. Finally, the process of simulating the energy behaviour of a building is outlined, focusing on the modelling steps, validation of the model, sensitivity analysis and optimisation of the proposed energy upgrade measures. Through interviews, experts in the field explain the limitations and possibilities of using BIM tools in the simulation of the energy performance of historic buildings (Gigliarelli et al. 2022). By the end of this module, it is expected that one will be familiar with various relevant tools, while simultaneously acquiring a methodology for the conservation and enhancement of the environmental strategies that can be applied to vernacular dwellings.

Module III addresses '*Traditional building materials and vernacular forms*'. In this module, the traditional materials used in vernacular architecture are presented, together with ways in which they may affect the morphology and environmental characteristics of the structures. More specifically, this module is divided into two parts and covers, (a) an introduction to climate parameters, with an emphasis on the Mediterranean climate, and (b) an overview of traditional building materials and construction techniques, including earthen, stone, and lightweight timber structures (Fig. 17.1). Different forms of roofing are shown, as well as contributions made by thermal mass elements in buildings. In the first part of this module, the inextricable links of vernacular architecture to local climatic factors are presented. In examining the island of Cyprus, for example, this climatic aspect presents several geomorphological peculiarities. It is expected that this may lead to differences in how vernacular architecture is expressed, depending on the region wherein it is considered (Philokyprou et al. 2017).

The second part of this module looks into traditional building materials, with three typical materials being examined: timber, earth and stone. Earthen structures, for example, are widespread throughout the world, while stone constructions are to be found both as masonry wall, roof and floor structural elements. Due to the high heat capacity of these building materials, heat is stored in the building shell during the day and released during the night. This enables internal temperatures to remain relatively constant with respect to the fluctuation of the external environment's temperature (Philokyprou et al. 2018; Michael et al. 2016). By the end of this module, it is expected that one will be able to identify and understand the role of traditional building materials in improving the environmental conditions inside vernacular dwellings.

Module IV, entitled '*Environmental strategies at urban and building scale*', covers the main environmental strategies that are found in vernacular architecture. A sample of the teaching material is shown in Fig. 17.2. This module is divided into two parts; the first includes a brief introduction to the environmental principles in question, and the second shifts focus to passive design strategies concerning the different scales of development (Fig. 17.1). More specifically, the environmental strategies concerning heating, cooling and managing the microclimate are presented. Starting from the urban scale, the layout of the settlements is discussed, paying particular attention to the continuous building system which is encountered in many traditional settlements, and which leads to compact building layouts that reduce a settlement's exposure to adverse climatic conditions (Meir and Roaf 2005). Reference is also made to underground (sunken) vernacular dwellings, a practice that further reduces the exposure of the building envelope to the environmental conditions, while taking advantage of the thermal mass of the soil (Cañas and Mazarrón 2009; Silvia and Ignacio 2005). Another important urban-scale cooling feature is the formation of narrow streets that provide shading to both the street area and building facades (Thravalou et al. 2024).

Moving on to the building scale, special emphasis is placed on the strategies applied to the building envelope itself, through specific typological and morphological choices. The cross arrangement of openings for example, enhances natural ventilation and cooling. Correspondingly, light structures with many openings projecting into the street, the so-called *sahnisi*, contribute to the improvement of natural ventilation and natural lighting (Thravalou and Philokyprou 2021; Michael et al. 2017; Oikonomou and Bougiatioti 2011). Examples of traditional structures, such as wind towers, are also presented, as these are used for passive cooling and are to be found in many hot regions (Bahadori et al. 2016; Jomehzadeh et al. 2020).

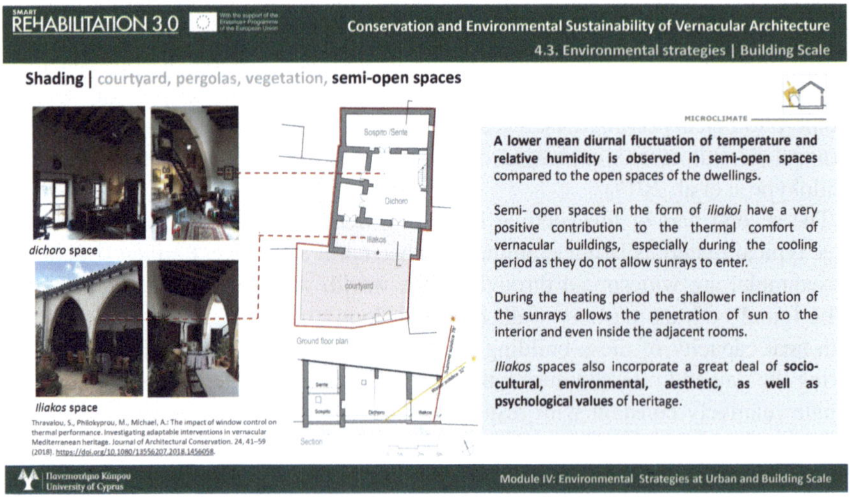

Fig. 17.2 Sample of the teaching material of module IV. *Source* Authors

Shading in private yards, using suitable structures such as pergolas, deciduous trees and semi-outdoor areas, limits solar gains during the summer due to solar height (Fig. 17.2). Accordingly, during the winter, when the solar azimuth is low, the sun penetrates the interior through the semi-outdoor areas and heats the main living spaces. During the night, the temperature difference between indoor and outdoor environments enhances night ventilation (Aldawoud and Clark 2008; Edwards 2006; Ghaffarianhoseini et al. 2015). Water features found in the courtyards of vernacular dwellings also contribute to cooling through evaporation.

In the last part of this module, reference is made to heating and the heat loss protection strategies inherent in vernacular dwellings. Finally, a brief reference is made to strategies dealing with dampness and protection from water, such as building houses on wooden stilts. By the end of module IV, it is expected that one will be able to identify the various different environmental cooling and heating strategies incorporated in vernacular dwellings.

Best Practices for Environmental Conservation

Finally, the last module of this course (module V), addresses '*Good practices for the environmental conservation of vernacular dwellings*'. More specifically, this module is divided into two parts and includes, (a) a detailed reference to the different strategies and principles of environmental conservation, and (b) examples of vernacular dwellings that were conserved in a way that preserves their bioclimatic characteristics (Fig. 17.1). The aim of this module is to propose a sustainable approach to the conservation of vernacular dwellings, by introducing environmental criteria into the conservation process. The investigation into the environmental features of vernacular dwellings showed that although they incorporate many environmental principles, there is often a need to further enhance these principles, in order to meet contemporary sustainability requirements (Philokyprou and Michael 2021; Troi and Bastian 2015; Troi 2001). An important factor with regards to the energy efficiency of traditional structures is the preservation of the high heat capacity of the envelope, that in turn creates rather stable indoor temperatures (Philokyprou et al. 2018; Michael et al. 2013). In order to maintain the important environmental role of the traditional courtyard, special care should be given to the preservation of its material palette and vegetation. The maintenance of semi-open spaces is considered an environmentally sustainable approach, as these spaces offer shading protection. The importance of the user's role in the thermal behaviour of vernacular dwellings is also shown to be important, especially when considering night ventilation (Thravalou et al. 2018; Demosthenous et al. 2015).

In the second part of module IV, selected examples of vernacular dwellings are analysed (see Fig. 17.3). During the conservation of these dwellings, their bioclimatic characteristics—which have been preserved—are showcased. The first example is a traditional house in the rural, semi-mountainous village of Kapedes in Cyprus

(Michael et al. 2014). This dwelling incorporates several cooling and heating strategies. During conservation, special attention was paid to the preservation of the original envelope, with an emphasis on the thermal mass of the masonry. The original solid timber planks at the openings were conserved. The cross arrangement of the openings that offer effective ventilation was preserved, as well as the timber balconies offering shading (Fig. 17.3). Another example of a vernacular dwelling in an urban centre is a private house in Kaimakli, i.e., a historic neighbourhood in Nicosia. The house features a tripartite arrangement and an internal courtyard to the rear of the plot. A glass sliding surface was installed in the arched opening of the semi-open space towards the yard. This space remains a closed-off area during the winter and can be opened up during the summer to avoid overheating and offer improved ventilation. An additional example is a house in the mountainous settlement of Askas. The spaces within the house are organised on two levels and incorporate a large number of heating strategies responding to the specific climatic conditions of the area. The conservation approach followed in a traditional urban building in Aglantzia in Nicosia, implemented as part of yet another research programme (HYBUILD), is also shown. In this building, photovoltaics were integrated on the roof.

In this module, the importance of a multicriteria assessment approach during conservation is emphasised. This assessment allows for the establishment of a contemporary conservation methodology, which proposes alternative environmental retrofitting solutions. A very important step in this process is the achievement of a balance between conserving the aesthetic appearance of the dwellings, and the retrofitting actions implemented. By completing this module, one will be able to understand the need to maintain and enhance the bioclimatic strategies of vernacular dwellings, by using easy-to-implement means of energy performance upgrading.

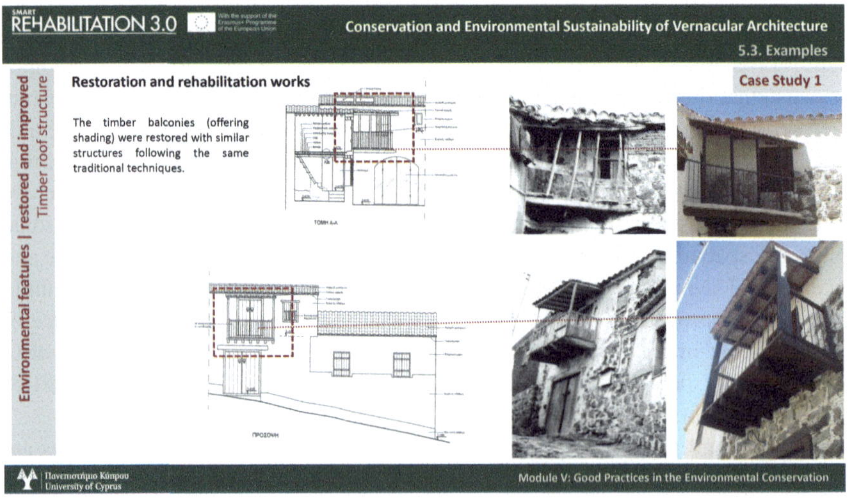

Fig. 17.3 Sample of module V showing one case-study building in Kapedes. *Source* Authors

Conclusions

There are a large number of traditional settlements across many European countries. The need to both protect and energy-upgrade this cultural heritage, in order to meet the contemporary demands of a sustainable approach to development, leads to the need for the study of vernacular architecture and the incorporation of this knowledge in university courses. By tackling vernacular dwellings and their environmentally sustainable conservation, students become aware of the values and the fundamental significance of strategies that address the built environment. A deep and meaningful engagement with vernacular architecture and its environmental elements, can inspire creative sustainable conservation solutions that contribute to the productive life of existing environments.

Continuous education that can be easily accessed and managed through online courses is especially useful for a society that aspires towards a more sustainable future. Sustainable development and the reuse of vernacular dwellings are inextricably linked, as two individual aspects aiming to address the preservation of non-renewable resources. It is thus important to determine the procedures to be followed for the rehabilitation of existing vernacular dwellings, since these offer the strongest 'living' examples of the architecture of the past. A detailed study of the environmental aspects of vernacular dwellings and the determination of the minimum changes that can be made to improve their energy efficiency is also significant.

The development of an online course in this area, such as the one described in this paper, can train both students and professionals in this industry sector. The interdisciplinary approach suggested in this MOOC is very important when studying vernacular architecture, as it helps to develop critical thinking. To achieve the sustainable reuse of vernacular dwellings, practical solutions are needed, together with successful design proposals that instigate social and economic impact. Critical thinking is promoted through the case studies included in this course, which in turn adopt various approaches corresponding to a different set of challenges for each case. Upon completing the course, the ultimate objective is to familiarise participants with the concepts of vernacular architecture and sustainable development, while providing them with useful, practical, and theoretical tools that may be used in this endeavour.

Acknowledgements This paper is based on the results of the Erasmus Plus Smart Rehabilitation 3.0, Innovating Professional Skills for Existing Building Sector (SEPIE) programme, under the coordination of Rehabimed (Spain). The partners of the programme are: Polytechnic University of Catalonia (Spain), Cesie (Italy), University of Palermo (Italy), University of Cyprus (Cyprus), Kaunas University of Technology (Lithuania), AEEBC (Ireland). Among the aims of the programme is the development of online training courses (MOOCs) and digital tools for accessing technological and innovative interventions in rehabilitation (through multiple databases accessible online https://smart-rehabilitation.eu/digital-library/).

References

3D PAST. 2016. 3D PAST Research Program: Living & Virtual Visiting European World Heritage. https://esg.pt/3dpast/.

Akande, O.K., D. Odeleye, A. Coday, and C. Jimenez-Bescos. 2016. Performance Evaluation of Operational Energy Use in Refurbishment, Reuse, and Conservation of Heritage Buildings for Optimum Sustainability. *Frontiers of Architectural Research* 5 (3): 371–382. https://doi.org/10.1016/j.foar.2016.06.002.

Aldawoud, Abdelsalam, and Ray Clark. 2008. Comparative Analysis of Energy Performance between Courtyard and Atrium in Buildings. *Energy and Buildings* 40 (3): 209–14. https://doi.org/10.1016/j.enbuild.2007.02.017.

Bahadori, Mehdi N., Alireza Dehghani-Sanij, and Ali Sayigh. 2016. *Wind Towers: Architecture*. Climate and Sustainability: Springer.

BEEP. 2019. BEEP Research Program: BIM for Energy Efficiency in the Public Sector. https://www.enicbcmed.eu/projects/beep.

BioCultural. 2013. BioCultural Research Project: Implementation of Sustainable Design Elements of Vernacular Architecture in the Rehabilitation of Traditional Buildings and in the Design of New Structures. http://www.biocultural.ac.cy/.

BioVernacular. 2012. BioVernacular Research Project: Innovative Methods for Protection and Conservation of Sustainable Design Elements of Vernacular Architecture in the Historic Centre of Nicosia. http://www.biovernacular.ac.cy/.

BIPV Meets History. 2018. BIPV Meets History Research Program. https://www.bipvmeetshistory.eu/en-gb/.

Cañas, Ignacio, and Fernando R. Mazarrón. 2009. The Effect of Traditional Wind Vents Called Zarceras on the Hygrothermal Behaviour of Underground Wine Cellars in Spain. *Building and Environment* 44 (9): 1818–1826. https://doi.org/10.1016/j.buildenv.2008.12.006.

Coch, Helena. 1998. Chapter 4—Bioclimatism in Vernacular Architecture. *Renewable and Sustainable Energy Reviews* 2 (1–2): 67–87. https://doi.org/10.1016/S1364-0321(98)00012-4.

Correia, Mariana, Letizia Dipasquale, and Saverio Mecca. 2014. In *VERSUS. Heritage for Tomorrow: Vernacular Knowledge for Sustainable Architecture*. Firenze: Firenze University Press.

Deligiorgi, Marissia, Stavroula Thravalou, Kristis Alexandrou, and Georgios Artopoulos. 2021. Data-Driven Integrated Practices for the Study and Protection of Built Heritage. In *Proceedings of the 5th Biennial of Urban and Architectural Heritage*. Nicosia.

Demosthenous, Despina, Amilios Michael, and Maria Philokyprou. 2015. Investigation of the Influence of Night Ventilation on Indoor Thermal Conditions in Rural Vernacular Architecture of Cyprus. In *International Conference on Sustainability in Architectural Cultural Heritage*, ed. M. Philokyprou, Aimilios Michael, and Andreas Savvides, 116–24. Nicosia.

Desogus, Giuseppe, Leonardo Giuseppe Felice. Cannas, and Antonello Sanna. 2016. Bioclimatic Lessons from Mediterranean Vernacular Architecture: The Sardinian Case Study. *Energy and Buildings* 129: 574–588. https://doi.org/10.1016/j.enbuild.2016.07.051.

Djordjevic, Aleksandra, Aleksandra Milovanović, Milica Milojevic, Ana Zorić, Mladen Pešić, Jelena Ristic Trajkovic, Ana Nikezić, and Vladan Djokić. 2022. Developing Methodological Framework for Addressing Sustainability and Heritage in Architectural Higher Education—Insights from HERSUS Project. *Sustainability* 14: 4597. https://doi.org/10.3390/su14084597.

Edwards, Brian (ed.). 2006. *Courtyard Housing: Past, Present, and Future*. Abingdon, England and New York: Taylor & Francis.

European Commission. 2019. The European Green Deal. https://eur-lex.europa.eu/legal-content/EN/TXT/?qid=1596443911913&uri=CELEX:52019DC0640#document2.

European Commission. 2020. Renovation Wave Strategy. https://ec.europa.eu/energy/topics/energy-efficiency/energy-efficient-buildings/renovation-wave_en.

European Commission. 2022. Strengthening Cultural Heritage Resilience for Climate Change: Where the European Green Deal Meets Cultural Heritage. Publications Office of the European Union. https://data.europa.eu/doi/https://doi.org/10.2766/44688.

European Parliament. 2010. Directive 2010/31/EU of the European Parliament and of the Council of 19 May 2010 on the Energy Performance of Buildings (Recast), vol. L 153/13. https://eur-lex.europa.eu/legal-content/en/TXT/?uri=CELEX:32010L0031.

European Parliament. 2018. Directive 2018/844/EU of the European Parliament and of the Council of 30 May 2018 Amending Directive 2010/31/EU on the Energy Performance of Buildings and Directive 2012/27/EU on Energy Efficiency (Text with EEA Relevance), vol. L 153/13. http://data.europa.eu/eli/dir/2018/844/oj/eng.

Ghaffarianhoseini, Amirhosein, Umberto Berardi, and Ali Ghaffarianhoseini. 2015. Thermal Performance Characteristics of Unshaded Courtyards in Hot and Humid Climates. *Building and Environment* 87: 154–168. https://doi.org/10.1016/j.buildenv.2015.02.001.

Gigliarelli, Elena, Filippo Calcerano, Letizia Martinelli, Georgios Artopoulos, Stavroula Thravalou, and Kristis Alexandrou. 2022. *Methodology for the Energy Renovation of Heritage Buildings Using BIM. Guidelines for the Development of an Energy Efficient Heritage Building Information Model (EE-HBIM)*, ed. Elena Gigliarelli, Filippo Calcerano, and Letizia Martinelli. Roma: Consiglio Nazionale delle Ricerche.

HERSUS. 2020. HERSUS Research Program: Enhancing of Heritage Awareness and Sustainability of Built Environment in Architectural and Urban Design Higher Education. https://hersus.org/about/.

HiBERatlas. n.d. *Historic Building Energy Retrofit Atlas*. https://www.hiberatlas.com/en/about-atlas-917.html.

Jomehzadeh, Fatemeh, Hasanen Mohammed Hussen, John Kaiser Calautit, Payam Nejat, and M. Salim Ferwati. 2020. Natural Ventilation by Windcatcher (Badgir): A Review on the Impacts of Geometry, Microclimate and Macroclimate. *Energy and Buildings* 226: 110396. https://doi.org/10.1016/j.enbuild.2020.110396.

Potts, Andrew. 2022. *European Cultural Heritage Green Paper "Putting Europe's Shared Heritage at the Heart of the European Green Deal"*. Europa Nostra.

Meir, A., Isaac, and Susan Roaf. 2005. The Future of the Vernacular. Towards New Methodologies for the Understanding and Optimization of the Performance of Vernacular Buildings. In *Vernacular Architecture in the 21st Century: Theory, Education and Practice*, 1st ed., ed. Lindsay Asquith and Marcel Vellinga, 215–30.

Michael, Aimilios, Chryso Heracleous, Stavroula Thravalou, and Maria Philokyprou. 2017. Lighting Performance of Urban Vernacular Architecture in the East-Mediterranean Area: Field Study and Simulation Analysis. *Indoor and Built Environment* 26 (4): 471–487. https://doi.org/10.1177/1420326X15621613.

Michael, Aimilios, Maria Philokyprou, and Chrystala Argiriou. 2014. Documentation and Evaluation of the Positive Contribution of Natural Ventilation in the Rural Vernacular Architecture of Cyprus. In *Proceedings of the International Conference on Cultural Heritage. Digital Heritage. Progress in Cultural Heritage Documentation, Preservation and Protection*, 310–19. Limassol, Cyprus.

Michael, Aimilios, Maria Philokyprou, Stavroula Thravalou, and Ioannis Ioannou. 2016. The Role of Adobes in the Thermal Performance of Vernacular Dwellings. In *Terra 2016: XIIth World Congress on Earthen Architectures*, 7. Lyon, France: CRAterre.

Michael, Amilios, Maria Philokyprou, Stavroula Thravalou, and Ioannis Ioannou. 2013. Contribution of Bioclimatic Elements to Thermal Comfort: Heritage Case Study in Nicosia, Cyprus. In *PLEA 2013: Sustainable Architecture for a Renewable Future*, vol. E-Book. Munich.

Oikonomou, Aineias, and Flora Bougiatioti. 2011. Architectural Structure and Environmental Performance of the Traditional Buildings in Florina, NW Greece. *Building and Environment* 46 (3): 669–689. https://doi.org/10.1016/j.buildenv.2010.09.012.

Onecha, B., M. Bosch, M. Bosch Prat, C. Cornadó, E. Crespo, and J. Martí. 2022. 4 X 4: Methodology to Design a Set of 4 MOOCs between 4 Entities under the Same Topic—Erasmus+ SMART

REHABILITATION 3.0 Project. In *EDULEARN22 Proceedings*, 2034–2040. Palma, Spain: IATED. https://doi.org/10.21125/edulearn.2022.0529.

Onecha, B., C. Cornadó, M. Bosch, M. Bosch Prat, E. Crespo, A. Pérez, R. Cabrera, J. R. Rosell, and X. Casanovas. 2021. Accompanied Self-Renovation: A Massive Online Course To Meet The Needs Of a Professional Rehabilitation Expert In People-Centered Renovations. In *ICERI2021 Proceedings*, 1863–1869. Online Conference: IATED. https://doi.org/10.21125/iceri.2021.0498.

Otto, Daniel, Sandra Caeiro, Paula Nicolau, Antje Disterheft, António Teixeira, Sara Becker, Alexander Bollmann, and Kirsten Sander. 2019. Can MOOCs Empower People to Critically Think about Climate Change? A Learning Outcome Based Comparison of Two MOOCs. *Journal of Cleaner Production* 222: 12–21. https://doi.org/10.1016/j.jclepro.2019.02.190.

Philokyprou, Maria. 2011. Teaching Conservation and Vernacular Architecture. *Journal of Architectural Conservation* 17 (2): 7–24. https://doi.org/10.1080/13556207.2011.10785086.

Philokyprou, Maria, and Aimilios Michael. 2021. Environmental Sustainability in the Conservation of Vernacular Architecture. The Case of Rural and Urban Traditional Settlements in Cyprus. *International Journal of Architectural Heritage* 15 (11): 1741–1763. https://doi.org/10.1080/15583058.2020.1719235.

Philokyprou, Maria, Aimilios Michael, Eleni Malaktou, and Andreas Savvides. 2017. Environmentally Responsive Design in Eastern Mediterranean. The Case of Vernacular Architecture in the Coastal, Lowland and Mountainous Regions of Cyprus. *Building and Environment* 111: 91–109. https://doi.org/10.1016/j.buildenv.2016.10.010.

Philokyprou, Maria, Aimilios Michael, Stavroula Thravalou, and Ioannis Ioannou. 2018. Thermal Performance Assessment of Vernacular Residential Semi-Open Spaces in Mediterranean Climate. *Indoor and Built Environment* 27 (8): 1050–1068. https://doi.org/10.1177/1420326X17699037.

Philokyprou, Maria, Amilios Michael, Stavroula Thravalou, and Ioannis Ioannou. 2013. Evaluation of Sustainable Design Elements in the Historic Centre of Nicosia, Cyprus. In *Vernacular Heritage and Earthen Architecture*, ed. Mariana Correia, Gilberto Carlos, and Sandra Rocha, 631–637. London: Taylor & Francis Group.

Philokyprou, Maria, and Stavroula Thravalou. 2022. Lessons Learned from the Development of Massive Open Courses on Innovative Strategies for the Rehabilitation of Built Heritage. In *ICERI2022 Proceedings*, 4095–4101. Seville, Spain: IATED. https://doi.org/10.21125/iceri.2022.0994.

RIBuild. 2015. RUBuild Research Program: Historic Buildings and Internal Insulation. https://www.ribuild.eu/home.

Senevirathne, Malith, Dilanthi Amaratunga, Richard Haigh, Dhimman Kumer, and Arturas Kaklauskas. 2022. A Common Framework for MOOC Curricular Development in Climate Change Education—Findings and Adaptations under the BECK Project for Higher Education Institutions in Europe and Asia. *Progress in Disaster Science* 14: 100222. https://doi.org/10.1016/j.pdisas.2022.100222.

Silvia, Martín Ocaña., and Cañas Guerrero. Ignacio. 2005. Comparison of Hygro-Thermal Conditions in Underground Wine Cellars from a Spanish Area. *Building and Environment* 40 (10): 1384–1394. https://doi.org/10.1016/j.buildenv.2004.11.005.

Thravalou, Stavroula, Kristis Alexandrou, and Georgios Artopoulos. 2023. 'A Simulation-Based Approach for Defining Energy Retrofit Strategies of Built Heritage Through the Use of H-Bim Tools'. In *Mediterranean Architecture and the Green-Digital Transition: Selected Papers from the World Renewable Energy Congress Med Green Forum* 2022, ed. A. Sayigh, 545–555. Springer International Publishing. https://doi.org/10.1007/978-3-031-33148-0_43.

Thravalou, Stavroula, Petros Mouzourides, Aimilios Michael, Maria Philokyprou, and Marina K. -A Neophytou. 2024. 'A Field Study on the Environmental Conditions of Street Canyons in Dense, Historic, Urban Centres in the Mediterranean. The Case of Nicosia, Cyprus'. In *Recent Advances in Environmental Science from the Euro-Mediterranean and Surrounding Regions (3rd Edition). EMCEI 2021*, eds. Mohamed Ksibi, Abdelazim Negm, Olfa Hentati, Achraf Ghorbal, Arturo

Sousa, Jesus Rodrigo-Comino, Sandeep Panda, José Lopes Velho, Ahmed M. El-Kenawy, and Nicola Perilli, 637–40. Advances in Science, Technology & Innovation. Springer. https://doi.org/10.1007/978-3-031-43922-3_143.

Thravalou, Stavroula, and Maria Philokyprou. 2021. Urban Design Considerations in the Environmental Assessment of Vernacular Buildings with Timber Projections (Sachnisi): The Case of Nicosia's Historic Center. *Frontiers of Architectural Research* 10 (1): 176–189. https://doi.org/10.1016/j.foar.2020.11.001.

Thravalou, Stavroula, Maria Philokyprou, and Amilios Michael. 2018. The Impact of Window Control on Thermal Performance. Investigating Adaptable Interventions in Vernacular Mediterranean Heritage. *Journal of Architectural Conservation* 24 (1): 41–59. https://doi.org/10.1080/13556207.2018.1456058.

Troi, Alexandra. 2001. Historic Buildings and City Centres—The Potential Impact of Conservation Compatible Energy Refurbishment on Climate Protection and Living Conditions, 10.

Troi, Alexandra, and Zeno Bastian. 2015. *Energy Efficiency Solutions for Historic Buildings. A Handbook*. Basel: Birkhauser.

VERSUS. 2012. VERSUS Research Program: Lessons from Vernacular Heritage to Sustainable Architecture. https://www.esg.pt/versus/outcomes.html.

Chapter 18
Building Technologies for Sustainable Heritage Treatment: Concepts, Design Strategies and Tools

Ana Radivojević, Dušan Ignjatović, Nataša Ćuković Ignjatović, Bojana Zeković, and Nevena Lukić

Keywords Building technology · Design strategies · Design tools · Building assessment

Introduction

During the last decade, the issues of climate change, global vulnerability regarding the resource supply and socioeconomic uncertainties have transferred from speculations about the future development to very tangible present. This means that the focus is shifting, from raising awareness and formulating general directions, to more direct and immediate actions. In building sector, such actions demand both shift in design approach and technology in order to provide sustainable solutions for new and existing buildings alike.

The motives for extending a building's lifespan may vary from preservation of its architectural, cultural, historical, or symbolical values to preservation of the materials, energy, work, and any other resources that are already embedded in its present form. There are, however, general denominators regarding building technology that may help to enable sustainable heritage treatment. When thoroughly integrated into contemporary design approaches, they provide the path for coping with current and future challenges.

From the basic analysis of buildings, starting from applied materials and structural characteristics to the complex design interventions and their impact on future building

A. Radivojević · D. Ignjatović · N. Ćuković Ignjatović (✉) · B. Zeković · N. Lukić
Department for Architectural Technologies, University of Belgrade - Faculty of Architecture, Belgrade, Serbia
e-mail: natasa@arh.bg.ac.rs

M. Loren-Méndez et al. (eds.), *Sustainable Heritage Management - Design: Notions, Methods and Techniques*, https://doi.org/10.1007/978-3-031-81970-4_18

performance, design strategies and tools related to technological issues are constantly being developed. These tools are used throughout the design process and serve for testing and validation of various concepts. On the other hand, many strategies and approaches integrate technological issues in the core steps of building design, starting from the initial concepts. Usually used as an adjective for the word design (environmentally responsive, energy conscious, climate-sensitive, carbon neutral, etc.) these numerous concepts carry an added value, which mainly contributes to the qualities of environmental protection and/or comfort issues. While these diverse tendencies should be encouraged, there is also a rising need for all of those values to become integral to the design process. This approach is to some extent being encouraged by legislation, and to a larger extent by market uptake (mainly through labelling), but still, all of the mentioned values are too often being regarded as an added feature, mainly relying on technological solutions. **Mapping the overlaps** in the different approaches and actions which all have a similar technological background is the first step in defining a common methodology for sustainable heritage treatment.

Concepts

Originating concepts dating from the mid-twentieth century have largely evolved, from pioneering days of solar architecture, all the way to contemporary ecological imperatives. The main theoretical approaches include **solar**, **bioclimatic**, **green**, **sustainable**, and **resilient** architecture (Fig. 18.1). As professional knowledge and general awareness grew, the postulates became more complex: bioclimatic has added the climate and interaction with the environment to the basis of solar architecture (Olgyay 1963; Herzog 1996); green architecture has expanded to the environmental issues in the wider sense as well as to human health and comfort (Klaus 1997; Szokolay 2008). Sustainability has introduced socio-economic factors, simultaneously looking further into the future while the ever-growing uncertainty has put the spotlight on the issues of resilience.

The initial concepts emerged in the 1950s, practically as soon as the technology empowered architecture to "disconnect" from its environment, providing fresh air, light and convenient room temperature through building systems rather than through building's architectural features. Victor Olgyay published "Design with Climate" in 1963, encompassing his research and publications that started with "The Temperate House" (1951), followed by papers "Bioclimatic approach to Architecture" and "Solar Control and Orientation to Meet Bioclimatic Requirements" (1952). Olgyay's interest in environmental context was triggered by his engagement in preparing experimental building designs (Olgyay 1963), so much of his work resulted in general principles and their applicability in architectural design and still stands as a good starting point both for students and practitioners. Architecture's response to the sun, wind and water bodies in relation to human comfort, was, of, course, recognized ever since the ancient times—Socrates depicts a house which offers a pleasant retreat at all seasons, Vitruvius writes about relations toward the sun and

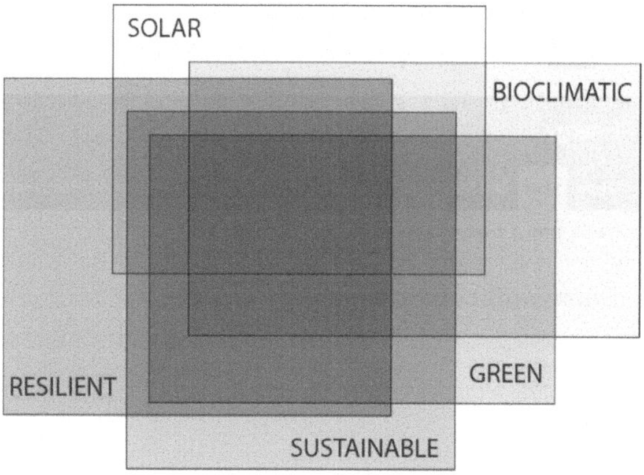

Fig. 18.1 Evolution and overlaps of design concepts

the resources in "De Architectura" (Vitruvius 1990), Alberti in "De re aedificatoria" (Alberti 1988), Le Corbusier in "La Ville radieuse", Athens charter and most of his work (Le Corbusier 1935). Olgyay brings a contemporary, systematic, and multidisciplinary scientific approach, enabling development of innovative design strategies and tools with metrics that remains relevant to date. In Szokolay's "Introduction to Architectural Science" (Szokolay 2008) the correlations of physical phenomena and architectural responses are further explored and explained, providing the path for exploring design solutions and verifying their effects.

Since the 1970s, growing energy demand has fostered resources depletion, excessive CO_2 emissions, and many geopolitical turbulences on a global and local scale, reflecting the environmental and social relevance of energy, so several concepts that are focus on the energy performance of buildings became dominant and continue to shape the current architectural practice (Hegger et al. 2008). In mid-1990's, this approach became more visible through European Charter for Solar Energy in Architecture and Urban Planning, initially drafted by Thomas Herzog, and signed by 27 leading European architects, including Norman Foster, Richard Rogers, Renzo Piano, Frei Otto, Nickolas Grimshaw, Alexandros Tombazis (Herzog 1996). As Norman Foster wrote in his foreword to *Sol Power*: "*It is not about fashion but survival, given the enormity of challenge.* "Starting from **energy-efficient** through **passive**, **nearly-zero**, **net-zero** to **energy positive**, these energy-related concepts are becoming an integral part of contemporary practice, and buildings' energy performance is perceived as an important feature by legislation, the public, and consequently the real estate market. The current overview of major topics—legislation, goals, strategies, and initiatives related to energy-efficient buildings by the European Commission (Fig. 18.2) indicates the importance of improving energy performance of existing buildings.

Energy, however, is not the sole component of sustainability. Environmental impact of material flows, resources depletion and side-effects of extraction also became visible, as did the unsustainability of conventional "take-make-waste" concept of linear economy. Considering the whole life cycle impacts the way the

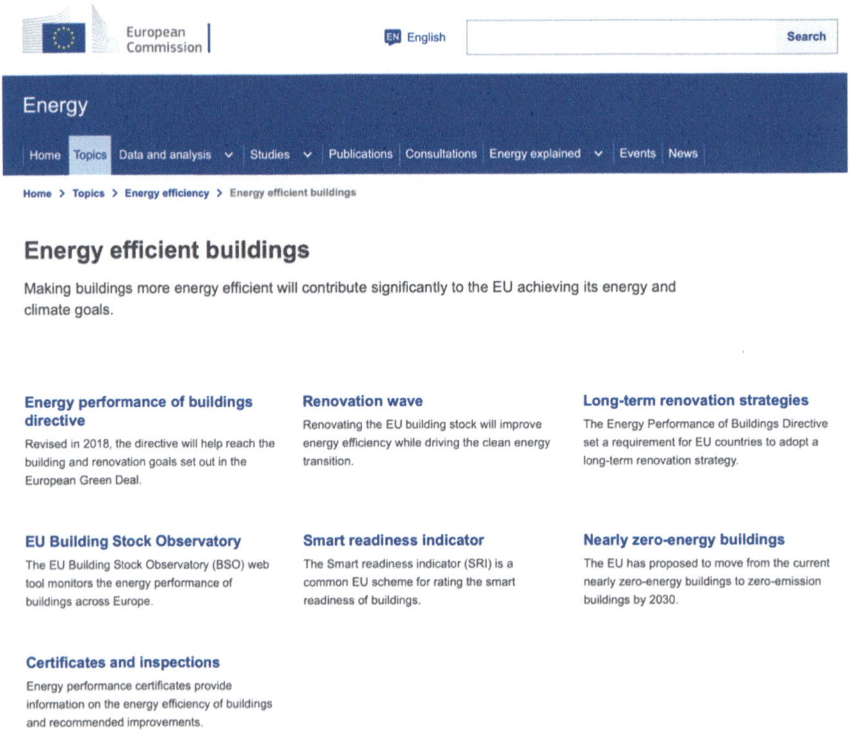

Fig. 18.2 European commission: energy-efficient buildings (https://energy.ec.europa.eu/topics/energy-efficiency/energy-efficient-buildings_en, accessed 8.3.2023.)

buildings are designed, the materials that are chosen, but, above all, places the use and adaptive reuse of built heritage at the very top, as the most desirable option. The life cycle assessment (LCA) thus finds its way in all design stages and scales—from the conceptual design to detailing and procurement, from building level to specific products and materials. The concept of circular economy emerged as a response to the evident shortcomings of traditional approach and it is changing the way we address architectural design, research, and practice, becoming an inherent part of education and formation of future architects (Ioannou et al. 2022).

Technological advancements and use of new materials also changed the way the buildings may affect the health and wellbeing of its users, leading to sick building syndrome (SBS WHO 1983). The concept "Cradle-to-Cradle" applies a holistic approach, encompassing circularity and impact on human health in pursuit of truly sustainable architecture and product design, eliminating the very concept of waste. Initially defined through The Hannover Principles (McDonough and Braungart 1992) and their manifesto "Cradle to Cradle: Remaking the Way We Make

Things" (McDonough and Braungart 2002 and 2008), it has evolved into a certification system and well-recognized label. The concept was presented at 15th Biennale di Venezia, taking its place at the Central Pavilion.

All the above-mentioned concepts are aligned with a number of UN Sustainability Goals, adopted by all UN Member States in 2015, as part of the 2030 Agenda for Sustainable Development. Those concepts most often directly address Goal 11 (Sustainable cities and communities), 12 (Responsible consumption and production), 13 (Climate action), while, in fact, the implications can be related to all 17 goals.

Design Strategies

Design strategies can be defined as paths that transfer concepts into tangible design features and technologies. Some are more relevant in the early design stages, as the consequences of design choices often can't be reverted, or remediation requires excessive amounts of time, money, and resources. Basic strategies in the preliminary design of new buildings mainly refer to placement, orientation, and thermal zoning. Together with incorporation of passive solar measures and natural ventilation, these strategies are considered the base principles of bioclimatic architecture (Hausladen et al. 2012). When dealing with the existing buildings, this means that the initial site assessments and deep understanding of the relevant potentials and shortcomings present crucial input for the basis of design. Aside from their cultural and historical values, the existing buildings are also a significant resource, considering the embedded materials, energy, and work, so preservation and refurbishment are imminent to the viable sustainable approach. The thorough understanding of building as a resource is best reflected in the schematic diagram of design phases, for new and existing buildings, presenting relevant design strategies, and introducing the postulates of (re)shaping and (re)materialization, followed by modernization and optimization of technical systems (Fig. 18.3). In the last step, a building may even generate a positive environmental impact, by producing more energy than needed, emitting clean air into the environment, water suitable for multiple uses, and even enabling food production.

Thermal Zoning

Thermal zoning principle is based on the functional organization that takes into account different requirements of spaces in terms of thermal comfort. This means that spaces that have higher demands in terms of thermal comfort, based on their use (like higher temperatures in living/working spaces contrary to auxiliary, service spaces) are to be placed and grouped in a manner to achieve better protection from negative climate impacts (low outside temperatures or dominant winds) and better use of thermal gains (passive solar or internal gains) (Hegger et al. 2008). This strategy

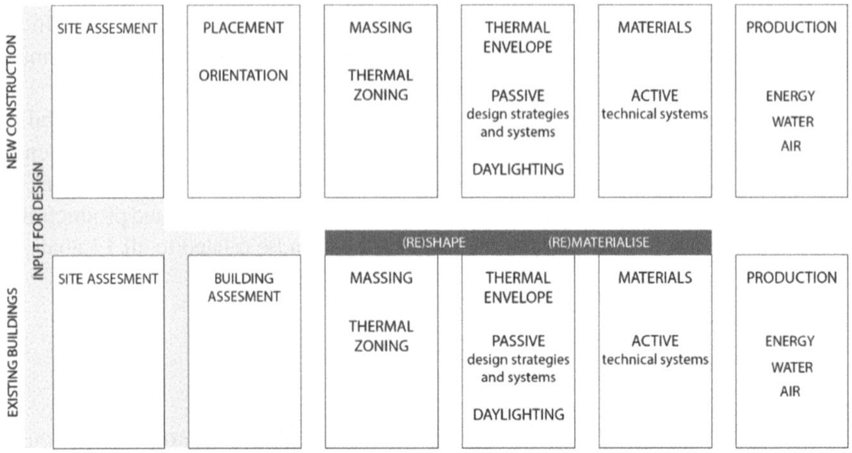

Fig. 18.3 Design strategies for new and existing buildings

is directly connected to other basic design strategies like orientation and utilization of passive solar gains.

When dealing with existing buildings, one of the key starting points may be analysis of thermal behaviour inside the buildings, according to its function, and if thermal zoning is inappropriate for a given climate, reconsider functional reorganization that would further determine other design strategies.

Thermal Massing

Using thermal mass as a bioclimatic design strategy is based on the heat storage capacity of used materials, predominantly bricks, concrete, stone, and rammed earth. Building elements (partitions, floors, roof slabs, exterior walls etc.) built with these materials, in sufficient thickness (Daniels 1997) and used as heat storage, combined with other passive design measures (passive solar gains and improved natural ventilation), can improve thermal comfort by **lowering** the need for heating and/or cooling. In cold and moderate climates this principle is used for lowering the need for heating, by **improving** use of **solar** gains (internal or passive solar), while in hot arid climates, combined with improved natural ventilation, it lowers the overheating issues. Also, combined with thermo-mechanical systems, this principle can be used for building core activation.

This design principle is hard to incorporate when dealing with existing buildings if they are not built using materials with high thermal capacity. However, since for a majority of heritage buildings their large potential for refurbishment lies in exactly the massive material of primary construction, its activation as a thermal mass storage may be a key passive bioclimatic strategy in its redesign.

Thermal Envelope

Thermal envelope consists of those building elements that separate conditioned areas (heated, cooled) to exterior space (air, soil) or unconditioned areas. It consists of opaque and transparent elements. Most common elements of thermal envelope in existing buildings are exterior walls, roofs, windows, partitions to unheated areas such as corridors, attic space, basement space, as well as walls and floors in contact with soil. The main characteristics of opaque elements is their heat transfer coefficient (U-value), while for transparent elements (windows) solar heat gain coefficients are also important. Heat losses that are related to elements of thermal envelope are mainly transmission heat losses, while transparent elements are also crucial to ventilation heat losses. Largest share of transmission heat losses in existing buildings (for example, kindergartens) is related to exterior walls, windows and roofs or ceilings to unheated attic (Ćuković et al. 2020). Shading elements of transparent elements are important to consider in order to prevent overheating.

Due to mostly poor conditions of thermal envelope elements in existing buildings, and their underperformance in terms of thermal protection according to current standards, their improvement is a necessity. Demands for improvement levels of thermal envelope in existing buildings may even be higher, due to some restraints in shape factor or building orientation which can prevent higher levels of energy savings (Atanasiu et al. 2011). In buildings that are under some regime of heritage protection, usually exterior components of thermal envelope (walls, windows) have restrictions in terms of available refurbishment options.

Passive Solar Measures

Passive solar design measures are recognized as relatively simple solutions which use conventional building elements (such as windows), transform them into solar receptors and act in combination with other strategies such as thermal mass in further use of captured heat. While ranging from such basic ones to overly complex, even those simple, affordable, easy to install and maintain are considered effective in lowering energy use and contributing to more sustainable solutions (Morrissey et al. 2011). Their function is based on the heat transfer mechanism, which can be direct, indirect, or isolated (Pucar 2006), based on the connection between the solar receptors, heat storage and heated space. The most common passive solar systems are large windows facing the direction of maximum solar gains and different types of conservatories, ranging from glazed balconies or loggias to the glazed volumes and atriums. The more complex system that includes a glazed massive exterior wall with heat transfer mechanism in-between is known as Trombe-Michel Wall. What is common to all types of passive solar spaces is that they function as a thermal buffer and lower transmission heat losses while providing additional solar gains. This is why they are considered the most effective strategy in achieving high energy efficiency (Whang and Kim 2014).

Passive solar design measures are also easy to incorporate when dealing with existing buildings and their refurbishment, especially when dealing with heritage buildings. Interventions on existing buildings in the form of addition of multifunctional glazed spaces have become quite common with the progress of technological solutions in their materialization and function (slim profiles, triple glazing, integrated solutions for water drainage etc.). These elements can have a strong visual and aesthetic statement while providing benefits for energy consumption and thermal function of the refurbished building.

Daylighting

Daylighting strategies are tightly connected to other design strategies, mainly regarding the treatment of transparent elements of the building envelope and to achieve visual comfort inside the buildings. The maximum use of natural light in order to achieve optimal conditions of light comfort, while reducing the use of artificial light, is a recommendation of all standards that deal with this issue (Radivojević and Đukanović 2018). Although provision of passive solar heat gains through these elements is considered effective in reducing building energy need for heating and providing abundance of daylight (which also leads to lower energy consumption for artificial lighting) it can also contribute to the issue of overheating and glare, which has a negative effect on visual comfort. Design options for introducing Sun-shading elements, while providing sufficient daylight are numerous, most common being light-shelves. Various Sun-shading elements which use various kinds of sensors and innovative technologies are constantly emerging and change the image of contemporary facades.

Provision of sufficient daylight in treatment of existing buildings is an important issue not only from the aspect of energy efficiency through limited use of artificial light, but also from the aspect of health benefits of future users. In heritage buildings built in massive construction techniques, a small window-to-wall ratio in terms of providing sufficient levels of daylight may be an issue. This needs to be adequately addressed in the early stages of design, either by reshaping the facades, or by introducing some elements for maximization of daylight within the existing windows.

Design Tools

Technological tools for assessment of the achieved design qualities are being developed in parallel with the emergence of new concepts and range from simple tools to complex systems. Those that could be considered appropriate from the building technology perspective are: building life cycle analysis, rating systems for green building assessment, building energy performance simulation, and in-situ measurements of building performance.

Building Life Cycle Analysis (Building LCA)

If a building is understood as a specific system, then the Life Cycle Thinking concept applied on the building represents a comprehensive understanding of impacts that buildings, together with constituent materials and products have on the environment. Therefore, we can discuss the **BUILDING LIFE CYCLE ANALYSIS** which can answer numerous questions and help reduce the impact building has on the environment during the different stages of a building's life: design, construction, operation, as well as demolition and disposal processes. The usual model of the life cycle of a building is a linear process, composed of four phases: design, construction, operation, and maintenance, or use stage, and end-of-life-stage. Based on the way of ending the life of a building, which can be manifested as disposal, recycling, or reuse, one can speak of a linear (ending with disposal), or a preferable circular life cycle of a building (ending with recycling or reuse of building components).

Building LCA is a process that identifies the flow of materials, energy, and waste originating from materials or a building as a whole during its entire life cycle so that the impact on the environment can be predicted in advance. Problems arising from the interaction of materials with the environment can be observed in the domain of: used energy and consequent pollution, raw materials and waste, or health risks/toxicity of the materials themselves. Therefore, the output of an LCA analysis can include different aspects of the environmental footprint of a building (energy use, CO_2 emissions, global warming potentials, etc.).

Green Building Certification Systems

Achieving the ambitious goals related to sustainability is rather challenging and so is the verification that the job is well done. Green building rating systems aim to quantify various aspects of a sustainable building, resulting in a label/certificate that clearly communicates the environmental quality of a particular building. They allow the comprehensive assessment of a building as an entire system (Ebert et al. 2011). In order to achieve that, they define a series of relevant building performance indicators (e.g., energy consumption, CO_2 emission, waste reduction, application of green materials, indoor air quality, etc.), providing the methodology and metrics for verifying each indicator. They constantly evolve, "recalibrating" according to the shifts in practice, legislation, new discoveries and offer solid guidelines for various stages of design and construction. All certification systems initially originated as a product designed and developed for a certain market, but the major ones soon became international, with adjustments for various conditions. They all offer options that are developed for existing buildings, either for the case that major construction works are planned or for the evaluation of building's operation and maintenance, interior design etc. The most common certification systems used internationally include LEED (USA), BREEAM (UK), DGNB (Germany), HQE (France). Although technically not a rating system, European common framework Level(s) can also be used for guidance in certain areas, since it offers 16 indicators related to 6 macro-objectives that describe how the building performance can be aligned with the strategic EU policy objectives (European Commission 2021).

Building Energy Performance Simulation

Building energy performance simulation primarily takes place during the building design process and can be a valuable method of control of different building characteristics and their expected performance. Enormous potential for this type of analysis has been recognized due to the development and increasing application of BIM technologies. In practice, several different software tools are available to simulate either the overall picture of building performance or some individual parameters (thermal characteristics, indoor air quality, building surroundings, daylight, acoustics, etc.).

Thermal simulation of a building represents a dynamic analysis of building's energy performance by using computer models and simulation tools. Currently there are over 400 applications which can be used for analysing building energy and thermal simulation (Bahar et al. 2013). Among these we differ simulation tools with visual communication (so called, *frontend*), such as *DesignBuilder*, *Ecotect* and *IES Virtual*, where through creation of a 3D building model and assigning relevant properties to modelled elements all relevant parameters are defined, and simulation-calculation tools, where all necessary data are inputted in textual/numerical format (simulation tools such as *EnergyPlus*, *DOE*, *HTB*, etc.). For the latter, this means that all preceding analysis of geometry and thermal zoning are done in some other modelling software, or frontend simulation software. Interoperability of both types of simulation tools with existing 3D modelling software's (such as *ArhiCAD*, *Revit*, *SketchUp*, etc.) represents a developing field, with the purpose of integrating energy modelling with the design process as early as possible in the design development. The main use of energy modelling is comparison between several design options in terms of energy performance, and also, certification of certain design when compared to a determined base case model. Also, thermal simulations can be used for various research and practical purposes (Ignjatović et al. 2018).

One of the aspects that significantly influence the results of the dynamic simulation is modelling of occupant behaviour. This may not be the crucial issue when comparing several designs based on the same occupancy behaviour pattern, but it may be very significant if we are trying to predict future energy consumption or compare the model and the real building, in the case of refurbishment of existing buildings.

In-Situ Measurements of Building Performance

In-situ measurements of building performance are quantitative indicators that examine and verify different building characteristics. Different measurement tools can be used in different situations: (a) after the building construction to make sure that intended performances are achieved, (b) during the exploitation of the facility to determine whether the facility functions adequately, and (c) during a renovation to determine whether and to what extent the conditions have improved. Some of the most commonly used non-invasive *in-situ* measurement tools are thermal imaging, blower door testing and measurements of thermal characteristics (heat flux) of elements of thermal envelope.

Measurements of heat flux in-situ are conducted with specialized equipment, consisting of different sensors (temperature and heat-flux sensors) and data logging devices. There are some limitations and rules regarding the measurement procedure: ventilated building structures cannot be measured (structures where final cladding is ventilated); the temperature difference must be at least 10 degrees; temperature variations should be minimal, which implies that indoor and outdoor temperature should be constant; measurements should be performed for at least 72 h, while optimal duration is about two weeks due to temperature stabilization; influences of direct sunlight, wind and moisture should be kept minimal. After the measured values of inside and outside air and/or wall temperatures and heat flux are recorded and reviewed, the reference period for further calculations is determined as the one where most stable heat flux is occurring. Based on these values, further calculations of U-values and conclusion on the state of thermal properties of the measured element of thermal envelope are derived. This method is widely used in practice, to determine the cause of various defects in building elements (moisture, leaking, bad thermal performance etc.) but also in scientific purposes, to determine characteristics of many heritage building types which have not been previously studied in detail (Lovec et al. 2018).

Blower door test or air permeability test is a method of measuring the air flow through the envelope of a building at a certain difference between the internal and external static pressure. This method can be used to measure the airtightness of an individual building component, or whole building envelope. It is predominantly used in the existing buildings. (Abishdid and Bitsuamlak 2012) Test is performed by placing and sealing a frame panel with a fan used to achieve the pressure difference (usually 50 Pa) as well as a device that measures the air that flows through the fan after the pressure difference is achieved by compression/decompression in the object.

The numerical value defines the amount of air necessary to maintain the pressure level that infiltrates/exfiltrates through the cracks of the object at a given pressure difference. Blower Door testing only supplies the numerical value of infiltration rate, and this method cannot locate the anomaly of a thermal envelope of the building—the place where the excess air flow occurs. Therefore, it needs other verification methods to make the insights complete and measurements more exact. (Lukic et al. 2020) The application of these methods can be significant in energy rehabilitation processes, analysing and evaluating the quality of building construction and components, checking the technical condition of the building before exploitation, for the certification of buildings, when buying real estate, etc.

Infrared imaging is one of the most common non-invasive methods for estimation of thermal performance of the building envelopes. Apparent ease of application, usually requiring just a camera of certain resolution and sensitivity, does not mean that the analysis process can be performed instantly and accurately. As other performance inspections, it is largely influenced by the environmental parameters therefore a specific measuring conditions have to be met: temperature difference between outside and inside of min 10 degrees in the period of minimum 4 h before inspection, absence of direct sunlight, limited air movements (wind), low perspiration are the most important ones. The idea behind this constraint is that recorded radiation in the infrared band of spectrum coming from the building envelope is resulting from the thermal behaviour of the structure and not from external sources and that readings are not deviated by them. Also, the nature of the materials has to be understood as different materials emit radiation according to their structure and surface treatment, which if not fully understood can lead to errors. The emissivity. i.e. ability of the material to emit radiation is the crucial parameter which has to be analysed and estimated.

In the performance analysis infrared inspection is commonly used in two ways: as a qualitative and quantitative method. For qualitative analysis, a thermal image is acquired and analysed focusing on the patterns of thermal readings analysing them according to the designed envelope structure and identifying the deviations areas. Variations in readings can be expected and appear as the result of the very nature and structure of the envelope (as designed) but can appear as a consequence of the faults in the construction process. If we are looking for the specific, numerical, estimation of performance oscillation, we have to perform a quantitative analysis where usually infrared inspection is combined with flux measurements or compared to the numerical two or three-dimensional heat transfer modelling. In this way, we can compare the readings and quantify the deviations precisely. (Ignjatovic 2016).

Infrared inspection has its limitations, apart from the environmental conditions, appearing mainly due to the structure of the envelope. Ventilated facades are "masking" or even blocking the radiation and in this type of construction it is advised that the measurements are performed prior to finalization. Also, highly reflective surfaces require the addition of a contact layer (black tape or special easy removable paint) which will increase emissivity and make the readings more realistic.

Conclusion

Architectural heritage is facing new challenges, yielding new approaches, with many technological aspects overlapping with the ones initially conceived for new buildings and contemporary design solutions. The environmental crisis puts an added value to the existing buildings, treating them as a valuable resource for the future, making us rethink the material and design features, both in the context of the current condition and anticipated future use. The concepts are evolving, putting into place new design strategies and tools for evaluation and verification of the environmental effects related to specific design solutions. The problem is complex, so are the paths to the desirable solution—good architecture, as always, doesn't emerge from linear thinking and strict guidance. The concepts presented in this paper produce a variety of design strategies and tools, but the connection is never linear—they overlap and intertwine, and should be considered primarily as a basic starting point for each individual case, rather than a set of instructions on how to achieve sustainable buildings.

References

Abishdid, C., and G.T. Bitsuamlak. 2012. Air infiltration through building envelopes: A review. *Journal of Building Physics* 35 (3): 267–302.

Alberti, L. B. 1988. *On the Art of Building in Ten Books* (N. Leach & R. Tavernor, Trans.) The MIT Press.

Atanasiu, B. 2011. *Principles for Nearly Zero-Energy Buildings: Paving the Way for Effective Implementation of Policy Requirements.* Buildings Performance Institute Europe (BPIE).

Bahar, Y.N., C. Pere, J. Landrieu, and C. Nicolle. 2013. A thermal simulation tool for building and its interoperability through the building information modelling (BIM) platform. *Buildings* 3: 380–398.

Le Corbusier. 1935. *La Ville radieuse.* Boulogne-sur-Seine: Éditions de l'Architecture d'Aujourd'hui, Collection de l'équipement de la civilisation machinist.

Ćuković Ignjatović, N., D. Ignjatović, and B. Zeković. 2020. Improving energy efficiency of kindergartens in Serbia: Challenges and potentials. *Thermal Science* 24 (6): 3521–3532.

Daniels, Klaus. 1997. *The Technology of Ecological Building: The Fundamentals and Approaches, Examples and Ideas.* Birkhäuser, Basel.

Ebert, Thilo, Nathalie Eßig, and Gerd Hauser. 2011. *Green Building Certification Systems: Assessing Sustainability—International System Comparison—Economic Impact of Certifications.* München: Institut für Internationale Architektur-Dokumentation.

European Commission, Directorate-General for Environment. 2021. *Level(s), What's in it for architects, designers, engineers and quantity surveyors?* Publications Office of the European Union. https://doi.org/10.2779/550618. Accessed 8.3.2023.

Hausladen, G., P. Liedl, and M. de Saldanha. 2012. *Building to Suit the Climate—A Handbook.* Birkhauser.

Hegger, M., M. Fuchs, T. Stark, M. Zeumer. 2008. *Energy Manual—Sustainable Architecture.* Edition DETAIL. Institut für Internationale Architekturdokumentation / Birkhäuser Verlag AG, München / Basel, Berlin, Boston.

Herzog, Thomas. 1996. *Solar Energy in Architecture and Urban Planning.* Munich: Prestel.

Ignjatović, Dušan, M. 2016. *Estimation of residential building energy performance using the infrared thermography method* [in Serbian], Doctoral thesis, University of Belgrade—Faculty of Architecture https://hdl.handle.net/21.15107/rcub_nardus_6119

Ignjatović, D., A. Rajčić, B. Stanković. 2018. Building Simulations and Modelling: Energy, in Konstantinou T., Ćukovic Ignjatović N., Zbašnik-Senegačnik M. (Eds.) *Energy resources and building performance, thematic book series Reviews of Sustainability and Resilience of the Built Environment for Education, Research and Design as results of the Erasmus+ project, Creating the Network of Knowledge Labs for Sustainable and Resilient Environments (KLABS).* TU Delft Open, pp. 129–144.

Ioannou, Olga, Bob Geldermans, Tillmann Klein, and Alexander Wandl. 2020. Planning for change: A methodological framework for integrating circularity into TU delft's faculty of architecture and the built environment's curricula. *SAJ-Serbian Architectural Journal* 12 (3): 234–269. https://doi.org/10.5937/saj2003234I.

Lovec, V., M. Jovanović-Popović, and B. Živković. 2018. The thermal behavior of rammed earth wall in traditional house in Vojvodina: Thermal mass as a key element for thermal comfort. *Thermal Science* 22 (4): 1143–1155.

Lukić, N., Lj. Đukanović, A. Radivojević. 2020. Implication of infiltration on building energy demand—a review of verification methods. *Proceedings of the international Conference on Contemporary Theory and Practice in Construction XIV STEPGRAD, Online, Banja Luka, Bosnia and Herzegovina, 11–12 June 2020,* Borković, A., Malinoić, M., eds., University of Banja Luka. Banja Luka, Bosnia and Herzegovina. pp. 235–249.

McDonough, William and Braungart, Michael. 1992. *The Hannover Principles: Design for Sustainability.* Commissioned by the City of Hannover, Germany, as design principles for Expo 2000, The World's Fair. Available at https://mcdonough.com/writings/the-hannover-principles/. Accessed 8.3.2023.

McDonough, William, and Michael Braungart. 2002. *Cradle to Cradle: Remaking the Way We Make Things.* New York: North Point Press.

McDonough, William, and Michael Braungart. 2008. *Cradle to Cradle: Remaking the Way We Make Things.* London: Vintage Books.

Morrissey, J., T. Moore, and R.E. Horne. 2011. Affordable passive solar design in a temperate climate: An experiment in residential building orientation. *Renewable Energy* 36: 568–577.

Olgay, Victor. 1963. *Design with Climate: Bioclimatic Approach to Architectural Regionalism.* Princeton: Princeton University Press.

Pucar, Mila. 2006. *Bioclimatic Architecture: Glazed Spaces and Passive Solar Systems.* Institute for Architecture and Urbanism. Belgrade.

Radivojević A., Lj. Đukanović. 2018. Material aspects of Energy Performance and Thermal Comfort in Buildings, in Konstantinou T., Ćukovic Ignjatović N., Zbašnik-Senegačnik M. (Eds.) *Energy resources and building performance, thematic book series Reviews of Sustainability and Resilience of the Built Environment for Education, Research and Design as results of the Erasmus+ project, Creating the Network of Knowledge Labs for Sustainable and Resilient Environments (KLABS).* TU Delft Open, pp. 61–85.

Szokolay, Steven V. 2008. *Introduction to Architectural Science: Basis of Sustainable Design.* Oxford: Elsevier/Architectural Press.

Vitruvius, Pollio, Marcus. 1990. *Ten Books on Architecture* (in Serbian, translated by Matija Lopac, 1990). Sarajevo: Svjetlost.

Whang, S.-W., and S. Kim. 2014. Determining sustainable design management using passive design elements for a zero emission house during the schematic design. *Energy and Buildings* 77: 304–312.

WHO—European Centre for Environment and Health. 1983. *WHO Guidelines for Indoor Air Quality: Selected Pollutants.* EURO Reports and Studies, no 78. Bonn Germany Office: WHO Regional Office for Europe (Copenhagen). https://energy.ec.europa.eu/topics/energy-efficiency/energy-efficient-buildings_en. Accessed 8.3.2023.

Chapter 19
Creative and Accessible Cartographic Approaches to Cultural Heritage. CACHt_cart. Two Characterisations of Contemporary Landscape

Enrique Larive-López and José Peral-López

Keywords Landscape · Cultural heritage · Spatial data infrastructure · Cartographies · Accessibility

Introduction

The knowledge of the territory and its transmission has been an occupation of different societies since the beginning of their configuration. Sharing this knowledge is considered for this paper as part of the communication between people. Evidently, this is not a novelty since other authors from various disciplines have already proposed it. From fields of knowledge such as semantics, geography and of course from architecture, cartography has been considered as a language of closed meanings but also as a field of open knowledge. In this double condition of the representation of the territory, it is the task of contemporaneity to introduce movement, sensorial aspects and the social component. With the consideration of these aspects we will be in a position to approach the contemporary representation of landscape. For this we propose two ways; from creativity and from accessibility.

Representing the territory is also a relevant cultural fact, an instrumental action that, together with transcribing its main characteristics, data and processes, allows

E. Larive-López (✉) · J. Peral-López
Universidad de Sevilla, Seville, Spain
e-mail: elarive@us.es

J. Peral-López
e-mail: jperal@us.es

M. Loren-Méndez et al. (eds.), *Sustainable Heritage Management - Design: Notions, Methods and Techniques*, https://doi.org/10.1007/978-3-031-81970-4_19

us to reveal its condition as a landscape. In this approach, we are interested in situating the value of representation as an opportunity to explore the contemporary, to navigate landscapes and weave new fields, projections and displacements, broadening the spectrum in which different realities and scales can decant: visible and invisible, cultural and natural, individual and collective, scientific and phenomenological, rational and affective. The aim is to transfer a polyphasic and overlapping geographical reality, in constant change and simultaneity, into a multi-layered information operation, an intelligent and creative cartography that allows us to open a fluid and active gaze on the landscapes of contemporaneity.

In order to understand the option of accessible cartographies, the starting point is the use of a teaching and research methodology. This case was an exhibition held on the Guadalquivir River and its relationship with society in history. From the set of contents, the interaction of the visitor with the contents is emphasised, by uploading experiences in relation to the thematic areas. In this context, accessibility is emphasised from the perspective of knowledge, i.e. providing everyone with the keys to the territory.

Concepts, Ideas and Tools

The relationship between man and his environment, and the understanding of the keys that have characterised it, has required exploring the complexity of its historical construction and therefore its current configuration from different fields of study. Knowledge of the territory, since the first movements and settlements, has been one of the main needs of society as a civilisation. Both nomadic and sedentary attitudes have prioritised their understanding as part of their survival as a species, in such a way that their transmission ensures continuity over time, either by marking the locations of resources in the case of nomadism or to ensure the protection of populations. In any case, the description of the environment and its representation form part of the task of communicating.

In this regard and in the city of Bologna, Italy, an intellectual environment is generated where communication and cartographic representation converge in the works of Eco (2013) and Farinelli (2016). Geographer Lladó (2013) takes up Eco's proposal to define two types of semantic representation in order to understand Farinelli's way of understanding and studying maps. In the first of these, language can be studied as a finite set of expressions giving rise to a formal and controlled language where each expression has a literal and univocal meaning. For the second, there is no limit to the number of expressions, where meanings depend on the context with no intention of limiting the language by assigning a single use to words. Thus, in this semantic representation, references to knowledge are considered to be open knowledge as the meanings are no longer universal. Taking these languages to cartography, the first would refer to maps in a two-dimensional projection and without establishing a relationship with the observer. The second, defined by a biunivocal correspondence, allows as many meanings as contexts, making it possible to associate the image of the labyrinth as the most appropriate to represent the contemporary territory and its functioning.

Until the last decades of the twentieth century, in the theories that developed the construction of space by man and how it acquired the condition of place, movement was barely considered. From Giedion (1977) and above all with Norberg-Schulz (1979), the concept of topologies has been updated, differentiating between existential space and architectural space. The incorporation of the temporal dimension associated with space introduces movement through the senses. Architect Holl (1996) proposes the term phenomenal zones to define those areas that, being identified with a field of perception, play a key role in the materialisation of architecture. On the other hand, and with an identical starting point, geographer Nogué (2011) is interested in deepening the conceptualisation of landscape from the relationship between territory and the theory of communication. Between territory and society, the communicative elements are an ultimate but also immediate link that could simplify the complex issues of both and their relations, but at the same time, they are a true manifestation of the landscape as an identification of contemporary culture itself. Together with the incorporation of movement into space, the incorporation of the social component, the social construction of space, has contemplated not only its categorisation but also its production. The spatial appropriation linked to its use will share the results of relating to experience and perceptions resulting in the spatial practices that Harvey (1991) proposes on the basis of Lefebvre's (1992) theories.

Cartographic Preparations. A Grammar of the Landscape

Mapping a territory has been a sustained cultural practice to identify and know its attributes and properties, and at the same time, it has allowed the generation of a cultural identity about a certain place. Representing the territory is also a relevant cultural fact, an instrumental action that, together with transcribing its main characteristics, data and processes, allows its condition as a landscape to be revealed. Figure 19.1.

> ... Landscape is a phenomenon beyond immediate comprehension; it is not until we choose a perspective and a map of what we see, marking some aspects and ignoring others, that the landscape acquires meaning. (Corner 1991, 129) ... This is why mapping is never neutral, passive or without consequence; on the contrary, mapping is perhaps the most formative and creative act of any design process, firstly disclosing and then staging the conditions for the emergence of new realities. (Corner 2011, 91)

It is now interesting to reflect on this approach, which places the value of representation as an opportunity to explore the contemporary landscape, to navigate in its memory and weave new fields, projections and displacements, broadening the spectrum in which different realities and scales can be chosen, such as visible and invisible; cultural and natural; individual and collective; scientific and phenomenological. The aim is to transfer a polyphasic and overlapping geographical reality, in constant change and simultaneity, into a multi-layered information operation, an

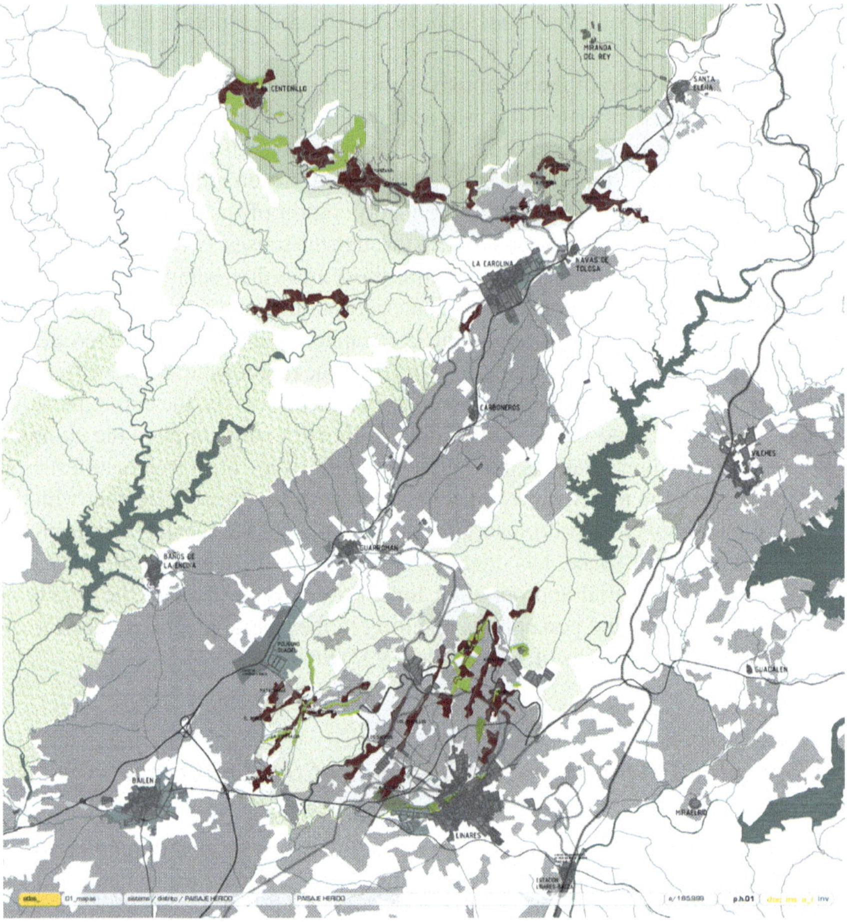

Fig. 19.1 Larive-Lopez, E. *Mapping a territory. Atlas Mining District Linares-La Carolina. Landscape Laboratory.* 2010

intelligent and creative cartography that allows us to open a fluid and active gaze on the landscapes of contemporaneity.

The evolution of representation has moved from the material and physical description of the ground to the description of invisible and often immaterial fields, forces and flows. This approach is supported by a wide availability of statistical, physical and cultural data, and the translation of this data into visual documentation and spatial infrastructure is now a ubiquitous component of landscape surveys, research and active management processes.

This research map responds to the explorations in techniques of observation and representation of the landscape through its intersections, confluences and crossroads developed in the Landscape Laboratory. A crossing between different approaches and creative events developed in the first decades of the twenty-first century in the Andalusian and American geographies. A displacement in time and space that has been constructing a grammar of the landscape that fuses spatial precision, creativity and cultural imagination and that seeks its arguments between the purely geographical and the freely abstract. Figure 19.2.

> ...The landscape laboratory is represented by the combination of geospatial technologies with the existing data in local and regional digital infrastructures; the approaches and distances to the place through phenomenological and collaborative experimentation; and the graphic and creative conceptualisation of the information through mappings, diagrams, panoramas and collages. The laboratory process, as play and experimentation, offers us a multiple reading of possible geographies, of landscapes of transition, landscapes of obsolescence, landscapes of exchange and landscapes of contamination. This extensive graphic and cartographic development aims to interfere between the personal experience of the reader-actor and the researcher-manager and to penetrate the conceptual and disciplinary boundaries that block the incorporation of border spaces into dialectical processes between present, past and future that favour their positive reuse in cities and in the territory... (Larive-López 2017, 407).

Approaches to the Field of Action

This research aims to investigate what happens when historical processes are thought about and described in spatial and local terms. To do so is to take seriously the unity of action, time and place. And as Karl Schlögel says:

> ... What happens if history and place are thought about together? ... What do we gain in perception and historical insight if we finally (again) take spaces and places seriously? (Schlögel 2007, 14)

But, of course, it goes without saying that there is a hidden path that is expressed in the five hypotheses of the research, almost in the form of days.

1. Thinking spatially. At the start of the twenty-first century there has been an important transition, if not transformation, in the ways of understanding, experiencing and studying the urban and territorial phenomenon and therefore the post-modern landscape. We are at the beginning of a revolutionary change in thinking, something different, spatial thinking or thinking of space.

> In the development of this critical spatial perspective, we find a wide referential world in Soja's writings (Soja 1996, 234) (Soja 2000, 162) where he analyses the post-modern perspective in depth. At least as far as the last two hundred years are concerned, never before has critical spatial thinking been so widespread, influential and relevant. What is happening is a transdisciplinary "spatial turn" that is affecting almost all social and human sciences.

2. The landscape as a disciplinary interstice. In recent decades, we have witnessed an extended discussion in the disciplines and fields that decide on, act on or study the relations between man and environment, and from this the landscape appeared as a new scale and/or another way of looking at reality. The emergence of the

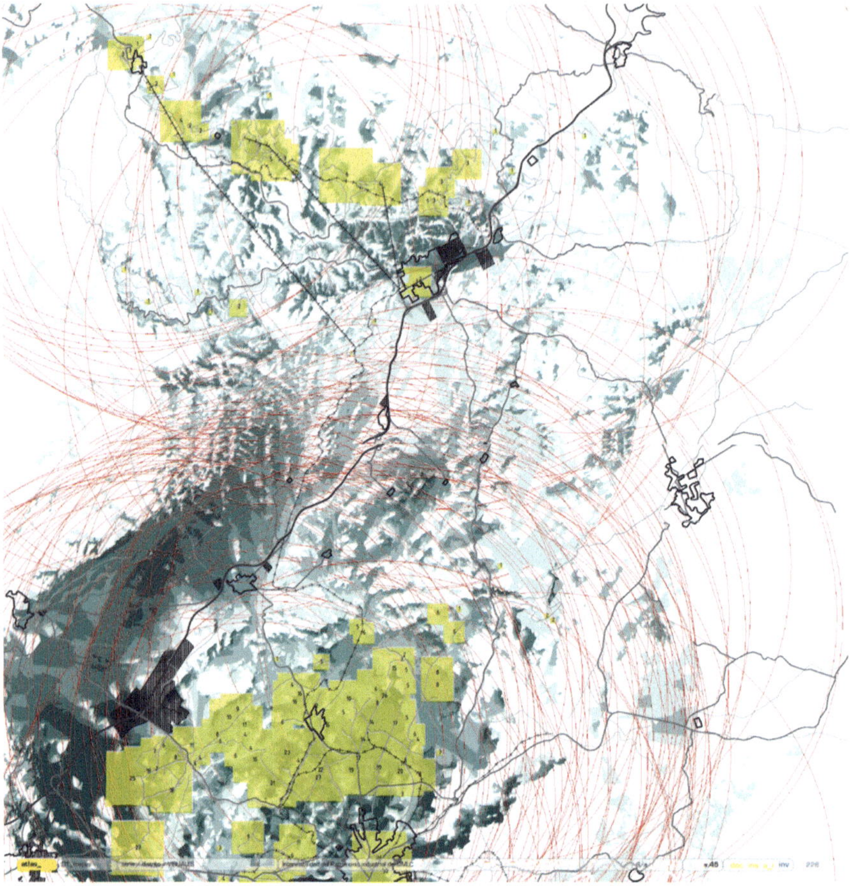

Fig. 19.2 Larive-Lopez, E. *Visible landscape. Atlas Mining District Linares-La Carolina. Landscape Laboratory*. 2010

concept is closely linked to the different forms of current nostalgia, reaction to changes in scale and speed, urbanisation, mobility, the emergence of globalisation and climate change. However, it is understood here that it is more interesting to highlight the emergence of the acceptance that the nature of reality is complex, interrelated and total, a consistency that the concept of landscape takes on and brings to us. In the interdisciplinary (dis)encounter, different conceptions, procedures and views of reality are observed, a polysemy of the concept that defines an extraordinarily rich framework/field. This richness also brings with it a need for definition, as within the same framework we find a situation of instability and "transhumance" in ideas that makes the field a slippery place. (Larive-López 2007, 89)

... The concept of Landscape as a way of seeing is extensively developed by Dennis Cosgrove in Social Formation and Symbolic Landscape. (Cosgrove 1998, 59, 66, 72, 73)

What can be done? What knowledge or attitudes can be most effective? How can the techniques of the researcher of historical landscapes of production be updated?

This research aims to go beyond narrow disciplinary frameworks and controversial terminological definitions to establish a bridge of pronounced transversality between architecture, archaeology, economics, technology, society and landscape and thus initiate a necessary debate between these fields of knowledge, often invisible to each other, allowing their confluence in the territory of production in which they take shape and become visible, in a disorderly manner, before those who study, observe or consume them. We therefore propose a construction of knowledge based on an active methodology that relates facts to ideas, bearing in mind the aim of contributing to the scientific and cultural progress of the students we are addressing.

... The construction of an analytical framework borrows tools from other landscape disciplines. The methodologies first of Patrick Geddes, then Ian L. McHarg (Design with Nature) and more recently, and in a more rigorous and practical way, those of Richard T.T. Forman (Land Mosaics)… (Ábalos 2005, 71).

To this end, we want to overcome disciplinary barriers and start from the premise that a paradigm is defined by its capacity to establish a new horizon in the process of knowledge by proposing objectives or tasks; putting into practice regulated procedures or methods; and establishing generalisable conclusions or theory. We want to achieve a qualitative and innovative leap marked by the social responsibility inherent in any field of knowledge. In this way, seeking the integration of diverse perspectives, we intend to clarify the confusing territory of contemporaneity, through discovery, observation, analysis and comparison. Landscape is presented as the medium through which culture, society and the individual interact with biophysical, meteorological and geological fluctuation or stability. Landscape is a conduit, an exaggerator, a proliferator, an inhibitor, an activator. Herein lies its timeless operational capacity. Landscape as infrastructure (Carlson 2013) and as an operational platform.

3. A new way of dealing with cultural heritage. The twenty-first century confronts us with new challenges regarding the research, protection and conservation of heritage. The increasingly strong links between natural and cultural heritage; between movable and immovable heritage; between tangible and intangible heritage; between object and context; between specialised technical management and citizen participation; between urban and rural; between local and global; between singular and generic; between concentrated and dispersed; lead us to a new heritage frontier.

A frontier where all the aforementioned circumstances come together in a disorderly manner. It is therefore necessary to establish a new heritage topology that goes beyond the previous narrow methodological and conceptual frameworks. We are facing new interpretations of cultural heritage and are faced with the need to design an approach that integrates the various currents of thought and the social demands that are being generated in our worlds from other fields of critical knowledge and sustainable action. In an instant, preservation, reuse and transformation come together, and issues of maintenance and conservation must be weighed against possible transformation or destruction through creative healing (Braae 2015, 13) that goes beyond a simple exercise of physical transformation of space or change of use.

4. Exploration of methodological processes. The research responds to a process of observation and exploration of methodological and instrumental systems of contemporary and creative approach to entropic territories, loaded with memory and waiting to be reactivated. The landscape can be considered as a privileged

observatory in which to discover, analyse and rethink the territorial, social and cultural conflicts that globalisation is currently maintaining with the permanence and recycling of these heritage contexts.

... What is an observatory? An observatory is a place where, through the mediation of technology and different techniques, a dialogue with nature can be established that translates the first and innocent experience of perception into knowledge. This is why we are particularly interested in this concept of Observatory, because it is a typology in which technique, nature and architecture interact to intensify the dialogue between man and the world, each adopting their respective roles in contemporary culture... (VVAA 2003, 19)

This spirit of restlessness and curiosity is a central element of the process, and we insist on the idea of the laboratory as a contemporary process and gesture to foster new thoughts and attitudes. The collaborative experience is an opportunity to show what technology is allowing us to achieve, especially in the production of data cartographies, active and transversal, connected to the already existing spatial data infrastructures that become the new *active Atlases of the landscape* (Larive-López 2017, 430). In other words, it becomes a cartographic language that will make meanings of different signs flow through a new geography of its own, woven on the basis of networks, de-hierarchised structures and multi-directional connections. Figure 19.3.

The user now becomes a protagonist and not a mere spectator, generating a common space. Interactivity is the precise term for this sphere and, in reality, it would be a further stage in the emphasis on the active role of the spectator (citizen) that the historical avant-gardes had already begun and which boosted new artistic behaviour at the end of the 1960s, although with a greater economic and social dimension, dissolving the classic opposition between object and process, author and spectator, producer and user in a large part of the dynamics activated, which allows us to speak of collective creation.

5. Active management of the landscape. We are now talking about active processes of interaction and continuous dialogue with the place. How can we decode the materials we have at our disposal? In traditional design epistemology, our attention would be placed in the future, while the emergent approach, which is experienced and explored in this research, moves between the present, the past, and the future. Thus, the difference between the two design epistemologies questions the fundamental relationship between space and time. Questions of authorship are diluted in the face of collaborative processes. New concepts appear and others are redefined. Disciplinary boundaries become more porous, giving way to transdisciplinarity, disciplinary interstices or disciplinary fluidity. It operates with instruments and methodologies that offer transversality of knowledge, participation and direct transfer. We are facing the theory of change, the *Transit Theory*. (Larive-López 2017, 420)

The term in transit implies that the landscape is reprogrammed, reused, recycled, resurfaced, or kept waiting in a slow-speed process, but that path requires entrepreneurship and creativity. The designer becomes a stalker (Tarkovsky 1979) of the border, improving permeability and facilitating transit. The work at the border is one of enquiry and search; of identification and recognition; of transfer and negotiation; and where time is important. Thus, transit is resilient to changing circumstances, avoiding being too open to succumb to future schemes that are better organised. (Berger 2006, 241)

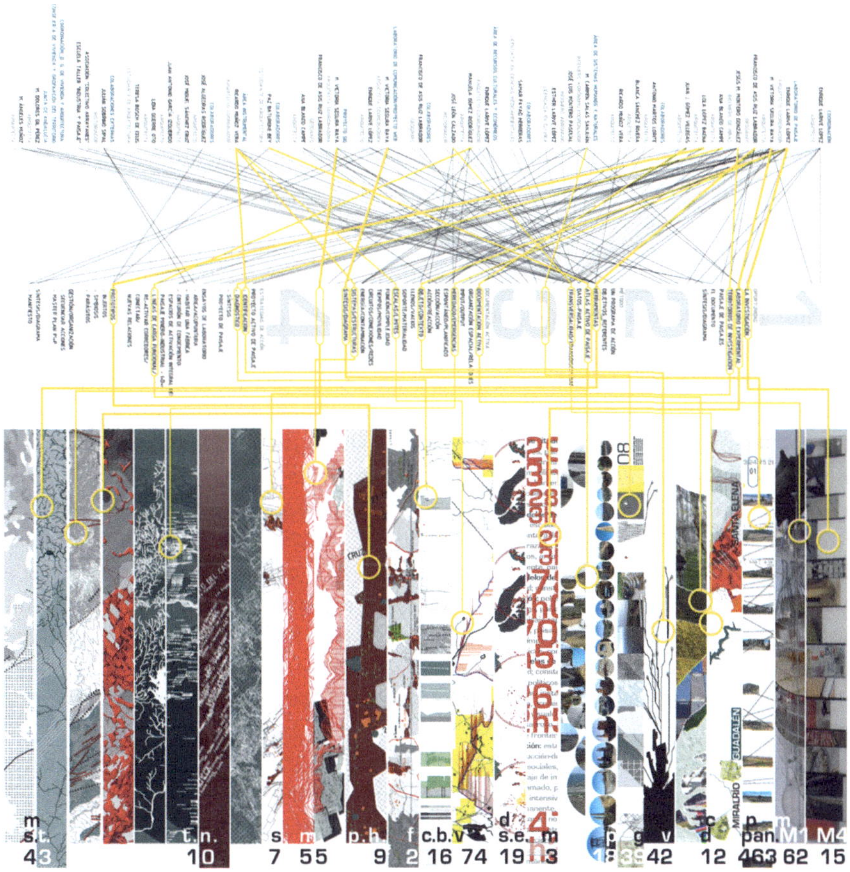

Fig. 19.3 Larive-Lopez, E. *Active Atlas of the landscape. Border Spaces. Landscapes in reclamation.* 2017

With *Transit Theory*, a new theoretical-methodological framework for approaching landscape is launched. The process is intensified at the margins rather than in the centre. This implies a shift and suggests a move away from the sole responsible planner or designer, towards a new entity or figure, the mediator, who encourages participation, negotiation and the laboratory, promotes the capacity for creative invention and entrepreneurship, and integrates the emergent and the "while" (Larive-López 2017, 421). These qualities are not sufficiently taught or rehearsed in traditional disciplines, but they are necessary in the interstices, to make borders porous.

Transit raises the need for vacant spaces as links between present, past and future (Braae 2015, 10), openly encourages the acceptance of uncertainty (Clèment 2007, 59) as a factor of development and banishes the voracious appetite for the ordering and manipulation of the territory as a methodology for the action of man on his environment. Faced with the seduction of the regulated, structured and closed, it invites us to be mere spectators of free, open and random evolution, directed exclusively by nature itself, understanding the Third Landscape as a power, as a rough diamond still in the process of consolidation.

Models and Sketches. Readings for Accessibility

Between October 2017 and March 2018, an exhibition was held at the General Archive of the Indies in Seville, commemorating the 90th anniversary of the creation of the water management body in Andalusia, the Guadalquivir Hydrographic Confederation. Entitled *Mapas y relatos de un río, imagen y miradas* (Maps and stories of a river, image and views), it focused on showing, through cartographies, documents, archaeological pieces and models, man's relationship with the Guadalquivir River from Phoenician times to the present day. This relationship was intended to be more than just a univocal one, with the spectator contemplating the objects on display and the explanations in the texts of the different sections. For this purpose, tablets were installed, one for each of the sections, where the public was asked to share their own experiences.[1] The ultimate aim of the exhibition was to create a new relationship between the exhibits, whose common characteristic was to convey a personal experience with the River, and the visitors' own experiences in real time. And all of this is with reference to a changing reality both physically and perceptively, as reference was made to hydraulic works over time and to the threats of increased flows or their divinisation. This exhibition was the opportunity to transfer to the general public a methodology developed in the teaching field, in the last semester of the studies of the School of Architecture at the University of Seville for its bachelor's degree. The course in particular is called *Paisaje, ciudad y arquitectura en Andalucía* (Landscape, city and architecture in Andalusia) and forms part of the optional offer of the degree with a total of 6 ECTS. Figure 19.4

The planning of architectural studies in higher education in Spain contemplates, in the section on specific competences, the acquisition of adequate knowledge of the relationship between cultural patterns and the social responsibilities of the future architect. However, in the area of Architectural Composition, how could heritage teaching be adapted to the new requirements? It was clear that the task was to incorporate concepts and skills and not to abandon the more traditional teachings of the History of Architecture, Aesthetics or Archaeology. To these, we added a perspective of accessibility and sustainability so that the results, in addition to meeting academic requirements, would be within the reach of a greater number of people, with a projection into the future in terms of regeneration. For the characterisation of a specific area, the following phase is proposed:

1. Observation. Considering an action and individual images as starting points, it is necessary to validate this experience and make it collective. In this way, the univocal relationship between the observer and their surroundings broadens its condition and becomes a two-way relationship, being used by other disciplines to approach the complexity of the landscape (Burne, Brasche). Thus, the first steps are aimed at establishing criteria that make it possible to define an area

[1] The sections into which the exhibition was divided were The River as a resource, The River as a threat, The River as a myth, The River tamed, The River as a project and The River managed. https://expobus.us.es/s/guadalquivir/page/inicio

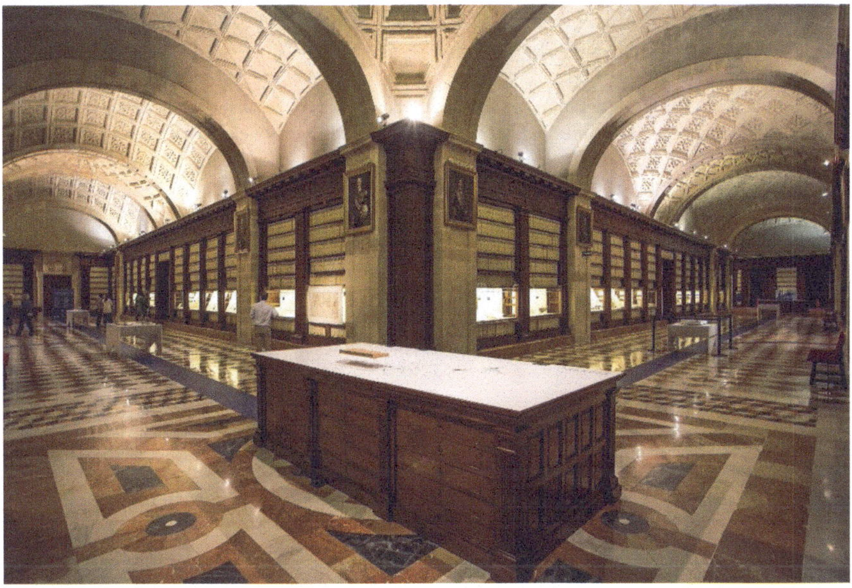

Fig. 19.4 Peral-López, J. *Maps and stories of a river, image and views* exhibition. 2017

with its own differentiating characteristics. Physical references such as orography and fluvial resources, as well as structures that have been maintained throughout history, are very important in the current configuration of the territory. Resources such as water and its conduction together with road infrastructures were, until the contemporary period, a highly important combination in the historical configuration of the landscape. From this point onwards, the modification/transformation produced by man with the passing of the study defines the following phases. Starting from the scale of the architectural monument, Manfredo Tafuri's appreciations allow us to transfer them to other scalar spheres:

> The observer is increasingly becoming a user who gives meaning to the object to the series, but who is more and more captivated and absorbed in this ambiguous situation of theirs. Inside and outside the work, at the same time, they are offered two alternatives: active participation and passive participation (although the latter is also a form of participation that changes the meaning of the work). (Tafuri 1972, 114)

2. Description (Thinking). With humanism and the revival of classical culture in Europe, the gradual disappearance of symbolism in the cartographies that had conveyed knowledge of the world in the Middle Ages began. Until the printing of the Ptolemaic descriptions, the vision of the world was ideological; a new conception of man's relationship with nature initiated the transition towards an abstract vision with the incorporation of flat and scaled representation. However, although the principle of veracity was gaining ground, it was recognised that the disappearance of symbolism in representation was not yet complete, as it was

practically impossible to produce a representation devoid of intentions and an objective image of the world.

In this phase it is proposed, within the univocal relationship of the observer, to take a journey through the different disciplines of knowledge, considering each one's own means of expression. Literary and narrative narration alternate with drawing to set the initiative at each moment, with one supporting the other and vice versa. The place, the moment and the development of the technique will be decisive in the configuration of the maps. Reason prevailed and identity gave way to objectivity, changing the diffuse limits of the known world for the limits of the cartographic sheets of the national maps, reaching the maximum of definition with ortho-photogrammetry. The final result, the map, had to be adapted and its layout justified:

... To draw is to select, to select is to interpret and to interpret is to propose (Solà and Parcerisa 1981, 12)

3. Gazes (Feeling). In the static and univocal relationship between the subject and the object (territory), history is considered as a critical element based on the results obtained with the different disciplines at the service of its elaboration. But how can images, whether written or drawn, be interpreted once they have been produced? It is time to incorporate communication and movement as a form of knowledge and its further transmission. It is now time to lower the viewpoint of the observer and bring it closer to the ground line and to consider the transmission of knowledge as an act of communication. By placing ourselves at the limits of truthfulness, we consider a map as a painting, as a work of art. The concept of the border is taken up again not only as its own, differentiated place but also as places between disciplines, in fields that have not been formulated. Figure 19.5.

To speak of representing movement is to speak of contemporaneity, and the reaction of models that identify them in specific areas requires creativity. Assuming Harvey's (1991) postulates that a spatial conception associated with time cannot be understood independently from a social action, he defined the concept of *spatio-temporal compression*:

I use this expression to refer to processes that generate a revolution of such magnitude in the objective qualities of space and time that they force us to modify, sometimes radically, our representation of the world. (Harvey 1991, 267)

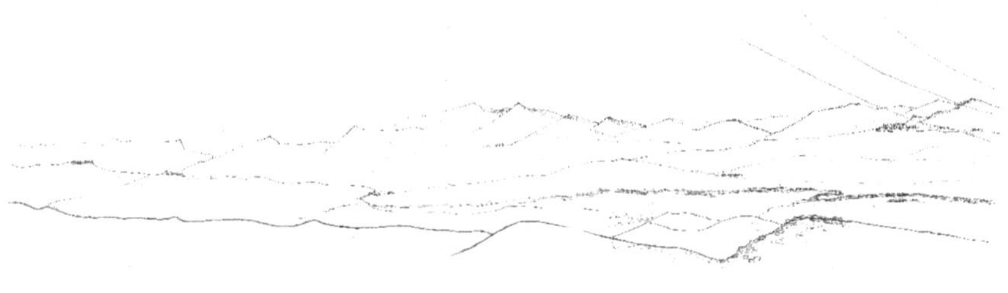

Fig. 19.5 Peral-López, J. *Crossing the River*. 2015

References

Ábalos, Iñaki. 2005. Atlas pintoresco. Vol. 1: el observatorio. Gustavo Gili: Barcelona.

Berger, Alan. 2006. Drosscape. Wasting Land in Urban America. Princeton Architectural Press.

Braae, Ellen. 2015. Beauty redeemed. Recycling post-industrial landscapes. Birkhäuser Architecture.

Carlson, Dane. 2013. *The Humanity of Infrastructure: Landscape as Operative Ground*. Scenario Journal. University of Pennsylvania. http://scenariojournal.com/article/humanity-of-infrastructure/

Clèment, Gilles. 2007. *Manifiesto del Tercer Paisaje*. Barcelona: Gustavo Gili.

Corner, James. 1991. Theory in crisis a discourse on theory II: Three tyrannies of contemporary theory and the alternative of hermeneutics. *Landscape Journal* 10 (2): 115–133. https://doi.org/10.3368/lj.10.2.115.

Corner, James. 2011. The Agency of Mapping: Speculation, Critique and Invention. Theories of Mapping Practice and Cartographic Representation. Editor(s): Martin Dodge, Rob Kitchin, Chris Perkins. John Wiley & Sons, Ltd: 89-101.

Cosgrove, Dennis. 1998. *Social Formation and Symbolic Landscape*. The University of Wisconsin Press.

Eco, Umberto. 2013. *The Book of Legendary Lands*. Rizzoli.

Farinelli, Franco. 2016. *Blinding Polyphemus. Geography and the Models of the World*. Seagull.

Giedion, Sigfried. 1977. *Space*. Time and Architecture: The Growth of a New Tradition. Harvard University Press.

Harvey, David. 1991. *The Condition of Postmodernity: An Enquiry into the Origins of Cultural Change*. Wiley-Blackwell.

Holl, Steven. 1996. *Anchoring*. Princeton Architectural Press.

Schlögel, Karl. 2016. *In Space We Read Time: On the History of Civilization and Geopolitics*. New York: Bard Graduate Center.

Larive-López, Enrique. 2007. Activar un paisaje herido. *Revista PH* 61: 82–91.

Larive López, Enrique. 2017. Border spaces. Landscapes in reclamation (Unpublished Doctoral Thesis). University of Seville, Seville. https://hdl.handle.net/11441/71040

Lefevbre, Henry. 1992. *The Production of Space*. Wiley-Blackwellz.

Lladó, Bernat. 2013. Franco Farinelli. Del mapa al laberinto. Barcelona: Icaria.

Peral-López, José (coord.) 2017. Guadalquivir. Mapas y relatos de un río. Imagen y mirada. Sevilla, Universidad de Sevilla. https://doi.org/10.12795/9788447221080

Nogué, Joan. 2011. The communicative dimension of landscape: A theoretical and applied proposal. *Revista De Geografía Norte Grande* 49: 7–24. https://doi.org/10.4067/S0718-34022011000200002.

Norberg-Schulz, Christian. 1979. *Genius Loci: Towards a Phenomenology of Architecture*. Rizzoli.

Soja, Edward W. 2000. *Postmetropolis: Critical Studies of Cities and Regions*. Los Ángeles: Blackwell Publishing.

Soja, Edward W. 1996. *Thirdspace. Journeys to Los Angeles and Other Real–and–Imagined Places*. Oxford: Blackwell.

Solà, Manuel, and Josep Parcerisa. 1989. La forma d'un pais. *Quaderns D'arquitectura i Urbanisme* 1: 3–13.

Tafuri, Manfredo. 1972. Teorías e historia de la arquitectura. Hacia una nueva concepción del espacio arquitectónico. Barcelona: Laia.

Tarkovsky, Andrei (Director). 1979. Stalker (Film). Mosfilm Production.

VVAA. 2003. Cuatro Observatorios de la Energía. Laboratorio de Técnicas y Paisajes Contemporáneos. Colegio Oficial de Arquitectos de Canarias: La Palma.

The manufacturer's authorised representative in the EU is Springer Nature Customer Service Centre GmbH, Europaplatz 3, 69115 Heidelberg, Germany. If you have any concerns regarding our products, please contact ProductSafety@springernature.com

Printed and bound by CPI Group (UK) Ltd, Croydon, CR0 4YY
01/10/2026
02225966-0012